THE MEMBRANES
OF CELLS

SECOND EDITION

THE MEMBRANES
OF CELLS

SECOND EDITION

Philip L. Yeagle
Department of Biochemistry
University of Buffalo
School of Medicine
Buffalo, New York

ACADEMIC PRESS, INC.
A Division of Harcourt Brace & Company
San Diego   New York   Boston
London   Sydney   Tokyo   Toronto

mc

Copyright © 1993, 1987 by ACADEMIC PRESS, INC.

All Rights Reserved.
No part of this publication may be reproduced or transmitted in any form or by any means, electronic or mechanical, including photocopy, recording, or any information storage and retrieval system, without permission in writing from the publisher.

Academic Press, Inc.
1250 Sixth Avenue, San Diego, California 92101-4311

United Kingdom Edition published by
Academic Press Limited
24–28 Oval Road, London NW1 7DX

Library of Congress Cataloging-in-Publication Data

Yeagle, Philip.
    The membranes of cells / Philip L. Yeagle. -- 2nd ed.
        p.   cm.
    Includes bibliographical references and index.
    ISBN 0-12-769041-7
    1. Cell membranes.   I. Title.
    [DNLM: 1. Cell Membrane.   QH 601 Y37m 1993]
    QH601.Y43   1993
    574.87'5--dc20
    DNLM/DLC
    for Library of Congress   93-17184
                                                                                CIP

PRINTED IN THE UNITED STATES OF AMERICA
93 94 95 96 97 98   BB   9 8 7 6 5 4 3 2 1

NWST
IA EB 8695

To Arlene, Jay, and David

Contents

Preface

This second edition was stimulated by rapid advances in some areas of the membrane field in the years since the first edition was published and by the gratifying responses from users of the first edition who needed an update. These responses and an evaluation of the status of investigation into biological membranes have led to some major changes in this new edition. One is the addition of a chapter on receptors, a field of great interest, a chapter which helps complete a survey of membranes within the context of one book. So much progress has been made in the field of membrane fusion that the chapter in the first edition was rendered almost completely obsolete and a new one has replaced it. Virtually all parts of the book have been reanalyzed and rewritten to provide a current and extensive survey of the study of the membranes of cells to benefit researchers and new students of the subject. In addition, at the suggestion of some of the users of the first edition, specific in-text referencing has been used instead of the more general referencing used previously. However, it should be noted that the references are representative, not exhaustive.

In addition to thanks to the people mentioned in the Preface to the previous edition, several should be particularly thanked for providing important information for the production of this edition. They include Drs. A. Albert, J. Bentz, E. Dennis, R. Epand, C. Huang, B. Litman, D. Siegel, T. Thompson, and B. Wattenberg. However, the responsibility for the representation of the content of this book remains with me. I want to thank Shirley Light of Academic Press and the production staff for making this book possible through much hard work. I also want to thank Dr. Tony Watts, the Biochemistry Department, and St. Hugh's College of Oxford University for providing a stimulating and refreshing sabbatical environment in which work on this book was completed.

Philip L. Yeagle

Preface to First Edition

This book is an outgrowth of a course on membranes that I have offered for a number of years. Thus, first and foremost, this is a textbook for students of cell membranes.

Additionally, this book is intended to serve researchers in the field of biological membranes and those scientists who wish to become better acquainted with this area.

To aid the reader in the understanding of this highly interdisciplinary subject, I have attempted to interpret the wealth of biophysical data available on biological membranes. A survey of much of the underlying biology and biochemistry also has been included.

It is my hope that this book, in its own way, will stimulate researchers and students alike, to pursue the many fascinating, vexing, and yet unanswered questions in this exciting field.

The creation of this book would not have been possible without the help of many scientific colleagues. A number of people read portions of the book and provided critical input, including A. Albert, C. Huang, P. Holloway, R. McElhaney, D. Papahadjopoulos, J. Bentz, I. C. P. Smith, G. Willsky, C. Pickert, D. Cadenhead, B. Cornell, and J. Silvius. Although I take responsibility for the content, the critical comments of these individuals were invaluable. B. Mierzwa edited the book to improve the style and construction, an effort for which I am very grateful. For much of the original inspiration in my career in membrane research which led to this effort, I would like to thank the "membrane group" at the University of Virginia.

Several investigators were kind enough to provide figures and information for the book, including G. Shull, D. MacLennon, R. Henderson, R. Kopito, R. McElhaney, W. Scherer, D. Cadenhead, B. Cornell, J. Griffin, W. Duax, T. Thompson, K. Jacobsen, and D. Engelman. I am especially grateful to S. W. Hui for the electron micrographs that appear in the book.

The preparation of the book was made possible by the efforts of Cindy Collins and particularly Janice Tokarczyk, who patiently prepared the many drafts of the manuscript. The help of the staff at Academic Press is also gratefully acknowledged.

Finally, the support, love, and understanding of my family is really what made it all possible.

Philip Yeagle

1

Introduction

The study of cell membranes is the study of vital structural elements that determine much of the cellular function. An understanding of the membranes of cells requires a synthesis of experimental insights from the fields of molecular biology, which include biophysical chemistry, molecular genetics, physiology, cell biology, and biochemistry. This synthesis is not normally possible in a discipline-oriented approach. The goal of this book is to obtain such a synthesis with membranes as the focal point of the study. A highly interdisciplinary approach is required. It will not be the purpose to become expert in the many disciplines, but rather to learn to appreciate and examine the membranes of cells from many points of view.

The pursuit of this goal will begin with an overview of the incredible diversity of roles cell membranes play in living cells. Gaining an understanding of how membrane structure and function lead to these roles will then be the province of the remainder of this book.

I. THE ORGANIZATION OF MEMBRANES IN CELLS

A. Prokaryotic Cell

The prokaryotic cell exhibits the most simple organization of cell membranes. It has no intracellular membranes. Since organelles are defined by intracellular membranes, there are no organelles in the prokaryotic cell.

The boundary of the cytoplasm of the prokaryotic cell is determined by the plasma membrane, or inner membrane. The plasma membrane forms a semipermeable sack allowing differentiation between the components and the composition inside and outside the cell. This plasma membrane is constructed of proteins and a lipid bilayer. The components of

1

the plasma membrane are held together by noncovalent forces. Thus the plasma membrane of prokaryotic (and other) cells is not a rigid structure. Although it effectively forms a sealed boundary, the plasma membrane is deformable. It gains rigidity from either intracellular membrane skeletons or extracellular matrices.

The plasma membrane of a simple cell is likely the most fundamental element of life in an evolutionary sense. Before life as we know it could progress, some distinction was required between the life processes and the soup in which those processes were occurring. The plasma membrane provided the necessary distinction.

Some prokaryotes express an extracellular matrix. The plasma membrane is surrounded by a relatively rigid structure held together largely by covalent bonds. This structure is called a cell wall. The cell wall is not a membrane. An important feature of the covalent structure of the cell wall is the peptidoglycan–carbohydrate linkages and amide linkages that cross-link the macromolecules of the cell wall and hold the structure together.

Outside the cell wall of bacteria is the outer membrane of the cell. The outer membrane is a structure that is not as rigid as the cell wall. It contains proteins and lipids, but is different from the plasma membrane in protein and lipid composition. In the case of gram-negative bacteria, the outer membrane contains some specialized glycolipids in great quantity, of which lipopolysaccharide is one. Among the proteins of the outer membrane are the porins, whose structure permits the formation of pores of specific size in the outer bacterial membrane. These pores permit passive transport of solutes across the outer membrane.

B. Eukaryotic Cell

The eukaryote is more complex than the prokaryote (see Fig. 1.1). Its greater complexity arises from the variety of membrane-bound compart-

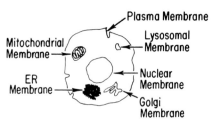

Fig. 1.1. Schematic diagram of a eukaryotic cell, indicating the plasma membrane and some of the membrane-bound intracellular organelles.

ments called organelles, which are found within the cell. These intracellular membranes subdivide cellular activities and permit more diverse and specialized functions than are possible in prokaryotes.

The plasma membrane is the boundary of the eukaryotic cell, as it is for the prokaryotic cell. Among the intracellular membrane-bound organelles are the nucleus, the mitochondria, the endoplasmic reticulum (ER), the Golgi, and the lysosome. These will be described below. Each of these organelle membranes exhibits a unique lipid and protein content crucial to their individual function. Each membrane is built around a lipid bilayer.

The membranes of eukaryotic plant cells follow the same structural and functional motifs exhibited by the more widely studied nonplant cells. For example, the plant cell membranes are built on a lipid bilayer, as are the membranes of nonplant cells. Membrane proteins add function to the plant membranes. The semipermeable nature of the biological membranes and the ability to create and utilize transmembrane chemical and electrical potentials lead to ATP synthesis in both mammalian and plant cells.

C. Virus

Enveloped viruses, although not cells, nonetheless are surrounded by a membrane with the same overall structural characteristics exhibited by cell membranes. The viral membranes will be described in more detail later in this chapter.

II. FUNCTIONS OF CELL MEMBRANES

A. General Motifs

Even though each membrane exhibits functions unique to that membrane, there are six features of cell membranes that are repeatedly used by a variety of membrane systems in cellular biology. One motif is the transmembrane gradient of the chemical species or charge. The cell utilizes a membrane to create, maintain, or use the energy stored in a concentration gradient from one side of the membrane to the other. Examples include ions and metabolites.

A second motif is the organization of enzymes into a complex. For example, in the inner mitochondrial membrane the enzymes involved in the electron transport chain are organized into functional groups.

A third motif is control of enzyme activity by membrane structure and by individual membrane components. The interactions between membrane proteins and the lipids of the cell membrane are multifaceted and offer many possible elements of control not available to nonmembrane enzymes.

A fourth motif is the membrane as substrate. One example is arachidonic acid, a precursor for prostaglandin biosynthesis, which is derived from membrane lipids.

A fifth motif is the transduction of molecular information from one side of a membrane to another. For example, plasma membrane receptors function on the cell surface to recognize extracellular signals and to alter intracellular behavior in response.

A sixth motif is compartmentalization: the physical separation of one compartment from another, with control over communication between them.

Through these six motifs, the unique (to biology) structures of the membranes of cells offer functions that are essential to cell viability, growth, and development. Appreciating the basic architecture of these motifs will lead to a better understanding of cell function.

B. Plasma Membranes

The plasma membrane of the cell defines the cell boundary. This membrane by weight is about half lipid and half protein. The plasma membrane of a cell exemplifies the basic compartmental function of membranes: it separates and delineates intracellular from extracellular domains.

It is a role of the plasma membrane to maintain the difference between the inside and the outside of the cell by controlling the entrance and exit of materials. All materials going in and out of the cell must encounter the plasma membrane. Specified cellular nutrients and products enter and leave the cell through the transport (and fusion) capabilities of the plasma membrane. Thus the plasma membrane plays an important regulatory role in the metabolism of a cell.

Consider the plasma membrane of the human erythrocyte as an example. The relatively simple structure of the human erythrocyte, with no intracellular organelles (in contrast to avian erythrocytes that are nucleated), has made this cell a useful model for other cell plasma membranes. The plasma membrane of the human erythrocyte contains a protein specific for the transmembrane movement of the cellular nutrient glucose. If this transport protein is blocked, glucose is much less able to penetrate the intact erythrocyte. The glucose transporter consists of a protein embedded in the plasma membrane. Membrane proteins responsible for transport usually are embedded in the membrane.

Another one of the major membrane proteins contained in the human erythrocyte plasma membrane is the protein responsible for anion transport. This protein functions as a small channel that specifically allows anions to move rapidly across the membrane. In erythrocytes rapid chloride–bicarbonate exchange across the plasma membrane is functionally integrated with the binding and absorption of oxygen by hemoglobin, and in CO_2 transport.

The inside and outside of the erythrocyte, as in most cells, is distinctly different in ion composition, thus illustrating another of the motifs described above. Sodium ion concentration is relatively low inside and potassium ion concentration is relatively high inside the cell. Na^+,K^+-ATPase, an enzyme of the plasma membrane, pumps sodium out of the cell and potassium into the cell, simultaneously. This ion pumping maintains the distinct difference in Na^+ and K^+ concentrations between the inside and the outside of the cell. The pump hydrolyzes ATP to provide energy for the transport process.

The Na^+ gradient across the plasma membrane generated by Na^+,K^+-ATPase is used for a variety of functions. One of the ways sodium can respond to the difference in chemical potential across the plasma membrane is to reenter the cell through a cotransport system. In a cotransport system a nutrient the cell requires is transported across the plasma membrane along with a sodium ion. Sodium is moving with its concentration gradient in this system. Therefore the energy stored in the sodium gradient can be used to drive a nutrient against its concentration gradient and into the cell. An example of this transport is found in intestinal epithelial cells.

To illustrate another motif, there are a number of receptor protein signaling systems in the plasma membrane involved in signal transduction. For example, the β-adrenergic receptor will respond to hormone outside the cell and subsequently activate the adenylate cyclase, located on the inside of the plasma membrane. Receptor-stimulated adenylate cyclase generates a signal or a second messenger, increasing cyclic AMP levels in the cell. Such an increase can cause a number of changes in the metabolic activity of the cell.

Another function of the plasma membrane of the cell is to control cell–cell interactions. During development, cells differentiate and organize into organs or organisms. This process depends on cells recognizing the right matrix and building the structure of the organism or organ according to the pattern dictated by its genome. This cell–cell recognition and cell–matrix recognition are mediated by the plasma membrane through receptors. The opposite extreme, when the cells do not recognize that they are encountering other cells or matrices, and consequently multiply without control, is characteristic of tumor cells.

A means of communication between cells is provided by the gap junction. This communication is particularly important in cell layers that function in a concerted fashion, as in involuntary muscle tissue. In the gap junction the plasma membranes of two cells are relatively close to each other but separated by 20–40 Å. Across this gap, connections between the two cells can form. These connections are apparently hollow protein tubes. Ions can pass through these tubes from one cell to another. This ion movement allows for cell communication on a primitive level, although at a rapid rate.

Cell plasma membranes can also form tight junctions. This is a region in which the electron microscope does not reveal any significant intercellular space between the two cell plasma membranes. The tight junction allows epithelial tissue, such as in the lining of the intestine and the kidney, to form sealed cell layers. As a consequence the two sides of the cell layer are not in direct communication with each other. The plasma membrane of the epithelial cells then controls movement of materials through the cell layer.

An example of specialization can be found in epithelial cell layers. The plasma membrane of such an epithelial cell is roughly divided into two regions. One is the basal lateral region where transported molecules are, in many cases, passively secreted. The other is the brush border membrane. The name, brush border, derives from the microvilli that decorate that surface of the cell. These numerous projections of the plasma membrane greatly increase the surface area, which facilitates the active transport carried out by the brush border membrane.

One other specialized membrane function is provided by the Schwann cell. During development the Schwann cell wraps itself around a nerve axon in a concentric and multitudinous fashion. It creates a rather simple membrane containing mostly lipid and a couple of proteins. It insulates the nerve electrically. When the insulation (myelin sheath) breaks down, the nerve axon is no longer properly insulated. Nerve conduction thus becomes faulty in demyelinating diseases.

Large quantities of carbohydrate are often found on the surface of the plasma membrane. The carbohydrate structure attached to glycoproteins and glycolipids in the plasma membrane plus material that is secreted makes a coat on the cell surface—the glycocalyx. This coat offers protection to the cell, and a means by which one cell can adhere to another.

Extending out of the plasma membranes of some cells, particularly unicellular organisms, are flagella pellicles. Flagella are long projections of the plasma membrane with a number of cytoskeletal components inside. These flagella are involved in locomotion of the microorganism. These projections constitute another specialized piece of plasma membrane.

This completes an overview of the functions of the plasma membrane of cells. The plasma membrane controls the passage of materials into and out of the cell through its transport functions. The plasma membrane is a focal point for regulation of cellular activity. The plasma membrane provides the means for cell–cell communication, and for cellular organization during development.

C. Intracellular Membranes

Organelles within eukaryotes are bound by either single or double membranes. These internal membranes not only enclose specialized regions within the cell, but also are involved in cellular processes such as biosynthesis, transport, recycling, energy metabolism, and degradation.

1. NUCLEAR MEMBRANE

The nuclear membrane is actually a double membrane. Thus one should properly refer to the inner and outer nuclear membranes. It would be anticipated that they would be functionally distinct and thus be different in composition. Unfortunately, almost no compositional information is available about the inner and outer membrane. Thin-section electron micrographs suggest that the nuclear membrane has patches or pores. However, these are not likely to be holes in the nuclear membrane, since the nucleus is not freely permeable. An interesting speculation is that the "pores" may be processing centers for mRNA. Chromosomal material may have some attachment to the nuclear membrane. Thus one can speculate on a direct role of the inner nuclear membrane in gene regulation. More likely is an indirect role, in which the nuclear membranes function somewhat analogously to the plasma membrane. The mechanism of such a role might involve receptors for hormones on the nuclear membrane, which when occupied, stimulate an intranuclear "second messenger" process, analogous to some plasma membrane receptors.

2. ENDOPLASMIC RETICULUM

The endoplasmic reticulum (ER) is an internal membrane system of the eukaryotic cell. The ER membrane system encloses a specialized region known as the lumen. Because the lumen of the ER is separate from the cytosol of the cell, the ER allows for separation of function and separation of materials. This membrane is the site of biosynthesis of many components. In particular many membrane components are synthesized on the ER, such as membrane proteins and lipids. The ER is subdivided into rough and smooth ER. The designation arises from the morphology observed in the electron microscope. Some parts of the ER have bumps, giving a rough appearance. The bumps are ribosomes that are protein manufacturing complexes bound to the ER.

The smooth ER represents regions where the ribosomes are not attached. These two regions of the ER apparently come in patches, thus enabling their distinction in the electron microscope.

The lumen of the ER contains proteins to be secreted. The lumen of the ER is related morphologically to the exterior of the cell. The lumenal contents of the ER can be transported to the Golgi. From the Golgi, the secreted proteins advance to the plasma membrane via vesicular transport. These membrane vesicles fuse with the plasma membrane and the inside of the vesicle becomes continuous with the outside of the cell. The proteins to be secreted are thus transferred to the exterior of the cell.

3. GOLGI

Golgi is a closed membrane system like the ER. Communication between the ER and the Golgi membranes appears to be via a vesicular pathway. Therefore the lumen of the Golgi is functionally connected with the lumen of the ER. The Golgi is also in communication with the plasma membrane of the cell, again by vesicular transport. The Golgi is involved in glycosylation of glycoproteins and glycolipids, and in acylation of membrane proteins.

Movement of material is a directional process in which the Golgi itself can be functionally subdivided. For example, there appears to be a progressive glycosylation as proteins move through the Golgi. The Golgi has been divided into cis, medial, and trans aspects to emphasize this compartmentalization. Communication among the Golgi aspects occurs through vesicular transport.

4. INTRACELLULAR TRANSPORT VESICLES

Figure 1.2 schematically outlines some of the roles of intracellular vesicles in the movement of movement of material and ''information'' through

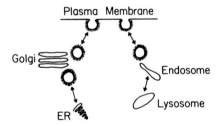

Fig. 1.2. Schematic diagram of some of the flow of membrane material within a eukaryotic cell involving vesicular transport. Vesicles provide communication between the Golgi and the ER and the ER and the plasma membrane. Coated vesicles are involved in the process of receptor-mediated endocytosis, which transports extracellular material through an endosome intermediary to lysosomes for catabolism.

the cell. These vesicles provide a membrane-bound compartment to transport membrane components, as well as trapped soluble components, from one organelle to another, or to and from the plasma membrane. As illustrated, some vesicles that form from the plasma membrane are coated with protein. This protein, which is called clathrin, is capable of making hollow baskets by itself without the membrane. The clathrin-coated vesicle eventually pinches off from the plasma membrane. This coated vesicle then can move to the Golgi or it can move to the lysosome, depending on function.

One of the challenges in the study of cell membranes is to understand the mechanisms of movement of material throughout the cell, considering that the target sites for this movement are all morphologically, and functionally, different. What governs this movement of material? If one considers the protein composition of the various intracellular membranes, one finds, for example, that there is little in common between the protein composition of the plasma membrane and the protein composition of the lysosome. Furthermore, there is little in common between the protein composition of the plasma membrane and the protein composition of the ER. How these differences in membrane composition are maintained in the face of interorganelle communication is a fascinating, unsolved mystery.

5. MITOCHONDRIAL MEMBRANES

Another organelle in the cell is the mitochondria, which has two membranes, the outer and the inner mitochondrial membranes. The inner mitochondrial membrane significantly differs from other membranes because of its high protein content. For example, the inner mitochondrial membrane has much less lipid than the plasma membrane and the ER membrane. It also contains a lipid, cardiolipin, that is not found in other membranes of most eukaryotes.

ATP synthesis is perhaps the most important function of the mitochondria. The pathway of ATP synthesis involves an intricate set of enzymes located in the inner mitochondrial membrane. This pathway is a good example of the motif that some cellular functions are uniquely dependent on enzyme complexes in membranes. The inner membrane is a highly convoluted membrane with lots of surface area and has many enzymes of the oxidative phosphorylation pathway incorporated in it in organized complexes. Oxidative phosphorylation is a series of biochemical steps in which electrons are transported and ATP is synthesized, while oxygen is reduced to water; this process involves a number of proteins. The complexes are organized to facilitate this controlled oxidative process.

Not only do all of these steps take place in and around the inner mitochondrial membrane, but oxidative phosphorylation would not take place without a membrane and its structural features. The process leading to ATP synthesis creates, through a mechanism involving the movement of electrons, a proton gradient across the inner mitochondrial membrane. Embedded in that membrane is an enzyme that is capable of ATP synthesis from ADP and inorganic phosphate. The enzyme obtains the energy for ATP synthesis from the proton gradient. In the process protons get pumped across the inner mitochondrial membrane and then flow back through this membrane via the protein responsible for ATP synthesis. It is that flow of protons through that protein which produces ATP. Such a pH gradient cannot exist without a membrane that is otherwise impermeable to protons. Therefore the ATP synthesis system functions only because it is a membrane system.

6. LYSOSOME

The lysosome is bounded by a single membrane. The lysosome is a degradative organelle. In fact in the lysosome, where the pH is lower than in the rest of the cell, there are enzymes responsible for degradation; these include enzymes for cleaving proteins, phospholipids, cholesterol esters, and carbohydrates into their respective subunits. Through intracellular vesicles, the lysosome is in communication with other organelles and the plasma membrane.

D. Viruses

Enveloped viruses constitute an important class of membranes separate from those of eukaryotic and prokaryotic cells. Viruses contain genetic material. Some contain RNA and some contain DNA. Viruses differ in the way that genetic material is packaged. It is usually packaged in connection with protein–nucleic acid interactions, which creates a complex of protein and RNA or protein and DNA. This complex is coiled and a tight structure is created that disassembles inside the cell upon infection and takes over the metabolic machinery of the cell. Some viruses consist only of protein and nucleic acid. Other viruses have an envelope, or a membrane, surrounding the protein–nucleic acid complex (nucleocapsid).

An example of the latter is the paramyxovirus Sendai (see Fig. 1.3). It contains a negative-strand RNA in its nucleocapsid. Sendai virus is roughly spherical, with spikes protruding from its enveloping membrane (envelope). These spikes are made of the surface glycoproteins of the virion, F and HN, which are the predominant proteins of the viral membrane. The F protein facilitates the fusion event required for viral entry. The stimulation of membrane fusion by F is exploited to promote artificially induced cell fusion. The other surface glycoprotein protein, HN,

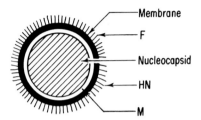

Membrane

F

Nucleocapsid

HN

M

Fig. 1.3. Schematic diagram of Sendai virus. The envelope membrane is represented by the thick black line.

has both hemagglutinin and neuraminadase activity. The hemagglutinating activity causes Sendai virus to stick to cell surfaces. The ability of Sendai virus to agglutinate red cells provides an assay for the activity of the HN protein. The lipids making the lipid bilayer of the Sendai envelope are derived from the plasma membrane of the infected cell.

III. THE HYDROPHOBIC EFFECT

Water is possibly the most important constituent contributing to the structure of biological membranes. However, water does not exist to any great extent within the biological membrane. Its contribution to the structure of membranes is described in the concept of the hydrophobic effect.[1] The hydrophobic effect describes how an aqueous medium deals with nonpolar moieties.

A good place to start this discussion is by describing what does not contribute to the hydrophobic effect. The concept of a hydrophobic bond is an idea that has been mistakenly confused with the hydrophobic effect. The hydrophobic bond implies that nonpolar species have an attraction for each other that excludes water from their environment. Consider the two interactions involved: hydrocarbon–hydrocarbon and hydrocarbon–water. One might believe that the hydrocarbon–hydrocarbon interaction would have the greater force of attraction. However, the forces of attraction have been measured and found to be similar. For this reason the term hydrophobic bond is incorrect. In contrast, water–water interactions do have a greater force compared to the other two interactions, due to hydrogen bonding between water molecules.

A. Water Structure

An analytical description of the structure of liquid water presents a largely intractable problem. However, a good place to begin is with the chemical structure of the water molecule. First consider the O–H bonds of water. They are polar bonds. They can readily form hydrogen bonds with appropriate species, such as with water itself.

These hydrogen bonds can be understood in terms of the polarity of the O–H bond. Because of the electronegativity difference between hydrogen and oxygen, the bonding electrons are asymmetrically distributed. They tend to associate more closely with the oxygen nucleus than with the hydrogen nucleus. From the point of view of the O–H bond, therefore, the oxygen end is partially negatively charged and the hydrogen is partially positively charged. From this an attractive interaction between the partial negative and partial positive charges can readily be envisioned. Therefore the H of one O–H bond is attracted to the O involved in another O–H bond on another water molecule. This attraction results in the proton of the hydrogen being shared between two oxygens of two water molecules. This interaction is termed the hydrogen bond. The water–water attraction is therefore much more energetically favorable because of this hydrogen bonding. The strength of these bonds can be seen in the structures of crystalline water or ice. See Fig. 1.4.

In liquid water a hydrogen bonding network likely extends throughout the liquid. When foreign species come in contact with this liquid, the network must react. Consider polar species that are either electrically neutral or carry charges. An example of a neutral species is ethanol. An example of a charged species is an amino acid. When ethanol comes in contact with water, it can accept a hydrogen bond from water, or donate a hydrogen bond to water. In fact, it does both. Therefore the interaction of ethanol with water is a relatively favorable interaction. This is because ethanol can accommodate itself with the hydrogen bonding network of

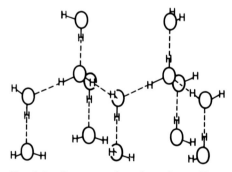

Fig. 1.4.   Representation of one form of ice.

water, without seriously disrupting that network. Thus ethanol and water are readily soluble in each other.

Charged species such as amino acids are capable of similar interactions with the water structure. The amino acid has a full charge on its polar groups near neutral pH. Therefore the COO^-, for example, is capable of accepting hydrogen bonds from water and the NH_3^+ is in a position to donate hydrogen bonds to water. Consequently the amino acid also fits in the water structure without serious disruption of that structure.

B. Nonpolar Hydrocarbon and Water

Now consider a pure hydrocarbon, for example, hexane. No groups on hexane can accept or donate hydrogen bonds to water. Hexane cannot insert itself into the water structure without disrupting that structure. Therefore a phase separation into droplets of pure hexane occurs when substantial amounts of hexane are present. However, a very small amount of hexane can be incorporated in water, and a single phase maintained. Introducing the hexane into the water structure requires a displacement of water to accommodate the hexane molecule. The hydrogen bonding network of the water must now be rebuilt to accommodate the hexane. The water network can be restructured by building a cage of water molecules around the hydrocarbon molecule. The construction of this cage causes a loss in the total number of ways in which hydrogen bonds can be built into the water structure. In a pure water phase there is a random distribution of water molecules, and a randomization of the hydrogen bonding network. When cages are built in response to the introduction of hexane, the organization of the hydrogen bond network is more ordered in the space surrounding the hexane molecule. Hydrogen bonds cannot form in the volume occupied by the hexane molecule. Therefore, the hydrogen bonding network cannot be random. Some water structures are prohibited by the presence of the hexane.

Entropy (S) is a measure of the randomness or the order of the system. The most significant fact characterizing the encounter of hydrocarbons and water is a decrease in entropy of the system, or an increase in ordering of the system. Therefore, ΔS is negative, and that makes a large positive contribution to ΔG, the change in free energy of the system,

$$\Delta G = \Delta H - T\Delta S, \qquad (1.1)$$

where T is the absolute temperature and ΔH is the enthalpy change. This corresponds to an energetically unfavorable process. The reason oil and water do not mix is then due primarily to the entropic parameter. This is the hydrophobic effect.

There is available some quantitative information on the unfavorable free energy cost of putting a hydrocarbon in water. This information is obtained from thermodynamics and some experimental observations. The thermodynamics comes about because one can write the chemical potential for the species, in this case a hydrocarbon, in water. The way that Tanford[1] chooses to do it is

$$\mu_w = \mu^{\circ}_w + RT (\ln X_w) + RT(\ln f_w), \qquad (1.2)$$

where X_w is the mole fraction of the hydrocarbon and f_w is the activity coefficient. μ°_w refers to a state of infinite dilution in water and μ_w is the chemical potential of the species in question in water. T is the temperature and R is the gas law constant. Correspondingly, for a hydrocarbon in hydrocarbon,

$$\mu_{HC} = \mu^{\circ}_{HC} + RT(\ln X_{HC}) + RT(\ln f_{HC}), \qquad (1.3)$$

where X_{HC} is the mole fraction of the hydrocarbon solute in the hydrocarbon solvent and f_{HC} is the corresponding activity coefficient. From Eqs. (1.2) and (1.3), one can derive the chemical potential difference between the unitary chemical potential of the hydrocarbon in hydrocarbon and the hydrocarbon in water.

Quantitative information was obtained by measuring the partitioning of a solute between an aqueous phase and a hydrocarbon phase. In so doing, a linear relationship was discovered between the free energy change involved in moving a hydrocarbon from water to hydrocarbon and the length of the hydrocarbon. This relationship holds as long as the hydrocarbons belong to a similar chemical series.

For example, consider the case of alcohols, such as propanol and butanol. The difference in the chemical potential is described in Fig. 1.5 and in the equation

$$\mu^{\circ}_{HC} - \mu^{0}_w = -2436 - 884n_C. \qquad (1.4)$$

This is a linear equation, with the variable, n_C, representing the number of carbon atoms in the alcohol. Therefore the term with the variable becomes more pronounced as the length of the hydrocarbon portion of the alcohol increases. Quantitatively, this corresponds to 884 calories per each additional methylene group. Equation (1.4) therefore says that to introduce butanol into water is 884 calories less favorable than putting propanol into water. Correspondingly, to remove butane from the aqueous phase and return it to the hydrocarbon phase is 884 calories more favorable than to remove propane from water in a similar fashion. As each methylene

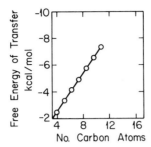

Fig. 1.5. Free energy transfer of aliphatic alcohols from aqueous solution to pure liquid alcohol. From C. Tanford, *The Hydrophobic Effect* (New York: Wiley, 1980), by permission of the publisher.

group is added, within a given series representing a single type of hydrocarbon, there is a uniform increment of free energy for each additional methylene group. Therefore as the alcohol hydrocarbon increases in length, it becomes less favorable to put that alcohol into water. See Fig. 1.5.

The same phenomenon is observed for alkanes. In fact, similar equations have been derived for dienes and carboxylic acids. Interestingly, the increment for each additional methylene group is similar for all of these species. The intercept is different, however, since the value of the intercept refers to the substituents in the structure that make one chemical series different from another chemical series. Thus an alcohol enters water more favorably than does an alkane, due to the polar hydroxyl on the alcohol, which can participate in the hydrogen bond network of the water. However, as the number of methylene groups is increased, the alcohol becomes more like a nonpolar hydrocarbon and consequently less soluble in water. These considerations provide a more quantitative understanding of the hydrophobic effect.

These results suggest an interesting hypothesis for us to consider: the hydrophobicity of a molecule is related to the hydrophobic surface area of that molecule. In other words, the larger the surface area of the hydrophobic molecule, the larger the water cage that must be built around that molecule, and the larger the unfavorable entropy contribution to the transfer of that molecule into an aqueous phase. Therefore the extent of the hydrophobic surface of the molecule determines the extent of the hydrophobic effect.

In the case of some very long molecules, one observes the following interesting phenomenon. The linear relationship between the number of methylene units and the free energy of transfer from hydrocarbon to water fails. In that case it is likely that the molecule folds on itself, thereby reducing the hydrophobic surface area that must be accommodated in the water phase. Alternatively, the molecule may be aggregated, thus reducing the hydrophobic surface area per molecule in contact with water (since some of the molecular surface is in contact with like molecules in the aggregate). For example, the partitioning behavior of cholesterol (between polar and hydrophobic phases) does not match well with the total molecular surface area. The solubility of cholesterol in water is higher than one would predict by the simple argument just described. This may result because cholesterol can form a dimer that would effectively reduce the hydrophobic surface area per molecule that must be accommodated in the aqueous phase.

Unsaturation can also have an effect on the relative hydrophobicity. Alkanes are more hydrophobic than alkenes of the same length and alkenes (one double bond) are more hydrophobic than dienes (two double bonds) of the same length. Thus, the more double bonds in the molecule, the less hydrophobic it becomes. These trends are useful in understanding phospholipid behavior.

As stated at the beginning of this discussion, the hydrophobic effect is perhaps the most important effect in determining the structure of membranes. The hydrophobic effect causes phospholipids to aggregate into the fundamental structural element of the biological membrane, the phospholipid bilayer. It is also because of the hydrophobic effect that membrane proteins are inserted into membranes. Through those proteins many of the functions of the biological membrane are expressed. Furthermore the structure of the membrane proteins themselves is also largely governed by the hydrophobic effect. All of these phenomena are due to the special hydrogen-bonding properties of water, which results in the entropy-driven hydrophobic effect.

IV. SUMMARY

Prokaryotic cells have a plasma membrane, but no intracellular membranes. They may also have a cell wall and an outer membrane. Eukaryotic cells, in addition to the plasma membrane, have a large variety of intracellular membranes that make up the organelles of the cell. These organelles increase the functional capability of the eukaryotic cell. In the case of plant cells, the eukaryotic cell may also have a cell wall. Some viruses are enclosed by a membrane. This membrane contains proteins essential to viral activity.

Biological membranes, because of their unique structure, impart at least six general properties to cells: the ability (1) to create and exploit transmembrane chemical and electrical gradients; (2) to organize enzymes for coordinated metabolic function; (3) to regulate enzyme activity; (4) to provide substrate for metabolism; (5) to provide signal transduction; and (6) to provide compartmentalization of cellular function.

The hydrophobic effect is one of the most important concepts necessary to an understanding of membrane structure. It is not an attractive force. Rather, it represents the relative inability of water to accommodate nonpolar species. This effect is due to the special hydrogen-bonding capabilities of water and the consequent tendency of water to form networks as part of the water structure. The hydrophobic effect arises predominantly from entropy considerations. The hydrophobic effects drive the structure of biology, including proteins and membranes, through a forced sequestration of hydrophobic chemical structures away from the aqueous environment.

REFERENCE

1. Tanford, C., *The Hydrophobic Effect* (New York: Wiley, 1980).

2

The Lipids of Cell Membranes

The most fundamental structure of the membranes of cells, the lipid bilayer, is formed from membrane lipids.[1] The membrane lipids generally have an amphipathic structure. One end of the lipid molecule has a polar group, such as a phosphate, an amine, and an alcohol. The remainder of the molecule is usually hydrophobic with extensive hydrocarbon character. These lipids are organized in an aqueous environment according to the hydrophobic effect, which often leads to the formation of the lipid bilayer. The formation of lipid bilayers will be discussed in Chapter 4. Here the chemical structures of the lipids will be examined, since it is the chemical structure that most fundamentally determines lipid behavior.

I. PHOSPHOLIPIDS

A. Chemical Structure

A good place to begin the story of the lipids is by looking at the best known and understood of the membrane lipids, the phospholipids.[2] Phospholipids derive their name from the phosphate group found in the polar headgroups of these lipids. The general structure is shown in Fig. 2.1. Two nonpolar hydrocarbon chains are esterified to a glycerol which in turn is esterified to the phosphate in the 3′ position. The phosphate, in its turn, is esterified to an alcohol. The name of the alcohol lends a further distinction to the nomenclature of the phospholipid. In the absence of any alcohol, the phospholipid is called phosphatidic acid. Phosphatidate, which refers to its normal, ionized form, is not found in large quantities in cell membranes, as it is largely a biosynthetic intermediate. This phospholipid will be known in this book by the abbreviation PA.

If the alcohol esterified to the phosphate is choline, the phospholipid is called phosphatidylcholine (or lecithin in an older nomenclature). Phos-

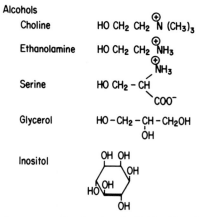

Fig. 2.1. Generic structure for the phospholipids. The alcohols listed in the lower left can be substituted at the position marked alcohol in the upper right structure to form the various classes of phospholipids.

phatidylcholine (often abbreviated as PC) is one of the most common of cell membrane phospholipids.

   If the alcohol esterified to the phosphate is ethanolamine, the resulting phospholipid is called phosphatidylethanolamine (or cephalin in an older nomenclature). Phosphatidylethanolamine (commonly abbreviated as PE) is another common cell membrane lipid. Note that PE contains a free amino group. This amino group can be stripped of a proton at high pH (9–10) to give an uncharged, primary amine.

   Other alcohols may be esterified to the phosphate. These include the L-amino acid serine, and sugars such as glycerol, and inositol. The phospholipids are then named, following the example of phosphatidylcholine, according to the corresponding alcohol. This nomenclature produces the names phosphatidylserine, phosphatidylglycerol, and phosphatidylinositol, respectively (commonly abbreviated as PS, PG, and PI, respectively).

The phospholipids so far listed include most of the common phospholipids found in cell membranes. All of them contain only one phosphate. Diphosphatidylglycerol (cardiolipin in an older nomenclature), which is found in mitochondrial membranes, is unique among common phospholipids in that it contains two diester phosphates linked by a glycerol (often abbreviated as DPG). The structures of some of these phospholipids appear in Fig. 2.2.

B. Electrical Charge

The chemical structure of the polar headgroup of these phospholipids determines what charge the phospholipid as a whole may carry. Phosphatidylcholine at physiological pH values carries a full negative charge on the phosphate and a full positive charge on the quaternary ammonium. It thus is zwitterionic, but electrically neutral. Phosphatidylethanolamine is constructed similarly. It carries a positive charge on the amine (which can be deprotonated as described above) as well as the negative charge on the phosphate.

Phosphatidylserine contains, in addition to the negatively charged phosphate, a positively charged amino group and a negatively charged carboxyl. Therefore this lipid exhibits an overall negative charge at neutral pH. The amino group of PS can also be deprotonated at high pH, whereas its carboxyl can be protonated at low pH.

The group of negatively charged lipids include phosphatidylglycerol and phosphatidylinositol. These phospholipids carry a net negative charge, because the sugar carries no positive charge to balance the negative charge of the phosphate. Diphosphatidylglycerol normally carries two negative charges, because of its two phosphates.

Interestingly, phosphatidylinositol sometimes appears in a phosphorylated form with extra phosphates esterified to hydroxyls on the inositol. It then is referred to as di- or triphosphatidylinositol (PIP and PIP_2, respectively), depending on the number of phosphates. Consequently phosphorylated PIs carry additional negative charge. The phosphate monoesters are often esterified to positions 4 and 5 of the inositol (although other combinations are possible), whereas the diester occurs at position 1 of the inositol.

The preceding discussion explains why some phospholipids carry net charges. Those charges are held at the surface by the organization of the membrane lipid bilayer. Phospholipids can therefore be important in determining the surface charge of the membrane.

As suggested by the variety of phospholipids just described, the phospholipid composition of cellular membranes can be complex. Individual

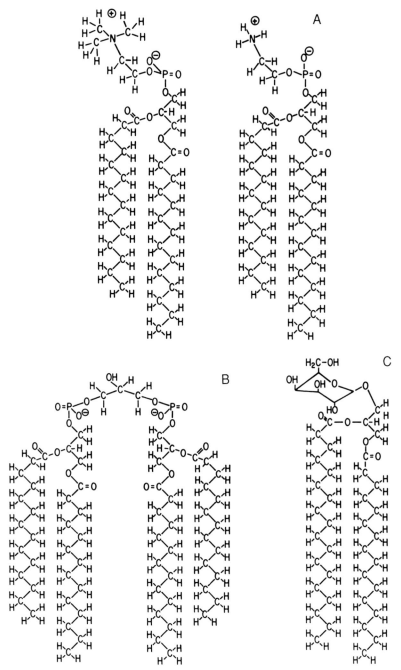

Fig. 2.2. (A) Chemical structure of phosphatidycholine (left) and phosphatidyethano-lamine (right). (B) Chemical structure of diphosphatidyglycerol (cardiolipin). (C) Chemical structure of phosphatidylinositol.

membranes often exhibit distinct phospholipid composition with respect to headgroup classes. For example, the phospholipid composition of several biological membranes are compared in Table 2.1. Such distinctions in membrane phospholipid composition may be important to membrane function. Some examples will appear in Chapter 7.

C. Fatty Acid Composition

The division of phospholipids into classes according to their headgroup structure is only one level of complexity for biological membrane lipid composition. Even greater complexity is observed in the hydrocarbon chain composition within each phospholipid class. In one common form of phospholipids the hydrocarbon chains are esterified fatty acids. Both saturated and unsaturated fatty acids are found esterified to the glycerol. The length of the commonly found fatty acids varies from as few as 12 carbons to as many as 26 carbons. The number of double bonds per fatty acid commonly ranges from none to as many as six. It is interesting to note that in the case of fatty acids with multiple double bonds, the double bonds are almost never conjugated. (In a rare case from a plant source, a multiply conjugated fatty acid has been isolated, called parinaric acid. Parinaric acid is fluorescent and is commonly used as a fluorescent probe for studies of lipids in membranes.)

Common names are given to many of the fatty acids and these common names are frequently encountered in the literature. Examples include myristic acid, with 14 carbon atoms and no double bonds, and oleic acid, with 18 carbon atoms and one double bond. In addition, the fatty acids are often referred to by a numbering scheme. In this scheme, the number before the colon describes the length of the fatty acid in carbon atoms, and the number after the colon refers to the number of carbon–carbon double bonds. Thus 16:0 refers to palmitic acid, a 16-carbon atom saturated fatty acid, and 18:1 refers to oleic acid, an 18-carbon atom unsaturated fatty acid containing one double bond. Fatty acids are numbered from the carboxyl terminal, with the carboxyl carbon labeled as number 1. The position of the double bond is indicated by the Δ nomenclature. As an example, Δ_9 refers to a double bond between carbons 9 and 10 of the hydrocarbon chain. Thus, 18:1, Δ_9 refers to oleic acid. Another nomenclature to indicate the position of the double bond is the $(n-)$ format of the IUPAC-IUB. Here the position of the first double bond from the terminal methyl is indicated. Thus 18:1 $(n$-9) designates oleic acid. A list of some of the common fatty acids and their common names appears in Table 2.2.

TABLE 2.1
Lipid Composition of Membrane Preparations

Source		Lipid composition (mole percentage)							
	Cholesterol	PC	SM	PE	PI	PS	PG	DPG	PA
Human erythrocyte	45	17	17	16	—	6			
Bovine rod outer segment disk	10	36		40	2	12			5
Escherichia coli	0	—		80			15		
Bacillus subtilis	0	0		30			12		
Sindbis virus	0	26	18	35		20			
Rabbit sarcoplasmic reticulum	10	63		17		10			
Rat liver									
Endoplasmic reticulum (rough)	6	55	3	16	8	3	—	—	—
Endoplasmic reticulum (smooth)	10	55	12	21	6.7			1.9	
Mitochondria (inner)	<3	45	2.5	25	6	1	2	18	0.7
Mitochondria (outer)	<5	50	5	23	13	2	2.5	3.5	1.3

TABLE 2.2
Some Naturally Occurring Fatty Acids

Symbol	Structure	Systematic name	Common name	m.p. (°C)
	A. Saturated fatty acids			
12:0	$CH_3(CH_2)_{10}COOH$	n-Dodecanoic	Lauric	44.2
14:0	$CH_3(CH_2)_{12}COOH$	n-Tetradecanoic	Myristic	53.9
16:0	$CH_3(CH_2)_{14}COOH$	n-Hexadecanoic	Palmitic	63.1
18:0	$CH_3(CH_2)_{16}COOH$	n-Octadecanoic	Stearic	69.6
20:0	$CH_3(CH_2)_{18}COOH$	n-Eicosanoic	Arachidic	76.5
24:0	$CH_3(CH_2)_{22}COOH$	n-Tetracosanoic	Lignoceric	86.0
	B. Unsaturated fatty acids			
16:1	$CH_3(CH_2)_5CH=CH(CH_2)_7COOH$		Palmitoleic	-0.5
18:1	$CH_3(CH_2)_7CH=CH(CH_2)_7COOH$		Oleic	13.4
18:2	$CH_3(CH_2)_4CH=CHCH_2CH=CH(CH_2)_7COOH$		Linoleic	-5
18:3	$CH_3CH_2CH=CHCH_2CH=CHCH_2CH=CH(CH_2)_7COOH$		Linolenic	-11
20:4	$CH_3(CH_2)_4(CH=CHCH_2)_3CH=CH(CH_2)_3COOH$		Arachidonic	-49.5

TABLE 2.3
**Fatty Acids Found in Egg Phosphatidylcholine and Human
Erythrocyte Phosphatidylethanolamine**

A. Fatty Acid Composition of Phospholipids[a]

Fatty acid	Mol% in egg PC	Mol% in erythrocyte PE
16:0	33	19
16:1	2	—
18:0	15	13
18:1	32	22
18:2	17.8	7
20:4	4.3	19
22:4	—	5
22:6	1.7	4

B. Fatty Acid Combination Found in Egg PC[b]

R₁	R₂	Mol%
16:0	18:1	45
16:0	18:2	31
18:0	18:2	12
18:0	18:1	10
18:0	20:4	8

[a] From P. R. Cullis and B. De Kruijff, *Biochim Biophys.
Acta* **513** (1978): 31–42.
[b] From N. A. Porter, R. A. Wolf, and J. R. Nixon, *Lipids*
14 (1979): 20–24.

The distribution of fatty acids in membrane phospholipids is peculiar
to the class of phospholipid and the membrane type. As an example,
Table 2.3 lists the fatty acids found in egg phosphatidylcholine and human
erythrocyte phosphatidylethanolamine.[3] The mixtures are complex. Although the importance of a particular blend of fatty acids to membrane
function is not fully understood, maintenance of a suitable membrane
environment by the fatty acid composition of the membrane phospholipids
does appear to be important. Some specific ideas on this subject will be
discussed in Chapters 4 and 7. Furthermore, arachidonic acid (20:4) is
important metabolically as a precursor to prostaglandins.

Although a distribution of fatty acids is apparent in the examples given,
these fatty acids are not randomly attached to positions 1′ and 2′ of the
glycerol. Unsaturated fatty acids are esterified predominantly at position
2′ and the saturated fatty acids occur mostly at position 1′ (Table 2.3).[4]
This distribution of fatty acids arises from the specificity of enzymes
involved in the biosynthesis of the phospholipids.

Two isomers are possible for the double bonds found in the fatty acids. The isomers are referred to as cis and trans. As can be seen in Fig. 2.3, the two isomers can produce quite different looking molecules, especially if the double bond is in the middle of the fatty acid. The cis isomer introduces a kink or bend in the molecule. This is a permanent structural feature of the fatty acid. On the other hand, the trans double bond does not introduce a strong structural perturbation into the fatty acid. As will be seen later, fatty acids with a trans double bond behave similarly to saturated fatty acids of the same length. The isomer most commonly found in biological membranes is the cis isomer. As a result, fatty acids with double bonds tend to perturb the membrane structure more than their saturated cousins because the double bonds necessarily force a kink to form in the hydrocarbon chain. In contrast, saturated fatty acids are capable of a straight-chain all-trans configuration about the carbon–carbon bonds. This is one of the ways in which the fatty acid content of the membrane can affect membrane structure and properties.

Although many positional isomers exist for fatty acids containing carbon–carbon double bonds, there are a relatively few isomers that are most common. When a hydrocarbon chain contains one double bond, it usually occurs, in mammalian cell membranes, between carbons 9 and 10. If there are two double bonds, they usually occur (in the longer fatty acids) between positions 9 and 10 and positions 12 and 13. Thus linoleic acid (18:2) is *cis*-octadeca-9,12-dienoic acid. Oleic acid (18:1) is *cis*-octadeca-9-enoic acid. When more than two double bonds are included in a fatty acid, this pattern does not necessarily hold. For example, arachidonic acid (20:4) is *cis*-eicosa-5,8,11,14-tetraenoic acid. The positions

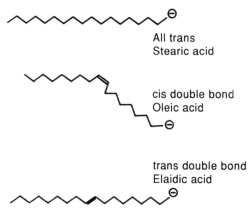

All trans
Stearic acid

cis double bond
Oleic acid

trans double bond
Elaidic acid

Fig. 2.3.   Comparison of the cis and trans carbon–carbon double bond in a fatty acid.

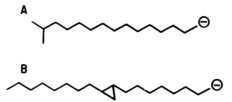

Fig. 2.4. Examples of branched-chain fatty acids. (A) Isopalmitate, (B) *cis*-11,12-methyleneoctadecanoic acid.

of these double bonds are determined during their biosynthesis by the desaturating and elongating enzyme systems.

Not all the phospholipids contain hydrocarbon chains with the above structure. Some branched-chain fatty acids are found such as isopalmitate and isomyristate. Another variation is the inclusion of a cyclopropane ring in the fatty acid. An example is seen in *cis*-11,12-methyleneoctadecanoic acid, shown in Fig. 2.4. Fatty acids containing cyclopropane can be found in bacteria like *Escherichia coli* when in an arrested state of their cell cycle.

D. Structurally Related Phosphorus-Containing Lipids

A further variation in lipid structure is found in a close relative of the phospholipid, the plasmalogen. The structure is quite analogous to phospholipids, with the exception occurring at position 2' of the glycerol. The hydrocarbon chain is attached through an unsaturated ether linkage. An example with ethanolamine as a headgroup appears in Fig. 2.5. Plasmalogens can constitute a major class of polar lipids in some membranes, including those of heart and brain in some species. Ethanolamine and choline headgroups are found on plasmalogens.

Platelet-activating factor (PAF) is a particularly potent biological lipid, which also is of the general structure 1-alkyl-2-acyl-glycerolphospholipid. The acyl group in this case is only an acetate. The structure of PAF is found in Fig. 2.5.

Phosphonolipids, common in protozoa such as *Tetrahymena pyriformis*,[5] exhibit a different headgroup structure. Here the bonding between the phosphate and the ethanolamine, found in PE, is replaced with a direct carbon–phosphorus bond (see Fig. 2.6). The 1-alkyl-2-acyl-glycerolphosphonolipid is commonly found, in which the alkyl group is connected to the glycerol by an ether linkage.

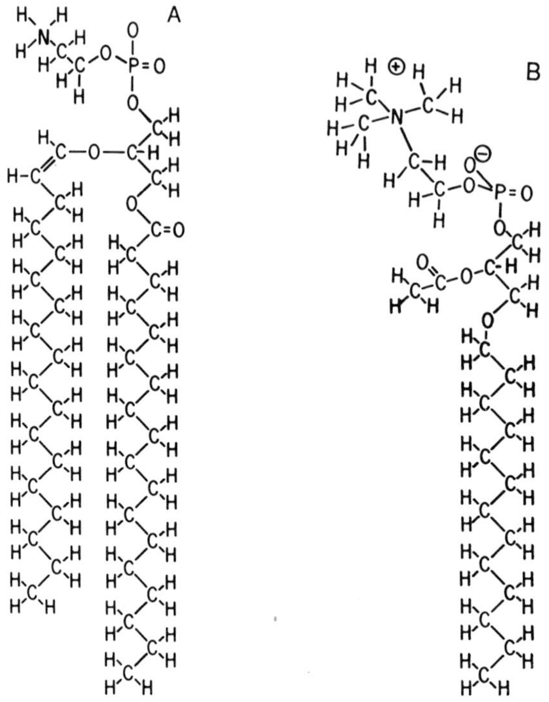

Fig. 2.5. (A) An example of plasmalogen. (B). Platelet-activating factor.

E. Characterization

The structures of the common membrane phospholipids have been elucidated through several approaches to structural analysis. A discussion of
this topic provides the opportunity to get acquainted with common methods of analysis in lipid biochemistry.[6] Because of the differences in headgroup structure, the various classes of phospholipids run differently when
analyzed by thin-layer chromatography on silicic acid. An example of
such an analysis is shown in Fig. 2.7.[7] This is a two-dimensional thin-
layer analysis, which is acidic in one direction and basic in the other.
Spots appear after visualization with 50% (v/v) sulfuric acid spray followed
by heating. (Aminophospholipids, such as phosphatidylserine and phosphatidylethanolamine, can be developed using a ninhydrin spray, before
the acid charring.) As can be seen, phospholipids with different headgroups are separated, providing a means for analyzing the phospholipid
content of a membrane.

In order to obtain the phospholipids for such an analysis, the membrane
preparation is subjected to a chloroform–methanol–water (8 : 4 : 1 (v/v))
extraction. In this two-phase system the phospholipids and neutral, hy-

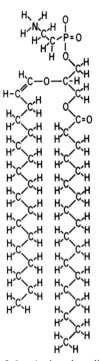

Fig. 2.6. A phosphonolipid.

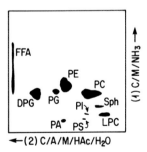

Fig. 2.7. An example of thin-layer chromatography on silica gel plates. Dimension 1 is run in chloroform/methanol/ammonia (65/25/5, v/v), and dimension 2 is run in chloroform/acetone/methanol/acetic acid/water (3/4/1/1/0.5, v/v). Reproduced with permission from G. Rouser, S. Fleisher, and A. Yamamoto, *Lipids* **5** (1970): 494.

drophobic lipids are extracted into the lower chloroform-rich phase. Very polar lipids, like some glycolipids, are not effectively extracted with this procedure and remain in the aqueous layer. Proteins, including membrane proteins, precipitate at the interface between the two phases.

As discussed above, a wide variety of fatty acids are esterified to the phospholipids. This is an important observation for the properties of cell membranes. How is information obtained on the fatty acid content of the phospholipid? To analyze the fatty acid content of a particular phospholipid from a membrane preparation requires three steps. First the phospholipid must be extracted and purified. Then the fatty acids must be hydrolyzed in KOH and methanol from the glycerol and converted to their methyl esters. This done, the fatty acid methyl esters are ready for separation by gas chromatography. Such an analysis for phosphatidylcholine 16:0, 18:1 is presented in Fig. 2.8. By using standards, the original fatty acids can be both identified and quantitated. Recently, lipid analysis has progressed further with the application of high-performance liquid chromatography (HPLC), which permits the separation of each headgroup class of phospholipids. Furthermore, a single headgroup class of the natural phospholipids may be separated into subclasses, each with different fatty acid compositions.[8]

F. Phospholipases

Phospholipids are substrate for a class of enzymes called phospholipases, which can hydrolyze the phospholipid at several different places.

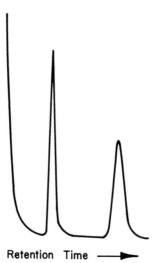

Retention   Time ⟶

Fig. 2.8.   Example of a gas chromatograph run with 16:0 fatty acid (second peak) derived by hydrolysis from 16:0, 18:1 PC.

Consequently, these enzymes can be useful tools in phospholipid structural analysis and in studies on biological membranes. Phospholipases also play roles in lipid metabolism. For example, the release of arachidonic acid for prostaglandin biosynthesis and the release of diacylglycerol for protein kinase activation result from the action of phospholipases. Furthermore, chemists have used phospholipases to assist in the synthesis of phospholipids.

Examine the activities of these enzymes more closely. Phospholipase A_2 hydrolyzes at position 2′ of the glycerol. These enzymes produce a derivative of phospholipids found in small quantities in membranes. For example, the action of phospholipase A_2 converts phosphatidylcholine to lysophosphatidylcholine, a phospholipid with only one hydrocarbon chain in position 1′ of the glycerol (see Fig. 2.9). This creates a molecule with properties considerably different from those of the original phospholipid. For example, lysophosphatidylcholine runs differently on a thin-layer silicic acid chromatography plate in some solvent systems than does the parent phosphatidylcholine. Furthermore, lysophosphatidylcholine has detergent-like properties, because of its tendency to form micelles (see

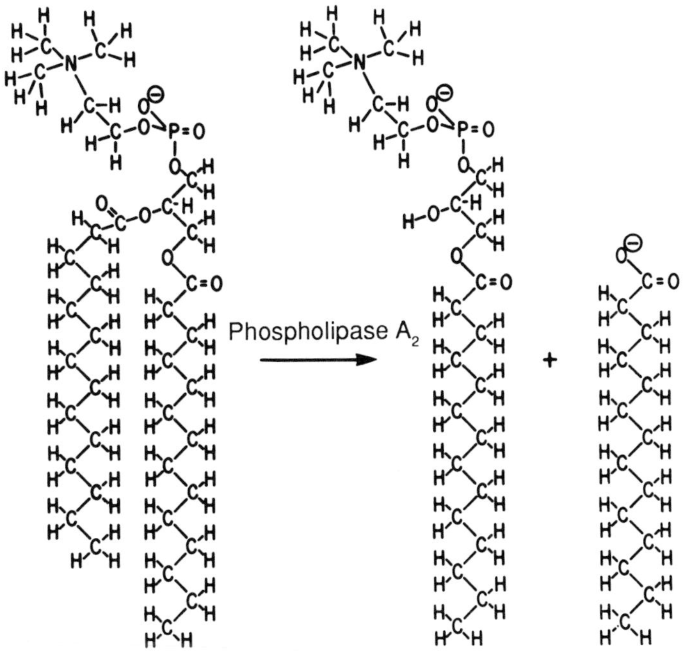

Fig. 2.9. Representation of the action of phospholipase A on phosphatidylcholine, producing lysophosphatidylcholine and a fatty acid.

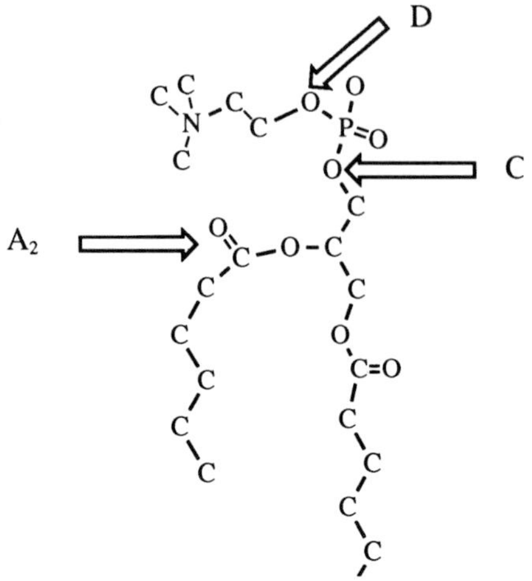

Fig. 2.10. Representation of the sites of cleavage for the common phospholipases.

Chapter 4). Receptor-stimulated phospholipase A_2 leads to the mobilization of arachidonic acid for prostaglandin synthesis.

Two other classes of phospholipase are commonly encountered. Phospholipase C cleaves the phospholipid between the phosphate and the 3' position of the glycerol, removing, for example, from phosphatidylcholine the phosphorylcholine group and diacylglycerol. In the case of phosphatidylinositol bisphosphate, phospholipase C releases inositol trisphosphate, an important second messenger involved in intracellular calcium release. The other product of phospholipase C hydrolysis, diacylglycerol, can stimulate protein kinase C activity.

Phospholipase D cleaves the phospholipid between the phosphate and the alcohol and leaves phosphatidic acid. Referring again to phosphatidylcholine, phospholipase D removes the choline alcohol (see Fig. 2.10). This reaction can play a role in phospholipid metabolism.

To act on the phospholipid substrate residing in a biological membrane, the phospholipase must gain access to the cleavage site on the lipid. This may entail modest penetration of the membrane bilayer surface and/or a limited extraction of the phospholipid substrate out of the bilayer surface. Figure 2.11 shows a representation of an interaction between phospholipase A_2 and substrate. Major molecular contact is observed between the phospholipid headgroup and the enzyme.

Figure 2.11 suggests there might be some modest specificity of certain phospholipases for particular phospholipid headgroup classes. This expectation is fulfilled for some phospholipases. Together the phospholipases offer the biochemist a variety of ways to degrade selectively phospholip-

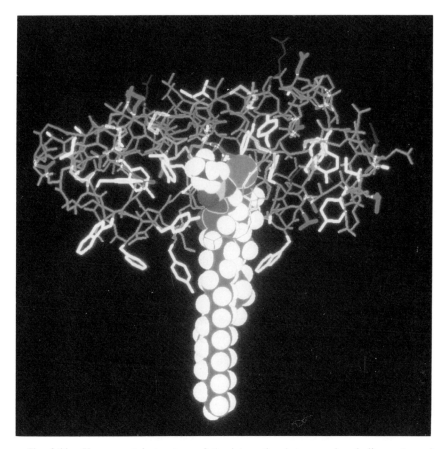

Fig. 2.11. X-ray crystal structure of the interaction between phospholipase A$_2$ and substrate. Major molecular contact is observed between the phospholipid headgroup and the enzyme. Figure courtesy of Edward A. Dennis, Brent Segelke, David Fremont, Nguyen-Huu Xuong, and Ian A. Wilson.

ids.[9] This can be useful, for instance, when it is necessary to determine the accessibility of phospholipids from the aqueous medium, or to confirm the influence of particular phospholipids on certain membrane behavior.

II. SPHINGOLIPIDS

Sphingomyelin is a sphingolipid and has a structure that is related to the structure of phospholipids. Sphingomyelin is found in some plasma membranes, including the human erythrocyte membrane where it appears in approximately equal frequency to phosphatidylcholine.

The structure of sphingomyelin appears in Fig. 2.12. This lipid contains the phosphorylcholine headgroup and thus is zwitterionic like phosphatidylcholine. It also contains two hydrocarbon chains like phosphatidylcholine. A closer look however reveals several differences from the diacylphosphatidylcholines. The backbone of sphingomyelin is sphingosine, whose structure is also shown in Fig. 2.12. The most common form of sphingosine contains 18 carbon atoms in the hydrocarbon chain. Some variation is possible in the length of the chain. The only double bond is near the polar end of the molecule and thus the hydrocarbon chain behaves in a manner similar to that of saturated fatty acid. The double bond is saturated in some sphingomyelins, creating dihydrosphingosine. Another difference from phospholipids can be seen in the free hydroxyl group on the sphingosine.

In sphingomyelin, the amino group of the sphingosine is esterified to fatty acids. These fatty acids exhibit considerable variability, as do the fatty acids in phospholipids, although the pattern of the composition is different. Multiple-unsaturated fatty acids are not commonly found in

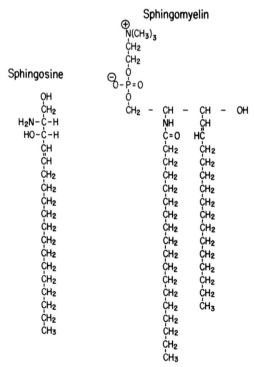

Fig. 2.12. Chemical structure of sphingosine and sphingomyelin.

TABLE 2.4
Fatty Acid Composition of Bovine
Brain Sphingomyelin[a]

Fatty acid	Mol%
16:0	3
18:0	36
20:0	1
22:0	4
23:0	4
24:0	10
25:0	5
24:1	31
25:1	4

[a] From C. F. Schmidt, Y. Barenholz, and T. E. Thompson, *Biochemistry* **16** (1977): 2649–2656.

sphingomyelin. Furthermore, the chain length is, on the average, somewhat longer than the fatty acids of egg phosphatidylcholine discussed earlier. The fatty acid composition of bovine brain sphingomyelin appears in Table 2.4.[10] (To obtain the fatty acids from sphingomyelin for analysis, acid hydrolysis must be used. The amide bond is reasonably stable to base hydrolysis.) Because of the cumulative effects of these structural differences from phosphatidylcholine, sphingomyelin runs on thin-layer chromatography in a manner different from that of phosphatidylcholine, even though both contain choline in their headgroups.

Other derivatives of sphingosine appear as lipids in membranes. Ceramides consist of a fatty acid esterified to sphingosine. Adding a sugar to the alcohol of the ceramide yields a cerebroside. Glucose and galactose are frequently found as the sugar in the cerebrosides. A further derivative of the cerebrosides adds a sulfate group to the sugar. These lipids are then called sulfatides and are commonly found in brain tissue.

Among the fatty acids of these sphingolipids are some with free hydroxyl groups α to the carboxyl. These α-hydroxy fatty acids exhibit longer retention times during gas chromatographic analysis than do their nonhydroxylated counterparts. Biosynthetically, these hydroxyls are added after the synthesis of the fatty acids.

Gangliosides are found at low levels in membranes (relative to phospholipids) but constitute an important component of the lipids of membranes. Gangliosides are derivatives of the ceramides with multiple additions of sugars, ending with N-acetylneuraminic acid (sialic acid), a sugar acid

carrying a negative charge. The roles of these gangliosides are not all
known. However, Tay-Sachs disease is known to be a lipid storage disease
involving one of the gangliosides. The binding of cholera toxin to membranes is believed to involve a ganglioside in the receptor-mediated interaction. An example of a ganglioside is given in Fig. 2.13.

III. GLYCOLIPIDS

The last classes of sphingolipids discussed can also be termed glycolipids, because their polar groups contain carbohydrate. Lipids with other
structures are also found in this class. Sugars can be attached to a glycerol
via a glycosidic linkage. If fatty acids are esterified to the other two
positions of the glycerol, glycosyldiglycerides result. The general structure
is given in Fig. 2.14. These lipids are more closely related to the phospholipid structure than the sphingolipids; a sugar replaces the phosphate and
alcohol. As with the gangliosides, a number of different derivatives can
be obtained through alterations in the structure of the polar headgroups

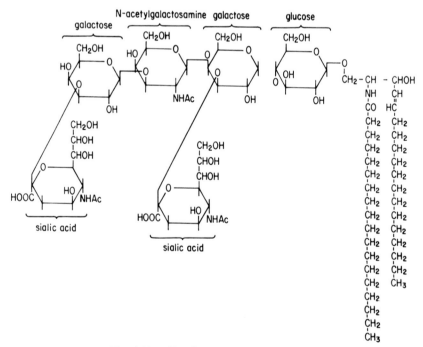

Fig. 2.13. Chemical structure of ganglioside.

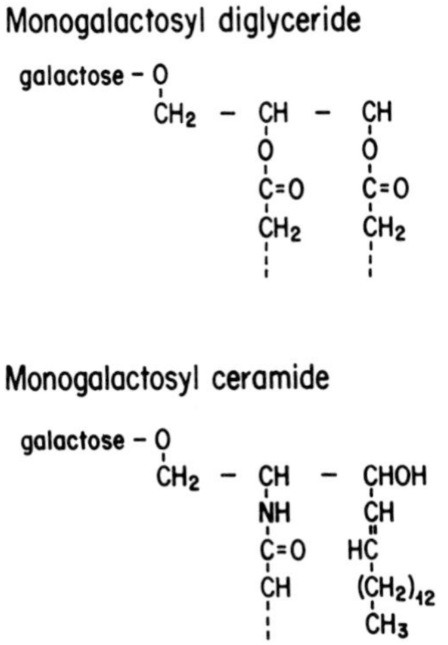

Monogalactosyl diglyceride

```
galactose - O
          |
          CH₂  -  CH   -   CH
                  |         |
                  O         O
                  |         |
                  C=O       C=O
                  |         |
                  CH₂       CH₂
                  |         |
                  |         |
```

Monogalactosyl ceramide

```
galactose - O
          |
          CH₂  -  CH   -   CHOH
                  |         |
                  NH        CH
                  |         ||
                  C=O       HC
                  |         |
                  CH       (CH₂)₄₂
                  |         |
                  |         CH₃
```

Fig. 2.14. Chemical structure of monogalactosyldiglyceride and monogalactosylceramide.

and in the fatty acids. Glycolipids of this sort are often found in bacterial membranes. The lipid composition of *Acholeplasma laidlawii* is given in Table 2.5.[11] Those lipids are also found in the thylakoid membranes. The lipid composition of the thylakoid membrane may be important to its structure because of interesting behavior of the individual membrane components.

IV. STEROLS

Cholesterol is the most common sterol of mammalian plasma membranes. The structure of cholesterol appears in Fig. 2.15. It consists of a

TABLE 2.5
**Lipid Composition of *Acholeplasma laidlawii* Membranes
Grown on 18:1 trans[a]**

Lipid	Mol%
Monoglucosyldiglyceride	18
Diglucosyldiglyceride	48
Phosphatidylglycerol	22
Glycerophosphorylmonoglycosyldiglyceride	2
Glycerophosphoryldiglucosyldiglyceride	10

[a] From A. Weislander, J. Ulmius, G. Lindblom, and K. Fontell, *Biochim. Biophys. Acta* **512** (1978): 241–253.

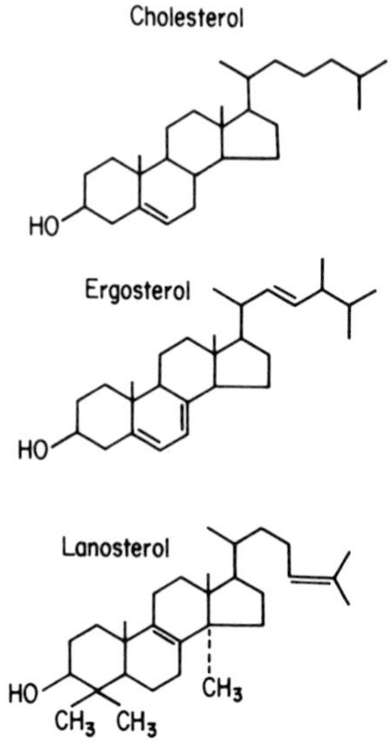

Fig. 2.15. Chemical structure of cholesterol, ergosterol, and lanosterol.

planar, fused-ring nucleus, a polar hydroxyl group, and a hydrocarbon tail. Cholesterol is quite different in structure from the other membrane lipids that have been discussed. However, as with the phospholipids, cholesterol is an amphipathic molecule, containing both a hydrophobic and a hydrophilic portion.

The physical properties of cholesterol have been well studied and are described in Chapter 5. Many of these properties derive from the rigid structure of cholesterol, conferred on the molecule by the fused ring system. Modifications to the chemical structure, like removal of the tail or addition of methyl groups on the ring system (thereby rendering it nonplanar), destroy the ability of the derivative to behave like cholesterol. This last observation led to the suggestion that enzymatic removal of the extra methyl on lanosterol, an intermediate in cholesterol biosynthesis, to produce cholesterol was obligatory for proper membrane assembly.

Mammalian cells cannot grow properly without including cholesterol in their membranes, suggesting that cholesterol plays a crucial role in mammalian cell biology. Cholesterol is not found in all cell membranes. In some membranes, close relatives of cholesterol, like ergosterol or stigmasterol, take its place. In *Tetrahymena* a sterol-like compound, tetrahymenal, takes the place and plays the role of sterol.

V. DETERGENTS

Detergents have far-ranging use in membrane studies. Because detergents are amphipathic molecules, like lipids, some of the same rules governing lipid behavior also apply to the detergents. Detergents come in two classes: nonionic and ionic detergents.

A. Nonionic Detergents

Nonionic detergents have polar headgroups that carry no charge. In recent years, the alkylglycosides have attracted much attention in membrane studies as nonionic detergents. Octylglucoside (OG) is one of the most highly used of these detergents. Its structure appears in Fig. 2.16. Relatives, such as laurylmaltoside, have also been used. One reason for their popularity is that a single molecular species can be obtained. Therefore, one only needs to worry about the stability of the detergent and its purity with respect to the alcohols from which it is formed.

Other nonionic detergents include the Triton and the Tween species. Both of these detergents suffer from the fact that they usually are obtained as a mixture of molecular species. This makes defining the behavior of the detergent rather difficult. There is, however, a series of similar compounds that are available in purer form. An example of these is $C_{12}E_8$. Structures of these detergents appear in Fig. 2.16.

B. Ionic Detergents

As their name indicates, ionic detergents carry a net charge as part of their polar moiety. Sodium dodecyl sulfate (SDS) may be the most familiar of the ionic detergents, since it is used extensively in polyacrylamide gel

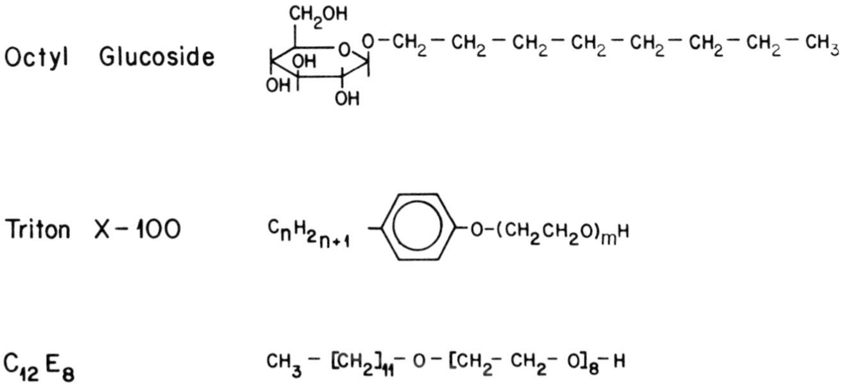

Fig. 2.16. Chemical structures of nonionic detergents.

electrophoresis. However, SDS suffers (as far as membrane studies are concerned) from a tendency to denature proteins irreversibly. Furthermore, SDS is difficult to remove from a solution. Interestingly, however, the Na^+,K^+-ATPase of plasma membranes can be effectively purified with the aid of low concentrations of SDS which help to remove proteins that are only loosely associated with the membranes. The structure of SDS is given in Fig. 2.17.

Other ionic detergents more commonly used in membrane studies include cholate, deoxycholate (DOC), 3-[(3-cholamidopropyl)dimethylam-monio]-1-propane sulfonate (CHAPS), and dodecyltrimethylammonium bromide (DTAB). The structures of these detergents are shown in Fig. 2.17. Phospholipid analogs that act as detergents have also been developed.

C. Chaotropic Agents

Chaotropic agents that change the properties of the aqueous phase faced by the membrane lipids are also useful in disrupting membrane structures. Examples of such agents include lithium diiodosalicylate (LIS) and guani-

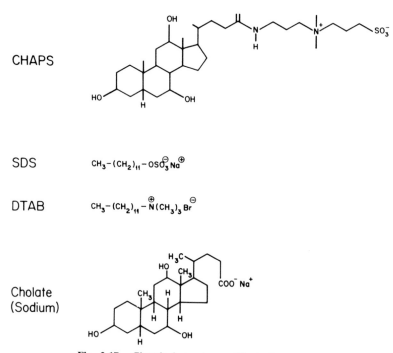

Fig. 2.17. Chemical structures of ionic detergents.

dine hydrochloride. These chemicals sufficiently change the thermodynamics of partitioning of the hydrophobic membrane components between the membrane phase and the aqueous phase so that some membrane components can be removed from the membrane. LIS, for example, is used to extract glycoproteins from the human erythrocyte membrane. Because of the large carbohydrate (i.e., polar) moiety of these glycoproteins, they are more easily solubilized than are other intrinsic membrane proteins of the erythrocyte, such as band 3. Thus this procedure achieves a selective solubilization. Proteins like band 3 require a more stringent treatment, such as that achieved by an ionic detergent.

VI. SUMMARY

A wide variety of lipids are found in biological membranes. The variety in structures leads to a collection of thousands of unique lipid chemical species in cell membranes. Phospholipids are among the most common. Phospholipids are classed according to their polar headgroup structure. In addition they have considerable variability in the structure of the hydrocarbon chains esterified to the glycerol of the phospholipid (or in some cases linked by nonester chemistry). The chemical structures of the lipids control the behavior of these lipids. The phospholipids can be degraded by phospholipases. Sphingolipids, built on the sphingosine base, and glycolipids are also important components of biological membranes. Cholesterol is an essential constituent of mammalian cell membranes. Given the structural variety, thousands of lipid species are found in biological membranes. As research continues, the roles of these many lipid species are gradually being defined. Detergents are useful tools for the study of membranes, since they assist in the purification of membrane proteins. Detergents can be classed as ionic and nonionic.

REFERENCES

1. Fox, C. F., and A. D. Keith, *Membrane Molecular Biology* (Stamford, Connecticut: Sinauer, 1972).
2. Chapman, D., *An Introduction to Lipids* (New York: McGraw–Hill, 1968).
3. Cullis, P. R., and B. de Kruijff, "The polymorphic phase behaviour of phosphatidylethanolamines of natural and synthetic origin," *Biochim. Biophys. Acta* **513** (1978): 31–42.
4. Vance, D., and J. Vance, *The Biochemistry of Lipids* (New York: Benjamin, 1985).
5. Pieringer, J., and R. L. Conner, "Positional distribution of fatty acids in the glycerophospholipids of *Tetrahymena pyriformis*," *J. Lipid Res.* **20** (1979): 363–370.
6. Kates, M., *Techniques of Lipidology* (Am. Elsvier, Amsterdam: North-Holland, 1972).

7. Rouser, G., S. Fleisher, and A. Yamamoto, "Two dimensional thin layer chromato-graphic separation of polar lipids and determination of phospholipids by phosphorus analysis of spots," *Lipids* 5 (1970): 494.
8. Porter, N. A., R. A. Wolf, and J. R. Nixon, "Separation and purification of lecithins by high pressure liquid chromatography," *Lipids* 124 (1979): 20–24.
9. Sundler, R., A. W. Alberts, and P. R. Vagelos, "Phospholipases as probes of membrane sidedness," *J. Biol. Chem.* 253 (1978): 5299–5304.
10. Schmidt, C. F., Y. Barenholz, and T. E. Thompson, "A nuclear magnetic resonance study of sphingomyelin in bilayer systems," *Biochemistry* 16 (1977): 2649–2656.
11. Weislander, A., J. Ulmius, G. Lindblom, and K. Fontell, "Water binding and phase structures for different *Acholeplasma laidlawii* membrane lipids studied by deuteron nuclear magnetic resonance and x-ray diffraction," *Biochim. Biophys. Acta* 512 (1978): 241–253.

3

Membrane Models and Model Membranes

I. MEMBRANE MODELS

In natural history the membranes of cells are ancient. Even the most primitive of life forms requires at least one membrane to enclose the constitutive functions. Without such a boundary, there is no way to differentiate the functions of an organism from the "soup" in which the organism lives. This boundary role represents only the most primitive of needs and functions for a cell membrane, in particular for a plasma membrane. There are many other functions of which the plasma membrane is capable. But this example does serve to illustrate the fundamental role of membranes in life.

Biological membranes contain as part of their structure a lipid bilayer. That lipid bilayer is constructed of amphipathic lipids in accordance with the hydrophobic effect. The first suggestions that such a structure was a fundamental element of biological membranes came from the following observation. When the lipids of a biological membrane were extracted from the membrane and redispersed in aqueous media, they generally formed milky dispersions.[1, 2] It was known that these lipid dispersions were made up of lipid bilayers. Therefore, it seemed reasonable that the lipids formed a bilayer in the biological membrane also.

Consider Fig. 3.1. Lipids are amphipathic molecules. They have nonpolar, or hydrophobic, hydrocarbon chains as part of their structure. The ΔG of transfer of the hydrocarbon into water is positive due predominantly to the unfavorable entropy contributions (discussed in Chapter 1). In addition the lipid contains a polar headgroup that interacts far more favorably with water does than the hydrocarbon portion. Transfer of this polar portion of the lipid molecule into the aqueous phase is thus characterized by a substantially more favorable ΔG than for transfer of the hydrocarbon chains.

Water

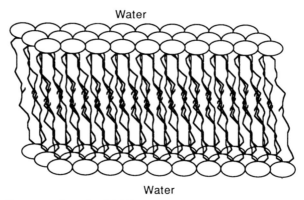

Water

Fig. 3.1. Representation of a lipid bilayer. The circles represent the lipid polar head-groups, and the wiggly lines represent the hydrophobic hydrocarbon chains of the lipids.

From this amphipathic lipid structure one would expect that the polar end of the lipid molecule would orient toward the aqueous phase, whereas the hydrophobic hydrocarbon region would organize to prevent contact with the aqueous phase. Reference to fig. 3.1 shows that the lipid bilayer is an excellent structure with which to achieve these goals. The principles behind the formation of the bilayer will be discussed further in Chapter 4.

In 1925, an interesting experiment was performed that further stimulated thinking about the lipid bilayer as an element of cell membrane structure. Gorter and Grendl extracted the lipid from the human erythrocyte and introduced this lipid to an air–water interface. At this interface, the lipid formed a monolayer. The lipid spread on this interface and then Gorter and Grendl measured the area of the interface covered by the lipid. The goal was to determine how much surface area was covered by the lipid derived from a single red blood cell. Having made that measurement to their satisfaction, they could then determine the relationship between that surface area and the surface area of the red blood cell. Their conclusion was that the lipids corresponded to twice as much surface area as the surface of the erythrocyte. From these observations they postulated that a lipid bilayer served as the plasma membrane of the erythrocyte.[3]

One of the next major points of historical interest for the development of models for biological membranes was provided by Davson and Danielli. They recognized that a membrane such as the erythrocyte contains proteins as well as lipid and therefore incorporated protein in their membrane model. The protein was suggested to coat the surface of the lipid bilayer, forming a trilamellar sandwich. The layering of the protein in their model was stimulated by the new (at that time) knowledge of β sheet structure. The protein was not allowed to penetrate the lipid bilayer in this model. These investigators were, perhaps, influenced by the earlier conclusion that the lipids of the erythrocyte completely covered the surface of the cell with a lipid bilayer.[4, 5]

An important new piece of information was provided subsequently by J. D. Robertson.[6] He stained a thin section of a glutaraldehyde-fixed human erythrocyte membrane preparation with the electron-dense material osmium tetroxide. A pair of dense lines 20 Å thick appeared in the image, separated by a light region 35 Å thick. In bilayers of unsaturated phospholipids, the osmium tetroxide likely binds more strongly to the double bonds of lipid hydrocarbon chains in membranes. Thus, in a lipid bilayer, one would expect two bands of staining, corresponding to the middle of the hydrocarbon chains. In more complex biological membranes, the osmium tetroxide may oxidatively react with other groups on the membrane constituents. The electron micrographs obtained by Robertson of the erythrocyte membrane showed two dark bands, separated by a light band, in a pattern described as a railroad track. These data were taken at the time as evidence for the presence of a lipid bilayer in biological membranes.

In the next period, the late 1960s and early 1970s, the experimental record began to gain momentum for the presence of a lipid bilayer in membranes.[7, 8] For example, a pure saturated phospholipid in a hydrated lipid bilayer was known to undergo a highly cooperative phase transition from a gel state to a liquid crystalline state (see Chapter 4).[9] This phase change was readily detected by differential scanning calorimetry. When *Acholeplasma* were encouraged to incorporate exogenous saturated fatty acid into the cell membranes, a similar phase transition from a gel state to a liquid crystalline state was detected in the biological membrane. Such data were consistent with a lipid bilayer in the cell membrane.[10]

Another physical technique was exploited around the same time to test the hypothesis of a lipid bilayer in biological membranes. X-ray diffraction measurements were used to develop an electron density profile across the membrane. The electron density profiles obtained were consistent with a phospholipid bilayer both in model membrane systems and in biological membranes.[11]

In an attempt to integrate protein structure into membranes, Green and colleagues proposed a model that was built around lipid–protein complexes as the fundamental unit of membrane structure.[12] This model buried part of the membrane protein structure into the membrane. Although this model did not give prominence to the lipid bilayer as a part of membrane structure, this model did get the proteins off the surface of the membrane, allowing for the concept of integral membrane proteins. Models of this sort also provide concepts useful to an understanding of membranes such as the mitochondrial membrane, in which proteins do appear to exist as complexes with phospholipids (in some cases, tight complexes).[13] Although it is now known that some of the mitochondrial lipids have the motional dynamics of lipids in bilayers, there is such a small (relatively) lipid content to that membrane that the concept of lipid–protein complexes is more useful to understanding that membrane than the fluid–mosaic model discussed below.

In an experiment with far-reaching implications, Frye and Edidin labeled a mouse and a human cell in tissue culture.[14] The mouse cell was labeled with one kind of fluorescent antibody. The human cell was labeled with another kind of fluorescent antibody. The emission from these two labels was different in color. Frye and Edidin then caused these two cells to fuse into one cell and observed the process under a fluorescent microscope. Immediately after fusion, the two membranes retained their original identity. However, with time, this identity was lost, as the labeled components of the two membranes mixed. These results suggested that membrane components could move laterally in the plane of the membrane.

Accumulation of data such as these stimulated Singer and Nicolson to describe a new model for membrane structure.[15] In this model the membranes were fluid, where fluid described the feature that components (both lipids and proteins) could diffuse laterally in the plane of the bilayer.

They also suggested that there were two classes of membrane proteins. One class was the peripheral membrane protein. Since this class of protein can be removed with salt treatment, or pH changes, peripheral membrane proteins appeared to be bound to the membrane surface. Cytochrome c was an example of a peripheral membrane protein. The other class of membrane proteins was the integral membrane protein. Integral membrane proteins were defined as being difficult to remove from the membrane without detergents or proteases. As a consequence, these proteins were envisioned with a considerable portion of their mass buried in the membrane. It was suggested that some may even span the membrane. Figure 3.2 is Singer and Nicolson's representation of their model.

This is a brief history of the development of membrane models. The Singer–Nicolson model, in particular, helped to stimulate an explosion in information about cell membranes since the early 1970s. The great amount of information obtained since the introduction of that model now permits considerable refinement of that model, which is the subject of this book.

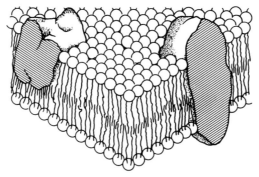

Fig. 3.2. Representation of the membrane model proposed by Singer and Nicolson. The large solid objects represent integral membrane proteins incorporated in the membrane. It should be noted that the lipid–protein mole ratio implied by this picture is much higher than found in almost all biological membranes. Reproduced with permission from S. J. Singer and G. L. Nicolson, *Science* **175** (1972): 720, AAAS.

II. MODEL MEMBRANE SYSTEMS

A. Structure of Model Membrane Systems

Knowledge of bilayer structure and properties has come largely from studies on model membrane systems. This discussion will examine what is known about five of these systems that have been extensively used for biophysical and biochemical studies. In each of the systems to be considered, except the last, the phospholipids are arranged in a bilayer. Thus these systems are not "modeling" the phospholipid bilayer, since they contain the bilayer. Although they do not exhibit all the complexity of a biological membrane, they effectively model some of the properties of biological membranes. However, because these pure lipid systems do not contain membrane proteins and because the lipid composition is usually simpler than is found in biological membranes, there are some properties of cell membranes that are not mimicked by these systems. Such distinctions are important when interpreting results from model membrane systems.

1. LIPOSOMES

To make phospholipid liposomes, pure phospholipids are dissolved in an organic solvent like chloroform where they are monodisperse or found in small aggregates. The phospholipids are dried down to a film on a

Fig. 3.3. Schematic representation of the multilamellar liposome. Because of artistic limitations, too few lamellae are shown here.

glass vessel. This film is then hydrated with a buffer solution. Vortexing produces a milky suspension of liposomes. Electron micrographs of these liposomes show that they consist of concentric bilayers of phospholipids, resembling an onion as seen in Figs. 3.3 and 3.4. They can be 10,000 Å or more in diameter. Between the bilayers is an aqueous space in which solutes can be trapped for transport or permeability studies. With this system a number of different questions of membrane structure can be addressed using a single pure phospholipid, or mixtures of phospholipids, or mixtures of phospholipids with other membrane lipids.[16] A summary of the properties of liposomes is listed in Table 3.1.[17]

There are a number of advantages of this model membrane system: it is easy to make, with reproducible properties, and can be made in large quantities and high concentrations. Since it is suitable for a wide variety of biophysical studies, it has been used extensively.

Drawbacks of this system include the unknown affects on the bilayer due to the close apposition of another bilayer, the limited aqueous space for trapping materials, the inhomogeneity in size and number of layers, and the inability to access both sides of the membrane with electrodes. Because of the concentric arrangement of the membranes, most of the

Fig. 3.4. (A) Negative stain electron micrograph of multilamellar liposomes of egg phosphatidylcholine. The stain was 2% ammonium molybdate. Bar: 100 nm. Micrograph kindly provided by Dr. S. W. Hui. (B) Freeze-fracture electron micrograph of multilamellar liposomes of egg PC and soybean PE (3/1). Bar: 100 nm. Micrograph courtesy of Dr. S. W. Hui.

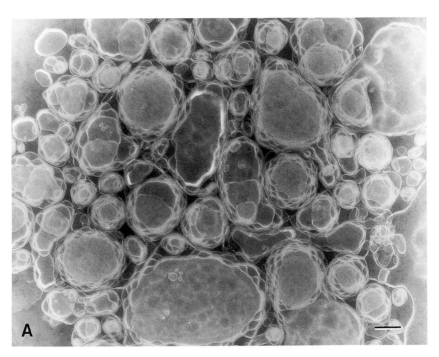

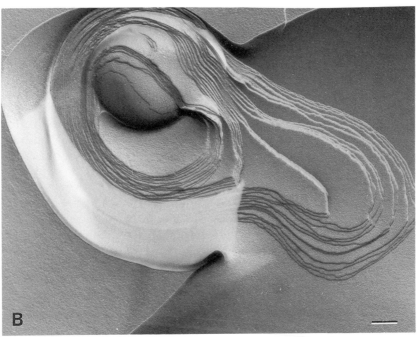

TABLE 3.1
Properties of Liposomes[a]

Properties	Liposomes
Size (diameter)	5–50 μm
Number of lamella	8–14
Fraction of lipid in the outermost monolayer	~8%
Transition characteristics for di-16:0-PC	
Temperature (T)	41.3°C
Enthalpy	9.1 kcal/mol
Width of transition at half height width	≤0.3°C
Cooperative unit	100–1000
Change in molal volume	4.0%

[a] M. K. Jain and R. C. Wagner, *Introduction to Biological Membranes* (New York: Wiley, 1980).

membrane surface is buried inside the liposome. Ten percent or less of the total phospholipid is found on the outside surface of the liposome.

2. SONICATED VESICLES

The problem of heterogeneity in size and number of layers can be overcome in the second model membrane system, the sonicated vesicles. These are also referred to as small, unilamellar vesicles (SUV). Sonicated vesicles are made by subjecting liposomes, formed as described in the previous section, to ultrasonic irradiation until the suspension appears to clarify. When that point is reached, the reduction in light scattering indicates that the average size of the vesicles has significantly decreased. In fact, the vesicles have a diameter well below 1000 Å. Such small vesicles do not scatter visible light as effectively as the much larger liposomes. Therefore the suspension appears nearly clear, with a slight bluish tinge (light scattering increases with a decrease in wavelength, so that UV wavelengths are still scattered effectively by the SUV). Then by chromatography on a gel like Sepharose 4B or by centrifugation, a population of vesicles remarkably homogeneous with respect to size can be isolated. Figure 3.5 shows the column profile obtained by monitoring the light scattering of the effluent from a vesicle preparation run on a Sepharose 4B column. The small vesicles appear in the included volume (fraction II), whereas the large liposomes appear in the void volume of the column.[18]

These vesicles have been carefully characterized, making them ideal for many studies.[19, 20] They are quite homogeneous with respect to size, which is useful for physical chemical studies.[21–26] Some of the properties of sonicated vesicles obtained with egg phosphatidylcholine are listed in

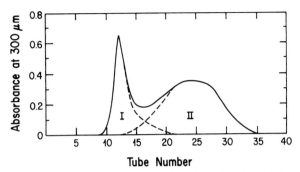

Fig. 3.5. Elution profile of a sonicated dispersion of egg phosphatidylcholine from Sepharose 4B. Fraction I consists of a heterogeneous mixture of large vesicles and liposomes. The second half of fraction II has been shown to be a homogeneous dispersion of small unilamellar vesicles. Reprinted with the permission from C. Huang *Biochemistry* **8** (1969): 344, American Chemical Society.

Table 3.2, and a schematic representation of the structure appears in Fig. 3.6. These vesicles consist of a hollow sphere, whose surface is a single phospholipid bilayer. Therefore, one side of all the phospholipid bilayers in this system is readily available from the aqueous medium. Because of the small radius of curvature, the outside surface of the phospholipid bilayer of these vesicles contains about 60 to 75% of the total phospholipid in the vesicle, depending upon the lipid composition, which can affect the size of the vesicle.

There are two significant drawbacks in the use of this system. One difficulty is the small size of the sonicated vesicle, which leads to a small radius of curvature compared to the curvature observed in many cell membranes. This leads to packing constraints in the lipid bilayer that do not mimic lipid packing in many biological membranes.[27] The second difficulty is that the internal aqueous space of the vesicle is very small, making the sonicated vesicles unsuitable for transport studies, and poor

TABLE 3.2
Sonicated Egg PC Vesicle Properties[a]

Molecular weight	1.88×10^6
Partial specific volume	0.9848 ml/g
Bilayer lipid ratio	2.1
Hydrated vesicle radius	105 Å
Outer hydration layer	6 Å

[a] C. Huang, *Biochemistry* **8** (1969): 344–352.

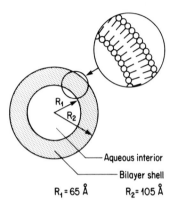

R$_1$ = 65 Å   R$_2$ = 105 Å

Fig. 3.6. Schematic representation of the small unimellar vesicle produced by sonication of egg phosphatidycholine dispersions (see Fig. 3.5). Reprinted with permission from T. E. Thompson, C-H. Huang, and B. J. Litman, In *The Cell Surface in Development* (A. A. Moscona, ed.) New York: (Wiley, 1974).

candidates for trapping anything but small solutes. Furthermore, in a suspension of SUV, the percentage of the total sample volume represented by the intravesicular space is quite small.

3. LARGE UNILAMELLAR VESICLES (LUV)

Several variations on the vesicle theme have been developed that produce medium or large size (relative to the small sonicated vesicles just discussed) vesicles for experimentation.[28–32] Development of these systems was stimulated by the limitations of the liposomes and the sonicated vesicles just discussed. One well-characterized system is obtained by detergent dialysis. To prepare LUV by detergent dialysis, phospholipid is dissolved in an octylglucoside solution at a defined detergent/phospholipid ratio. Dialysis of this solution removes the detergent, and vesicles form as the detergent is removed. The LUV are unilamellar and most exhibit diameters in the range of 1500 to 2500 Å. A summary of properties of LUV prepared by octylglucoside dialysis appears in Table 3.3.[33]

These vesicles are suitable for a number of studies, for which the model membrane systems described above are unsuitable. For example, they retain the advantage of the sonicated vesicles over the liposomes in that they are unilamellar. However, they are large enough to trap a significant amount of solute. The LUV are even large enough to do some modest transport studies. With SUV the intravesicular volume can only accommodate a few molecules at which point millimolar concentrations are reached. Therefore too few molecules can be transported to be sensed by a transport

TABLE 3.3
Properties of Large Unilamellar Vesicles[a]

Diameter	2300–2500 Å
Phospholipids/vesicles	4.19×10^5
Internal volume	5×10^{-15} ml
Cl^- permeability (rate constant)	2.1×10^{-5} sec^{-1}
Na^+ permeability	2.6×10^{-7} sec^{-1}

[a] L. T. Mimms, G. Zampighi, Y. Nozaki, C. Tanford, and J. A. Reynolds, *Biochemistry* **20** (1981): 833–840.

assay. The main disadvantage of this system is that residual detergent remains in the vesicles after dialysis.

Currently the most widely used method for preparation of LUV is extrusion, which avoids the problems of residual detergents.[34] In this method, phospholipid MLV are freeze-thawed and then repeatedly extruded through approximately 1 μm filters under pressure, until a limited range of vesicle size is observed and until the vesicles are predominantly unilamellar and bilamellar. With these vesicles, solutes can be readily trapped, and removed from the exterior through exclusion chromatography. Equipment is now available to accelerate this preparation utilizing much higher pressures. However, concerns have been expressed concerning distorted LUV shapes arising from the high-pressure extrusion.

Two other methods are available for preparing large unilamellar vesicles. One is the reverse-phase evaporation technique, which involves a controlled substitution of aqueous media for organic solvent. The other method involves injection of an ethanolic solution of phospholipid into aqueous media. Both methods produce unilamellar vesicles of a size that is much larger than the sonicated vesicles. Therefore they have the advantages described for the vesicles from detergent dialysis. Their major drawback is the presence of residual organic solvent in the membrane, which may affect membrane properties.

One other important development should be mentioned. Unlike sonicated vesicles, these large unilamellar vesicles cannot be fractionated on Sepharose 4B because they are too large. However, it is possible to separate much larger vesicles on Sephacryl 1000. Thus there are available a wide variety of sizes of unilamellar vesicles for membrane studies.

There are some interesting uses for these systems. Drugs can be trapped inside these structures for delivery via the bloodstream. Some investigators have suggested that with properly constructed surfaces, these drug-containing liposomes might be targeted for certain tissues within the body.

A great deal of energy and resources have been invested in the development of such drug delivery systems.

4. BLACK LIPID MEMBRANE (BLM)

The next model membrane system is unique in that it allows electrodes to be placed on both sides of a phospholipid bilayer for electrical conductivity measurements. This system is commonly referred to as the black lipid membrane developed by Mueller and Rudin.[35] The name results because of the optical behavior of the membrane when formed. The membrane is created by painting phospholipid in a solvent like hexane across an opening in a Teflon cup immersed in aqueous solution. Eventually the film across the opening thins down until it consists of a single bilayer of phospholipid. At this point the membrane becomes opaque due to destructive interference of reflected light from the front and back of the film; hence the name, black lipid membranes. Electrodes can be placed inside and outside the cup for electrical measurements, and the buffer composition on either side of the membrane can be modified. Thus the black lipid membrane model system is capable of measurements of the electrical properties of membranes that can be formed in this apparatus (unfortunately, it is not directly applicable to biological membranes). A diagram of this model membrane system appears in Fig. 3.7. (Patch clamping has now been developed as an alternative method for measuring electrical properties of membranes, which can be applied to cell membranes. It involves the extraction of a portion of cell membrane in a hollow electrode, the contents of which can be controlled for the electrical measurements of interest.)

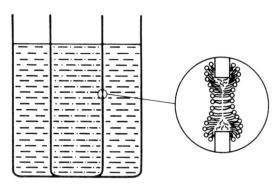

Fig. 3.7. Schematic representation of the black lipid membrane (BLM).

The principal drawback of the BLM is that solvent is left in the membrane, which can change its characteristics. This problem can be avoided, as described by Montal and Mueller, by creating the membrane from a surface monolayer of phospholipid.[36] In this technique a support is dipped into the lipid monolayer. Done properly, the support will be coated on each side with a phospholipid monolayer, thereby creating the bilayer desired.

Other problems include the instability of the black lipid film, and the difficulty of correlating results obtained from these system with those obtained from liposomes and vesicles. A comparison of the properties of the BLM and the solvent-free counterpart is given in Table 3.4.[37]

5. LIPID MONOLAYERS

Because of the amphipathic nature of the lipid molecules, they exhibit interesting surface activity. They can be spread in a monolayer at an air–water interface. The hydrophobic hydrocarbon chains orient with their primary exposure to the gas phase and usually perpendicular to the interface. The polar headgroups of the lipids in the monolayer immerse themselves in the aqueous phase. Therefore, the phospholipid monolayer resembles one-half of a phospholipid bilayer and has been used extensively for studies of phospholipid behavior. This is schematically represented in Fig. 3.8. From the aqueous phase, the monolayer surface resembles the surface of a membrane and has been used for studies of the behavior of molecules that are felt to be active at the membrane surface.

TABLE 3.4

Comparison of the Electrical Properties of Bimolecular Lipid Membranes in the Presence and Absence of Hydrocarbon Solvent with Those of Cell Membranes

Electrical property	Bilayer (with hydrocarbon solvent)	Bilayer from monolayers (no hydrocarbon solvent)	Cell membranes
Membrane resistance (ohm cm^2)	10^6–10^8	10^6–10^8	$<10^6$
Membrane capacity (μF cm^{-2})	0.45 ± 0.05	0.9 ± 0.1	Range: 0.8–1.2
Thickness of hydrocarbon region obtained from:			
a. Capacity data	42 Å	22 Å	16–24 Å
b. X-ray data	—	29 Å	—

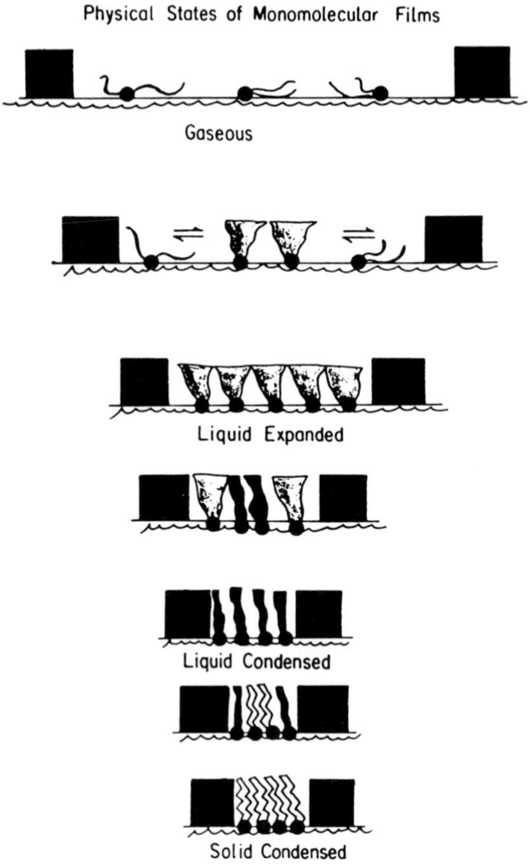

Physical States of Monomolecular Films

Gaseous

Liquid Expanded

Liquid Condensed

Solid Condensed

Fig. 3.8. Schematic representation of the states associated with lipids in monomolecular films. Figure courtesy of Dr. D. A. Cadenhead.

In Langmuir trough experiments, the lateral pressure on a phospholipid monolayer can be controlled. A known amount of lipid is applied to the surface, over which the lipid spreads as a monolayer. Since the total area occupied by the monolayer can be measured, the cross-sectional area occupied by each lipid molecule can be measured as a function of surface pressure. By so doing, several interesting pieces of information about lipids can be obtained. The surface area occupied per phospholipid molecule at a given pressure can be derived. For phosphatidylcholine, this is about 64 $Å^2$, in a pure phosphatidylcholine monolayer in the liquid expanded state, which resembles liquid crystal state of this phospholipid bilayer.[38] This number can decrease under greater lateral pressure.

A state similar to gel state in a membrane can also be induced in a monolayer by increasing the lateral pressure.[39, 40] The increase in lateral pressure reduces the cross-sectional area occupied by the hydrocarbon chains of the lipids. This is achieved by reducing the freedom of motion of those chains. If gauche–trans isomerizations are inhibited, the hydrocarbon chains adopt a more pure all-trans configuration (see Chapter 4). The minimum cross-sectional area a hydrocarbon chain can occupy is that described by a pure all-trans configuration of all the carbon–carbon bonds in the chain. This is analogous to the configuration adopted by phospholipids when the phospholipid bilayer is in a gel state.

In a pure monolayer, the transition between the liquid expanded state and the liquid condensed state is a sharp boundary, again analogous to the bilayer, in which there is a sharp boundary between the liquid crystal state and the gel state.

At higher pressures, a solid condensed state can be observed in the monolayer. Large increases in lateral pressure in this phase produce little further condensation of the cross-sectional area of the lipid, presumably because there is little else the lipid molecule can do to reduce the area it occupies. Rather, the film will collapse at higher pressures, as the lipids pile up on top of each other. The condensed solid phase in a monolayer may find an analogy in the subphase inducible in a phospholipid bilayer at low temperatures, where motion of the molecule is largely frozen and the headgroups are relatively dehydrated.

Cholesterol, in a pure cholesterol monolayer, occupies a surface area corresponding to about 35 Å. An interesting phenomenon occurs when cholesterol and phospholipid are mixed in a monolayer. At a given surface pressure, the area per molecule is not a weighted average of the two constituents, according to composition. Rather, the area per phosphatidylcholine molecule decreases as cholesterol content increases. This is the well-known "condensing" effect of cholesterol.[41]

For phosphatidylcholine, an increase in lateral pressure leads to a staged reduction in area per molecule, but cholesterol, in a monolayer, exhibits quite different properties. Because the molecule consists of a rigid steroid ring that is planar, the surface area occupied by cholesterol cannot be modified significantly by increases in lateral pressure. Therefore, area per molecule is not strongly dependent on lateral pressure, and the result is that pressure leads rather directly to collapse of the cholesterol monolayer.

The principal drawbacks of the lipid monolayer for modeling membranes are twofold. One is that the properties are those of a monolayer and not a bilayer. The second major problem is the difficulty in deciding what lateral surface pressure corresponds to a real membrane.

Some interesting studies on monolayers and bilayers may provide interesting hints concerning such questions. The ability of phospholipases from
various sources to cleave phospholipids was examined as a function of
the lateral pressure in a phospholipid monolayer. The data suggested that
different lipases were sensitive to different thresholds of lateral pressure
on phospholipid hydrolysis (i.e., too high a lateral pressure would inhibit
phospholipase activity). This effect presumably results because the lipase
must penetrate the surface of the monolayer to catalyze the hydrolysis
reaction. By comparison of monolayer studies and studies of phospholipase action on the erythrocyte membrane, it was concluded that a lateral
surface pressure of 31–35 dyn/cm described this biological membrane.[42]

It is interesting at this point to return to the Gorter and Grendl experiment in which the lipids of the erythrocyte membrane were spread on a
surface monolayer. Although their result was important to the development of the concept of a bilayer in biological membranes, their interpreta-
tion, was, in fact, incorrect. They chose the wrong lateral surface pressure
in the monolayer to model the erythrocyte membrane. The membrane
lipid does not account for all the erythrocyte surface area; protein also
makes a substantial contribution. One must be careful how one applies
monolayer data to conclusions about biological membrane structure.

B. Properties of Model Membrane Systems

The thickness of the bilayer is a function of many properties. It is
represented in X-ray diffraction experiments as the distance between the
phosphates of the phospholipids in the membrane because of the relatively
electron-dense phosphate. Since the headgroup conformation of phosphatidylcholine in membranes is parallel to the bilayer surface (see Chapter
4), the trans-membrane phosphate-to-phosphate distance corresponds well
with the width of the phospholipid bilayer. However, there is another
interesting way to view this piece of information. The thickness of the
bilayer does correspond to approximately the distance covered by two
phospholipid molecules placed end to end. Therefore this was also one of
the early clues to the existence of a phospholipid bilayer in cell membranes.

Liposome and SUV may behave differently when cycled through a
phase transition. The liposome is stable to a gel to liquid crystal phase
transition, and the system can be cycled through the transition region
repeatedly. The small sonicated vesicles are not always as stable.[43] Cycling
through the phase transition region can cause the small vesicles to fuse
into larger vesicles. However, in the liquid crystal state, the SUV can be
stable nearly indefinitely if properly cared for. The behavior of large
unilamellar vesicles under such conditions has not been extensively examined.

This raises an interesting question concerning the small sonicated vesicles. Are they in fact thermodynamically stable structures? The observations just chronicled do not fully support the concept that the vesicles are stable structures. Given the opportunity they can spontaneously form somewhat larger structures by fusion. Since they are formed under the influence of ultrasonic energy, the small vesicles may well represent a kinetically trapped species.

One further comparison should be made concerning the small sonicated vesicles and liposomes. The packing of individual molecules in the two model membrane systems cannot be identical. A check of the schematic representations of their respective structures reveals why. The liposomes appear as nearly flat surfaces in comparison to the sonicated vesicles because of the large difference in radius of curvature between the two membrane systems. In the small vesicles, the headgroups on the inside of the vesicle are packed more tightly than on the outside, whereas those in the liposomes are packed at a density that is probably similar to the outside surface of the small vesicle. The difference in packing of the phospholipid headgroups can be seen in the difference in ^{1}H NMR[24] and ^{31}P NMR[44] chemical shift arising from the phospholipids on the outside of the vesicle (downfield) and the phospholipids on the inside of the vesicle (upfield) as seen in Fig. 3.9.

The same structural constraints propagate their effects to the packing of the hydrocarbon chains in the bilayer interior. Physical methods have shown that the two halves of the small vesicle bilayer are distinguishable because of these packing differences. There is a difference between the ^{19}F NMR chemical shifts of specifically labeled phosphatidylcholine hydro-

40 Hz

Fig. 3.9. Effect of differences in packing of phospholipid headgroups on the inside and outside of small sonicated phospholipid vesicles on the high field ^{1}H NMR spectra of the N-methyl protons of PC. At 600 MHz, two peaks are observed, the upfield peak (right) corresponding to headgroups on the inside of the vesicle. The observation of the two peaks is indicative of different environments on the inside and outside of the vesicle for the lipid headgroups.

carbon chains between the inside and the outside of the vesicle.[45] These differences are important to keep in mind when reviewing data reflecting the properties of these model membrane systems.

Returning to the monolayer model system, a number of measurements have been made to determine the cross-sectional area occupied by each lipid molecule. Table 3.5 presents a summary of some of that data.[17] One of the interesting observations is at the top of Table 3.5. Introduction of a double bond in a saturated chain produces an increase in surface area occupied by the lipid. It is then interesting to consider whether the same effect is noted when the fatty acids are part of a phospholipid molecule. Table 3.6 provides the answer.[41] The introduction of the first double bond produces a significant change in surface area. Beyond that, the addition of double bonds also increases the surface area in smaller increments. Why does this happen? The introduction of a double bond into a hydrocarbon chain produces a permanent kink in the chain conformation. This forces the chain to occupy more cross-sectional area, which is reflected in the surface area per molecule.

TABLE 3.5
Molecular Areas of Some Amphipaths[a]

Lipid	Limiting area[b] (Å^2) at 20–25°C
Stearic acid (18:0)	20.6
Oleic acid (18:1, cis)	32
Cholesterol	39
Dipalmitoyl PC	44.5
Dioleoyl PC	72
Egg PC	62
Sphingomyelin	42
Phosphatidyl serine[c]	50–100
Egg PE	42
Ganglioside[d]	140 (head)
	31 (chain)

[a] From M. K. Jain and R. C. Wagner, *Introduction to Biological Membranes* (New York: Wiley, 1980).

[b] The values are averages of several values reported in the literature from monolayer and X-ray diffraction of bilayer studies.

[c] From beef brain. Surface area depends on pH and the presence of multivalent ions.

[d] Forms micelles.

TABLE 3.6
Surface Area per Phospholipid[a]

Fatty acid composition	Area per molecule (Å^2)
18:0, 18:0	48
18:0, 18:1	77
16:0, 18:2	83
16:0, 18:3	86
16:0, 20:4	90
16:0, 22:6	91
18:2, 18:2	97
18:3, 18:3	104

[a] From R. A. Demel, W. S. M. G. van Kessel, and L. L. M. van Deenen, *Biochim. Biophys. Acta* **266** (1972); 26–40.

The multilamellar liposomes have been used to identify further characteristics of the interactions between bilayers in this model system. When phosphatidylcholine is hydrated out of a film, it spontaneously forms the multilamellar liposome, as described earlier. However, when the liposomes form, they swell to include water between the lamellae, but, for example, in the case of PC, they do not swell indefinitely. Apparently there is a balance achieved between interbilayer attractive forces and interbilayer repulsive forces. The forces involved can be measured by providing external pressure on the system through osmotic gradients.

C. Hydration

It was noted earlier that water was one of the most important constituents contributing to the structure of biological membranes. This was not because water was incorporated into membranes, but because the structure of water led to the hydrophobic effect. The hydrophobic effect was not an attractive force, but rather a dominant entropy term in the energy of interaction between the hydrocarbon portion of the lipids (and as will be seen later, proteins) and water.

However, lipids are amphipathic. The characteristics of the interaction between the polar portions of the lipid molecules and water are very different than for the hydrocarbon portions. Water will bind to the polar headgroups of phospholipids. However, the role of water goes well beyond the binding of a discrete number of water molecules to the lipid headgroups.

This farther reaching interaction is evident in the phenomenon of hydration repulsion.[46-48] Consider the following experiment. Under the influence of osmotic pressure, water can be withdrawn from between the bilayers of multilamellar liposomes. The effect is to reduce the repeat distance (see Fig. 3.10) measured from X-ray diffraction. When the distance between the bilayers approaches the region of 10 to 20 Å (depending on the lipid composition and the medium), the osmotic pressure required to further extract water increases dramatically. This is hydration repulsion.

The problem is that it requires considerable energy to remove the layers of water that are close to a membrane surface. These short range forces are strong. The distance at which this phenomenon occurs depends in part on the structure of the bilayer surface formed by the phospholipids. Phosphatidylcholine creates a bilayer surface that is relatively difficult to dehydrate; PE, on the other hand, creates a bilayer surface that is less costly, in free energy, to dehydrate than the PC surface.

Surface properties such as this can be important to a number of phenomena. Among these is membrane fusion, where two membranes come close together at an early stage in the fusion event. The close approach of the membranes is linked to the expulsion of water. The phase state of the lipid bilayer is influenced by the degree of hydration of the surface. Interactions of some peripheral membrane proteins with membrane surfaces will be influenced by the hydration of the surface. Even some integral membrane proteins may experience contact between the extramembranous portion of the protein and the membrane surface. Finally it should be noted that the issue of hydration appears whenever a solute is to be transported across the membrane. Water soluble compounds must be at least partially dehydrated at the initial stage of the transport process.

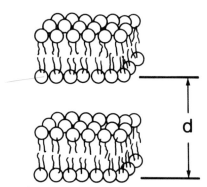

Fig. 3.10.   Representation of the repeat distance obtained from X-ray diffraction, *d*.

D. Ion Binding

Model membrane systems can bind cations on the surface. The cytosol of the cell and the outside of the cell contain cations. Many properties of the cell are controlled by the binding of such ions. Binding of cations can take place at the membrane surface. Some of the phospholipids in the membrane are negatively charged and can bind cations. Consequently, some of the ion binding sites on the surface of the membrane are phospholipid headgroups.

The binding of cations to membrane surfaces is not a simple phenomenon.[49] It involves biological membranes, with both lipid sites and protein sites available for cation binding. In the context of this text, only a brief overview can be given to ion binding to pure lipid model membrane systems, where cation binding is more easily understood.

Plotting the number of ions bound to the surface of a phospholipid bilayer as a function of a cation concentration does not produce a simple binding curve.[50] Rather, at very low cation occupancy of surface sites, the apparent affinity of the cations for the surface is relatively high. In contrast, at high concentrations of cations, and high occupancy of surface sites, the apparent cation affinity for the membrane surface is low. The affinity of cations for the surface varies according to the concentration of cations in the solution and consequently the occupancy of cation binding sites on the surface.

Consider a finite patch of surface with some negative charges on it and allow cations to interact sequentially with that surface. The surface has a negative potential of a certain value, i.e., a certain attractive force for cations. On binding of a cation, the surface changes. It does not carry as much of a negative charge because the cation has neutralized some of the negative charge. The second cation to approach does not bind as favorably as the first because the negative potential has been reduced. The affinity of the cation for the surface therefore decreases as the surface gets filled with bound cations.

When a number of cations are bound to the surface, the cations, in turn attract anions. A consequence is the build up of an electrical double layer made up of anions and cations from the solution.

Measurement of surface potential provides another means of looking at this problem of cation binding. Again consider a surface containing negative charge as part of the lipid structure. The surface potential is highly negative at the interface between the phospholipid headgroups and the aqueous phase. This potential drops exponentially from the surface, as represented in Fig. 3.11. The extent to which this surface potential extends from the surface of the membrane can be characterized by the

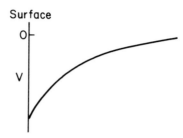

Fig. 3.11. Representation of a negative surface potential that falls to zero as the distance from the membrane surface increases (toward the right in the drawing).

Debye length i.e., the distance from the membrane surface at which for the surface potential drops to 37% its original value. For the surface of a phosphatidylcholine bilayer, in 0.1 M sodium chloride, the Debye length is about 9.7 Å. With the addition of 0.1 M calcium chloride, the Debye length drops to 4.9 Å. The surface potential is modified by the binding of cations to the surface of the membrane.

The effect of cations on the surface potential is particularly dramatic when observing the swelling of phospholipids in water. The lamellae swell to a limited extent in the case of phosphatidylcholine and incorporate a defined amount of water between the lamellae. The liposomes do not therefore swell indefinitely. The case of a negatively charged phospholipid like phosphatidylserine is different. In the absence of salt in solution, the lamellae will continue to swell indefinitely. This is because the negative surface potential on these membranes causes the surfaces to repel each other. In the presence of ions that repulsion is diminished. This effect has been measured in the following way. X-ray diffraction experiments provide information on the repeat distance in a multilamellar system. This distance is defined in Fig. 3.10. In the presence of low levels of sodium the repeat distance is about 107 Å. If the sodium concentration is increased to 0.5 M, the repeat distance is reduced to 60 Å. Thus the cations screen the negative charge on the membrane surface.

When calcium binds to the surface of phosphatidylserine bilayers, what kind of stoichiometry does one find? There are at least two possibilities. The first possibility is a 1 : 1 calcium PS complex. The second possibility is a 2 : 1 calcium PS complex. The 1 : 1 complex implies that calcium binds to a single phospholipid. The 2 : 1 complex implies that calcium bridges two phospholipids and ties them together. It is really not clear, at this point, which of the two models actually applies. Data have been interpreted in support of both models.

Where does the calcium bind? Again consider calcium binding to negatively charged PS. Phosphatidylserine has a free carboxyl group, which carries a negative charge. Evidence suggests that most of the binding is to the carboxyl. The phosphate group seems to be involved to a modest extent in binding the calcium. The remaining site for calcium binding involves chelation. One molecule provides more than one ligand for the coordination sphere of the calcium. For PS the carboxyl and amino groups form the chelate. In the case of phosphatidylcholine, calcium also binds, but since phosphatidylcholine does not carry a net negative charge, the binding is much weaker. The binding site for calcium on phosphatidylcholine is the negatively charged phosphate.[51]

III. SUMMARY

The present understanding of membrane structure is a reflection of how young the field of membrane studies is. The introduction of the fluid mosaic model stimulated a tremendous amount of research interest in membranes following its introduction in the early 1970s. Since then the shortcomings of that model have been identified and more modern models for membrane structures have been developed.

Model membrane systems have been exploited extensively to unmask the properties of lipids in membranes. One of the most commonly used model membrane system is the multilamellar liposome. It consists of many concentric layers of lipid bilayer resembling an onion in cross section. Unilamellar vesicle systems are perhaps the most popular of vesicle systems used today. These include small vesicles made by sonication and larger vesicles made by extrusion. These are useful because of their ability to encapsulate materials necessary for transport experiments, fusion experiments, and drug delivery. The third of the important model membrane systems is the lipid monolayer. Care, however, must be taken in extrapolating properties of lipids in monolayers to properties of lipids in bilayers.

The properties of these model membrane systems mimic in a number of ways the properties of biological membranes. This is due to the presence in the biological membranes of a lipid bilayer as a major structural element. Thus the study of the properties of model membrane systems often permits examination of questions for which the biological membranes present too complex a structure for experimentation.

Hydration repulsion, important to proteins structure, is a powerful force at the membrane surface. Dehydration of membrane surfaces requires energy. More energy is required to remove water from a PC surface than from a PE surface. Any process that requires two membrane surfaces, or one membrane surface and one protein surface, to approach each other closely will be influenced by hydration repulsion.

The binding of ions to phospholipid bilayers occurs at the phospholipid headgroups. Ion binding modifies the surface potential and occurs with an occupancy-dependent affinity.

REFERENCES

1. Bangham, A. D., "Membrane models with phospholipids," *Prog. Biophys. Mol. Biol.* **18** (1968): 29–95.
2. Israelachvili, J. N., D. J. Mitchell, and B. W. Ninham, "Theory of self-assembly of lipid bilayers and vesicles," *Biochim. Biophys. Acta* **470** (1977): 185–201.
3. Gorter, E., and F. Grendel, "On bimolecular layers of lipids on the chromocytes of the blood," *J. Exp. Med.* **41** (1925): 439–443.
4. Danielli, J. F., "Some properties of lipoid films in relation to the structure of the plasma membrane," *J. Cell. Comp. Physiol.* **7** (1936): 393.
5. Danielli, J. F., and H. Davson, "A contribution to the theory of permeability of thin films," *J. Cell. Comp. Physiol.* **5** (1935): 495.
6. Robertson, J. D., "The molecular structure and contact relationships of cell membranes," *Prog. Biophys. Biphysical Chem.* **10** (1960): 343–418.
7. Korn, E. D., "Structure of biological membranes," *Science* **153** (1966): 1491–1498.
8. Stoechenius, W., and D. M. Engelman, "Current models for the structure of biological membranes." *J. Cell Biol.* **42**:613–646 (1969).
9. Chapman, D., B. D. Ladbrooke, and R. M. Williams, "Physical studies of phospholipids. VI. Thermotropic and lyotropic mesomorphism of some 1,2-diacylphosphatidycholines (lecithins)," *Chem. Phys. Lipids* **1** (1967): 445–475.
10. Steim, J. M., M. E. Tourtellotte, J. C. Reinert, R. N. McElhaney, and R. L. Rader, "Calorimetric evidence for the liquid crystalline state of lipids in a biomembrane," *Proc. Natl. Acad. Sci. U.S.A.* **63** (1969): 104–109.
11. Luzzati, V., and F. Husson, "The structure of the liquid–crystalline phases of lipid–water systems," *J. Cell Biol.* **12** (1962): 207–210.
12. Green, D. E., and J. Perdue, "Membranes as expressions of repeating units," *Proc. Natl. Acad. Sci. U.S.A.* **55** (1966): 1295–1302.
13. Sjostrand, F., and L. Barajas, "A new model for mitochondrial membranes based on structural and on biochemical information," *J. Ultrastruct. Res.* **32** (1970): 293–306.
14. Frye, C. D., and M. Eddidin, "The rapid mixing of cell surface antigens after formation of mouse-human heterokaryons," *J. Cell Sci.* **7** (1970): 319–332.
15. Singer, S. J., and G. L. Nicholson, "The fluid mosaic model of the structure of cell membranes," *Science* **175** (1972): 720–731.
16. Bangham, A. D., M. M. Standish, and J. C. Watkins, "Diffusion of univalent ions across the lamellae of swollen phospholipids," *J. Mol. Biol.* **13** (1965): 238–252.
17. Jain, M. K., and R. C. Wagner, *Introduction to Biological Membranes* (New York: Wiley, 1980).
18. Huang, C., "Studies on phosphatidylcholine vesicles. Formation and physical characteristics," *Biochemistry* **8** (1969): 344–352.

19. Huang, C., and J. T. Mason, "Geometric packing constraints in egg phosphatidylcholine vesicles," *Proc. Natl. Acad. Sci. U.S.A.* **75** (1978): 308–310.
20. Mason, J. T., and C. Huang, "Hydrodynamic analysis of egg phosphatidylcholine vesicles," *Ann. N.Y. Acad. Sci.* **308** (1978): 29–49.
21. Berden, J. A., P. R. Cullis, D. I. Hoult, A. C. McLaughlin, G. K. Radda, and R. E. Richards, "Frequency dependence of P-31 NMR linewidths in sonicated phospholipid vesicles: Effects of chemical shift anisotropy," *FEBS Lett.* **46** (1974): 55–58.
22. Cullis, P. R., "Lateral diffusion rates of phosphatidylcholine in vesicle membranes: Effects of cholesterol and hydrocarbon phase transitions," *FEBS Lett.* **70** (1976): 223–228.
23. Schmidt, C. F., Y. Barenholz, C-H. Huang, and T. E. Thompson, "Phosphatidylcholine C-13 labelled carbonyls as a probe of bilayer structure," *Biochemistry* **16** (1977): 3948–3954.
24. Huang, C. H., J. P. Sipe, S. T. Chow, and R. B. Martin, "Differential interaction of cholesterol with phosphatidylcholine on the inner and outer surfaces of lipid bilayer vesicles," *Proc. Natl. Acad. Sci. U.S.A.* **71** (1974): 359–362.
25. Yeagle, P. L., W. C. Hutton, C-H. Huang, and R. B. Martin, "Phospholipid headgroup conformations. Intermolecular interactions and cholesterol effects," *Biochemistry* **16** (1977): 4344–4349.
26. Kornberg, R. D., M. G. McNamee, and H. M. McConnell, "Measurement of transmembrane potentials in phospholipid vesicles," *Proc. Natl. Acad. Sci. U.S.A.* **69** (1972): 1508–1513.
27. Cornell, B. A., J. Middlehurst, and F. Separovic, "The molecular packing and stability within highly curved phospholipid bilayers," *Biochim. Biophys. Acta* **598** (1980): 405–410.
28. Batzri, S., and E. D. Korn, "Single bilayer vesicles prepared without sonification," *Biochim. Biophys. Acta* **298** (1973): 1015–1019.
29. Enoch, H. G., and P. Strittmatter, "Formation and properties of 1000-Å-diameter, single bilayer phospholipid vesicles," *Proc. Natl. Acad. Sci. U.S.A.* **76** (1979): 145–149.
30. Milsmann, M. H. W., R. A. Schwendener, and H. G. Weder, "The preparation of large single bilayer liposomes by a fast and controlled dialysis," *Biochim. Biophys. Acta* **512** (1978): 147–155.
31. Nozaki, Y., D. D. Lasic, C. Tanford, and J. A. Reynolds, "Size analysis of phospholipid vesicle preparation," *Science* **217** (1982): 366–367.
32. Szoka, F., F. Olson, T. Heath, W. Vail, E. Mayhew, and D. Papahadjopoulos, "Preparation of unilamellar liposomes of intermediate size (0.1–0.2 μm) by a combination of reverse phase evaporation and extrusion through polycarbonate membranes," *Biochim. Biophys. Acta* **601** (1980): 559–571.
33. Mimms, L. T., G. Zampighi, Y. Nozaki, C. Tanford, and J. A. Reynolds, "Phospholipid vesicle formation and transmembrane protein incorporation using octyl glucoside," *Biochemistry* **20** (1981): 833–840.
34. Szoka, F., and D. Papahadjopoulos, "Procedure for preparation of liposomes with large internal aqueous space and high capture by reverse-phase evaporation," *Proc. Natl. Acad. Sci. U.S.A.* **75** (1978): 4194–4198.
35. Mueller, P., D. O. Rudin, H. T. Tien, W. C. Wescott, "Reconstitution of excitable cell membrane structure in vitro," *Circulation* **26**:1167–1170 (1962).
36. Montal, M., and P. Mueller, "Formation of bimolecular membranes from lipid monolayers and a study of their electrical properties," *Proc. Natl. Acad. Sci. U.S.A.* **69** (1972): 3561–3566.
37. Huang, C., L. Wheeldon, and T. E. Thompson, "The properties of lipid bilayer mem-

branes separating two aqeous phases: Formation of a membrane of a single composition," *J. Mol. Biol.* **8** (1964): 148–160.

38. Ghosh, D., M. A. Williams, and J. Tinoco, "The influence of lecithin structure on their monolayer behavior and interactions with cholesterol," *Biochim, Biophys. Acta* **291** (1973): 351–362.

39. MacDonald, R. C., and S. A. Simon, "Lipid monolayer states and their relationships to bilayers," *Proc. Natl. Acad. Sci. U.S.A.* **84** (1987): 4089–4093.

40. Nagle, J. F., "Theory of lipid monolayer and bilayer phase transitions: Effect of head-group interactions," *J. Membrane Biol.* **27** (1976): 233–250.

41. Demel, R. A., W. S. M. G. van Kessel, and L. L. M. van Deenen, "The properties of polyunsaturated lecithins in monolayers and liposomes and the interactions of these lecithins with cholesterol," *Biochim. Biophys. Acta* **266** (1972): 26–40.

42. Demel, R. A., W. S. M. Geurts van Kessel, R. F. A. Zwaal, B. Roelofsen, and L. L. M. van Deenen, "Relation between various phospholipase actions on human red cell membranes and the interfacial phospholipid pressure in monolayers." *Biochim. Biophys. Acta* **406** (1975): 97–107.

43. Schullery, S. E., C. F. Schmidt, P. Feigner, T. W. Tillack, and T. E. Thompson, "Fusion of dipalmitoylphosphatidylcholine vesicles," *Biochemistry* **19**, (1980): 3919–3923.

44. Bystrov, V. F., Y. E. Shapiro, A. V. Viktorov, L. I. Barsukov, and L. D. Bergelson, "P-31 NMR signals from inner and outer surfaces of phospholipid membranes," *FEBS Lett.* **25** (1972): 337–338.

45. Longmuir, K. J., and F. W. Dahlquist, "Direct spectroscopic observation of inner and outer hydrocarbon chains of lipid bilayer vesicles," *Proc. Natl. Acad. Sci. U.S.A.* **73** (1976): 2716–2719.

46. Leneveu, D. M., R. P. Rand, V. A. Parsegian, and D. Gingell, "Measurement and modification of forces between lecithin bilayers," *Biophys. J* **18** (1977): 209–230.

47. Rand, R. P., N. L. Fuller, and L. J. Lis, "Myelin swelling and measurement of forces between myelin membranes," *Nature (London)* **279** (1979): 258–260.

48. Rand, R. P., and V. A. Parsegian, "Hydration forces between phospholipid bilayers," *Biochim. Biophys. Acta* **988** (1989): 351–376.

49. McLaughlin, S., N. Mulrine, T. Gresalfi, G. Yaio, and A. McLaughlin, "Adsorption of divalent cations to bilayer membranes containing phosphatidylserine," *J. Gen. Physiol.* **77** (1981): 445–473.

50. McLaughlin, S., "The electrostatic properties of membranes," *Annu. Rev. Biophys. Chem.* **18** (1989): 113–136.

51. Altenbach, C., and J. Seelig, "Ca^{2+} binding to phosphatidylcholine bilayers as studied by deuterium magnetic resonance. Evidence for the formation of a Ca^{2+} complex with two phospholipid molecules," *Biochemistry* **23** (1984): 3913–3920.

4

Lipid Properties in Membranes

I. MEMBRANE STRUCTURE

Lipids form a variety of extended, noncovalent structures that include the fundamental lipid bilayer of the membranes of cells. Many of the examples in this chapter will be illustrated with phospholipids, a class of polar lipids. Phospholipids are amphipathic molecules, with a hydrophobic portion (the hydrocarbon chains) and a hydrophilic portion (the polar head group). These properties are exploited to establish a hydrophobic barrier to permeability in close proximity to an aqueous medium. The permeability barrier is maintained by the association of lipids in structures that sequester the hydrocarbon portions in hydrophobic regions away from the aqueous medium. Only the polar headgroups then encounter the polar phase. By so doing, the phospholipid molecule can satisfy the hydrophobic effect, which is the dominant "driving force" behind membrane assembly.

Interesting properties result from such organizations of amphipathic molecules. An appreciation of those properties is key to understanding the behavior of lipids in cell membranes. How the amphipathic lipid molecule elects to achieve these properties can be partially rationalized, although not uniquely, by an interplay between thermodynamics and the chemical properties imparted by the structure of the lipids. Several different interesting lipid assemblies can be formed. Each morphology is stabilized by a balance among favorable and unfavorable interactions. The nature of the interactions is determined by the lipid structures. Therefore, one can expect differences in lipid covalent structures to lead to differences in the morphologies of the assemblies of lipids that result.

A. Phase Structures

1. LAMELLAR (BILAYER) PHASE

The lamellar phase (lipid bilayers) is formed by the common lipids of biological membranes. In this structure, the polar heads of the lipids face the aqueous phase on both sides of the bilayer, and the hydrocarbon chains oppose each other inside the bilayer, as seen in Fig. 3.1. As an example, most common species of phosphatidylcholine form this phase.

One representation of the phospholipid bilayer comes from X-ray diffraction measurements. Phospholipids can be induced to form bilayers that stack, producing a repeating unit that can give rise to diffraction.[1]

X-ray diffraction data have been used to establish the electron density profile across a phospholipid bilayer. A representation of such an electron density profile appears in Fig. 4.1.[2] It is symmetrical because, in this case, the phospholipid bilayer is symmetrical in composition. When proteins are incorporated into the membrane, this symmetry is destroyed. The middle of the membrane has the lowest electron density. The membrane midplane contains the terminal methyl groups of the phospholipid hydrocarbon chains, which are experiencing considerable motional freedom. The points of highest electron density are near the outside edges of the membrane. This is the comparatively electron-dense region of the phosphate of the phospholipid. Thus it is possible to obtain the thickness

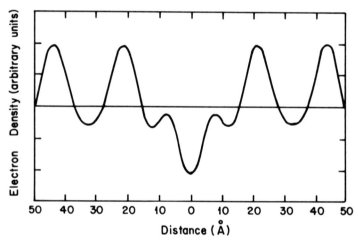

Fig. 4.1. Representation of an electron density profile derived from X-ray diffraction for dipalmitoylphosphatidylcholine. The location of the electron-dense phosphates can be seen clearly as the prominent positive peaks. Reprinted with permission from Y. K. Levine, *Prog. Surf. Sci.* **3**, (1973): 279.

of the membrane from the electron density profile by measuring the phosphate-to-phosphate distance. This is not far from the true membrane thickness because the polar headgroups of the phospholipids lie predominantly parallel to the membrane surface. The width so derived, from phosphate to phosphate across the membrane, is about 46Å in the gel state for dipalmitoylphosphatidylcholine.

One can see a change in this width if the length of the hydrocarbon chains of the lipids is altered. In other words, the width of the bilayer changes in direct proportion to changes in the effective length of the hydrocarbon chains. For example, the width of the bilayer decreases going from a gel state to a liquid crystal state, or from a gel state 18:0,18:0 PC to a gel state 14:0,14:0 PC. These observations lend further support to the concept of a bilayer of lipid in membranes.

A simple means of understanding the packing of lipids into a lipid bilayer can be obtained by considering the shape of the molecule. For phospholipids like phosphatidylcholine, the cross-sectional area of the headgroup is similar to the cross-sectional area of the hydrocarbon chains. Thus the shape of the molecule can be approximated as a cylinder. If one considers that these cylinders are to be packed to protect the hydrocarbon regions from contact with water, the packing of a lipid bilayer becomes inevitable (see Fig. 4.2). This can be represented more quantitatively by

$$P_r = \frac{A_h}{A_c},\tag{4.1}$$

where A_h is the effective cross-sectional area of the headgroup, and A_c is the effective cross-sectional area of the hydrocarbon chain region, and P_r is the packing ratio. A lipid bilayer should result when $P_r \approx 1$. As will be seen later in this chapter, modulation of the relationship between the cross-sectional area of the headgroups and the cross-sectional area of the hydrocarbon chains will lead to a modulation in the packing of the lipids.

Also interesting is the effect of unsaturation. The thickness of the mem-

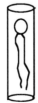
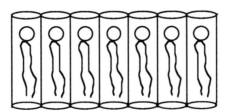

Fig. 4.2. Schematic representation of how an approximately cylindrical shape of an amphipathic lipid ($P_r = 1$) packs spontaneously into a bilayer configuration.

brane is decreased upon an increase in unsaturation, chain length otherwise being equal. The decrease in thickness reflects a reduction in the average chain length if a cis double bond is introduced. Furthermore an increase in unsaturation from no carbon–carbon double bonds to one double bond increases the cross-sectional area of the molecule. Thus with an increase in unsaturation, a phospholipid will increase the surface area it occupies and decrease its average length, thereby preserving its total molecular volume.

The bilayer is one of the basic elements of biological membrane architecture. As mentioned previously, most phospholipids in membranes inhabit a bilayer structure, and many of the properties of pure phospholipid bilayers mimic the properties of cell membranes. Consequently, this particular lipid structure will be the center of attention for much of the remainder of this chapter.

However, in some cases, variations in the structure of the lipid headgroup and its consequent hydration produce sufficient stress on the fabric of the bilayer structure that some lipids tend to adopt alternate morphologies. Prior to examining the lamellar phase in more detail, therefore, this survey of the phase structures accessible to amphipathic lipids will be completed.

2. INTERDIGITATED BILAYERS

A variation on the normal bilayer structure has recently been described. Under certain conditions, the lipid hydrocarbon chain termini are not all located in the bilayer midplane. Hydrocarbon chains from one leaflet of the bilayer may overlap hydrocarbon chains from the opposing leaflet of the same bilayer. This is called interdigitation. Some of the possible interdigitated structures are presented in Fig. 4.3.[3]

In Figs. 4.3a and 4.3b, two different, fully interdigitated forms are schematically represented.[4] One structure is made possible by a large difference in the effective length of the two hydrocarbon chains of the lipid. The effect is to locate one of the chain termini near the interface of the hydrophobic and hydrophilic regions of the bilayer. In the other fully interdigitated structure, both chain termini are near that interface. The thickness of the membrane in this latter structure approximates the length of one lipid molecule, rather than two as in the normal bilayer.

In both of these fully interdigitated bilayers, the lipid headgroups are required to guard more (than in a normal bilayer) of the bilayer surface against contact between the aqueous phase and the hydrophobic interior of the membrane. Such structures would thus appear to be favored by large, well-hydrated lipid headgroups. In this case, $P_r > 1$ (or the headgroup can be reasonably altered in conformation to make $P_r > 1$).

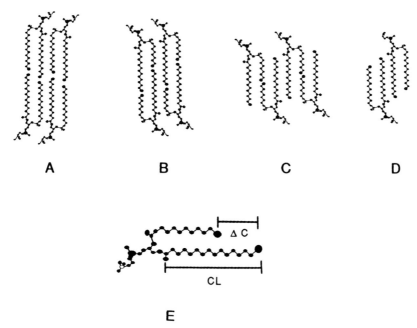

Fig. 4.3. Schematic representation of interdigitated bilayers. (A) Noninterdigitated bilayer; (B) partially interdigitated; (C) mixed interdigitated; (D) fully interdigitated. Figure courtesy of Dr. C.-H. Huang.

The partially interdigitated bilayer schematically presented in Fig. 4.3c does not depart as strongly from the normal bilayer as do the fully interdigitated forms. In this case the polar headgroups are not required to cover significantly greater surface area than in a normal bilayer. Again this form is favored by large differences in the effective length of the hydrocarbon chains. As in the other interdigitated forms, this partially interdigitated bilayer also would be expected to lead to a membrane that is thinner than the normal phospholipid bilayer.

To date, extended interdigitated bilayers have been found mostly in the gel state. Some of the strongest evidence for their formation has come from X-ray diffraction measurements of bilayer thickness. One example of a phospholipid that appears to form a partially interdigitated form in the liquid crystalline state is 18 : 0,10 : 0 PC, based on the large difference in effective chain lengths. Some of the sphingolipids, which in their natural form can exhibit large differences in effective chain length of the two hydrocarbon chains, would also be good candidates for interdigitation.

Transient interdigitation may be more widespread in the liquid crystalline state than is now appreciated. Two-dimensional nuclear Overhauser effect ^{1}H NMR experiments of liquid crystalline bilayers from several groups have provided evidence of dipolar interactions between phospholipid headgroups and terminal methyls of phospholipid hydrocarbon chains.[5] One way to explain such data is through transient interdigitation.

It is somewhat premature, given the state of the field, to speculate heavily on what roles bilayer interdigitation might play in membrane biology. However, it is worthwhile to keep in mind that membrane thickness has been demonstrated to modulate membrane enzyme activity (see Chapter 7). And the potent platelet-activating factor (PAF) is a candidate for interdigitation.

3. MICELLAR PHASE

In the case of lipids that contain well-hydrated headgroups and only a single hydrocarbon chain, packing problems in the hydrocarbon region (or bilayer interior) lead to sufficient stress on the lipid bilayer that an alternate structure forms. With only one hydrocarbon chain, these lipids would have difficulty filling all the volume of the interior of a bilayer, while accommodating the area per headgroup forced on the molecule by the hydration of the lipid headgroup. In this case, $P_r \gg 1$. This molecular shape can be approximated by a cone.

The geometric constraints on the packing of cones lead to the formation of a sphere. The resultant sphere would have a diameter roughly equal to the length of two lipids. The interior of this sphere is a hydrophobic domain. This spherical structure is called a micelle. The structure of a micelle is illustrated schematically in Fig. 4.4. Lysophosphatidylcholine and many common detergents are well-known examples of lipids that form micelles.

Lipid micelles represent a molecular assembly in which the individual components are thermodynamically in equilibrium with monomers of the same species in the surrounding medium. Even though part of the molecule is hydrophobic, these lipids do have a finite solubility in the aqueous phase. The extent of the solubility is, of course, determined by unfavorable entropy contribution due to the ordering of the water structure by the hydrophobic moiety. The greater the hydrophobic surface area exhibited by that moiety, the fewer molecules that can be dissolved in the aqueous phase, and the lower the solubility of that lipid. The solubility of the lipid is also influenced by the polar nature of the headgroup, which should help accommodate the lipid in the water structure through binding of water molecules. The balance between these interactions and the unfavorable entropy contributions from the hydrocarbon chain of the lysolipid in water, for example, determines the solubility. It is interesting that from these considerations, one can predict that even a phospholipid should have a

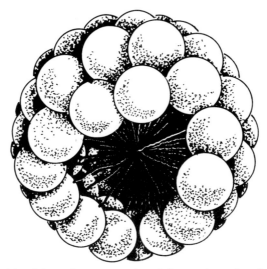

Fig. 4.4. Schematic representation of the structure of a micelle.

finite, if vanishingly small, solubility in water [for dipalmitoylphosphatidylcholine, the critical micelle concentration (CMC) = $10^{-10}M$]. This phenomenon is important in some kinds of lipid movement between membranes.

Now consider the behavior of a micelle-forming lipid, like lysophosphatidylcholine, as a function of lipid concentration. At very low concentrations of the lipid, only monomers are present in true solution. As the concentration of the lipid is increased, a point is reached at which the unfavorable entropy considerations, deriving from the hydrophobic end of the molecule, become dominant. At this point, the lipid hydrocarbon chains of a portion of the lipids must be sequestered away from the water. Therefore, the lipid starts aggregating into micelles. The concentration at which this occurs is referred to as the critical micelle concentration (CMC). Table 4.1 provides CMC data for a number of detergents. The larger the size of the hydrophobic moiety, the larger the micelle and the lower the CMC.

Above the CMC, addition of further lipid to the system results largely in an increase in micelle population. At this point, the system is more properly referred to as a lipid suspension with coexisting populations of monomeric detergent and detergent micelles. Unlimited increases in micelle population and/or size are not possible. Therefore, a maximum in the concentration of lipid in the suspension can be reached. As an example a phase diagram for SDS appears in Fig. 4.5.[6]

TABLE 4.1
Micellar Properties of Detergents

Detergent	CMC (mM)	n
SDS	8.1	120
SDS (30 mM NaCl)	3.1	72
SDS (50 mM NaCl)	2.3	84
SDS (100 mM NaCl)	1.3	85
DTAB	15	61
DTAB (50 mM NaCl)	5.7	72
DTAB (100 mM NaCl)	4.4	73
DOC	2	2
DOC (10 mM NaCl)	1.6	2
DOC (150 mM NaCl)	0.9	22
Cholate	10	2
Cholate (10 mM NaCl)	5.1	3
Cholate (150 mM NaCl)	3	5
Taurocholate	2.8	4
Palmitoyllysolecithin	0.01	—
Triton X-100	0.3	140
Octylglucoside	25	—
Laurylglucoside	0.2	—
$C_{12}E_8$	0.09	120

Micelles serve useful purposes in biology. Perhaps one of the most important is in the human digestive system. Bile salts are detergent species formulated by the liver from cholesterol and secreted into the intestine. These amphipathic molecules (an example of taurocholate is given in Fig. 4.6.) form micelles in aqueous media. In the bile, these bile salts are

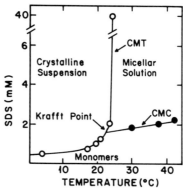

Fig. 4.5. Temperature–composition phase diagram for SDS in 100 mM sodium phosphate, pH 7.4. Reprinted with permission from R. Becker, A. Helenius, and K. Simons, *Biochemistry* **14** (1975): 1835, American Chemical Society.

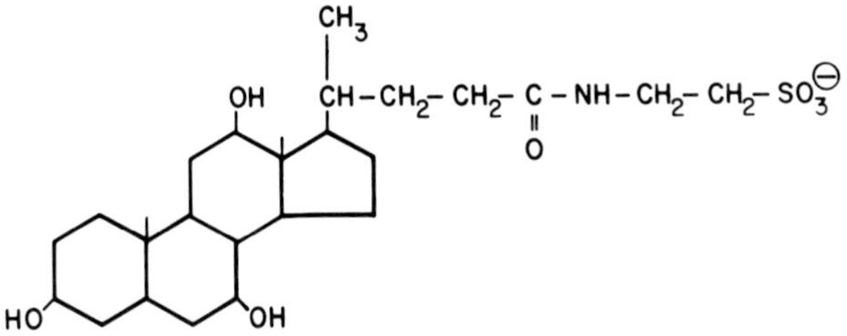

Fig. 4.6. Chemical structure of taurocholate.

distributed among micelles, phospholipid vesicles, and monomers in solution. The bile is concentrated and delivered to the gut.[7]

The bile is secreted into the gut to aid in digestion. Cholesterol and triglycerides ingested as part of the diet have very low solubilities in water. This is due to the hydrophobic effect. However, these neutral lipids are soluble in hydrocarbon solvents. Bile salt micelles provide, in their interior, hydrocarbon-like droplets that are effective at solubilizing triglycerides (as well as the products of lipase hydrolysis) and cholesterol. Cholesterol mimics the bile salt by placing its hydroxyl near the polar–nonpolar interface, and burying the fused ring and hydrocarbon tail in the hydrophobic part of the micelle. Oriented in these micelles in the gut, dietary lipids are in a form acceptable as substrate to a number of digestive enzymes (lipases, etc).

Since detergents form micelles, they are useful in dissolving oily residues such as those found in the kitchen sink. Some other classes of detergents have important uses in membrane studies. As is discussed later in Chapter 6, many proteins are tightly bound to membranes and are insoluble in water due to the hydrophobic effect. Detergents are frequently used to solubilize these membrane proteins (see Chapter 7, Section V). The detergent solubilization process separates the proteins from the host membrane and can be used to purify the protein for membrane studies. The detergent achieves its solubilizing effect by binding to the hydrophobic surface of the protein and creating a detergent micelle around that hydrophobic portion. This protects against the interaction of the hydrophobic protein surfaces with the aqueous phase. The detergent thus takes the place of the membrane lipid and creates a soluble detergent–protein complex, suitable for a variety of biochemical studies.

The CMC of detergents is of obvious importance to their ability to solubilize membrane constituents since the detergents must be in a detergent micelle to be effective at solubilization. Additionally the CMC of the detergent may also play a role in the removal of the detergent in the course of a reconstitution experiment. Detergents can sometimes be removed by dialysis, which relies on the ability of detergents to exist as monomers in solution at a finite concentration. Detergent monomers penetrate the dialysis membrane, whereas membrane vesicles and detergent micelles cannot.

The magnitude of the CMC controls the effectiveness of the dialysis technique. If the CMC is high, then a high concentration of monomers can exist in solution. This can lead to a large chemical potential difference for the detergent monomer on the two sides of the dialysis membrane if the dialysate is low in detergent concentration. The influence of a large difference in chemical potential can lead to a rapid removal of detergent by dialysis, if the kinetics of removal are a simple first-order process involving transfer of the detergent across the dialysis membrane. Conversely, if the CMC is low, the removal of detergent by this technique is not effective.

However, a high CMC is not necessarily a good criterion for choosing a detergent for solubilization or reconstitution. If the CMC is too high, then the concentration of micelles at accessible total detergent concentrations is low. The effectiveness of partitioning of the hydrophobic molecules from the membrane into the detergent micelles is reduced by decreasing the effective volume into which such partitioning can take place. However, a low CMC may be useful for studying detergent binding to proteins. Therefore, the choice of a detergent is not an easy task.

Other methods for detergent removal are available. Gel permeation column chromatography can be used, analogous to a desalting experiment. For some detergents like Triton X-100, the use of a product like Bio-Beads (Bio-Rad Richmond, CA) can be helpful by providing a surface into which the detergent can selectively partition.

4. HEXAGONAL I PHASE (H$_I$)

It was noted above that the micellar population in a detergent/water mixture cannot increase without limit as the detergent-to-water ratio increases. In fact, in the presence of low amounts of water, lipids that would normally form a micellar phase can form a larger aggregate. Long tubes of H$_I$ phase assemble, which can be thought of as many micelles fused together. These tubes have the polar headgroups facing out, and the hydrocarbon chains facing the interior. This phase structure is only seen under specialized conditions and is probably not particularly relevant to biological membranes.

5. HEXAGONAL II PHASE (H$_{II}$)

In the above consideration of the assembly of lipids into structures satisfying the requirements of the hydrophobic effect, the amphipathic chemical structure of the lipids and the shape of the lipid molecules, the region of $P_r \approx 1$, $P_r > 1$, and $P_r \gg 1$, were explored. Yet to be explored is the region of $P_r < 1$. In this region, one would expect that inverse cone shapes would lead to an inverse packing, such as seen in the hexagonal II (H_{II}) phase. However, as will be seen below, the formation of the nonlamellar H_{II} phase can be better understood in terms of the thermodynamics of hydration than on the basis of molecular shape.

Phosphatidylethanolamine is one of several common lipids derived from biological membranes that is capable of forming hexagonal II structures.[8] The hexagonal II phase has attracted interest because of the significant portion of lipids of cell membranes that favor the H_{II} phase when isolated from the other membrane components.[9] The question that has to be asked is why biological membranes would incorporate into their structure lipids that destabilize the bilayer, which itself is essential to the structure and proper function of the cell membrane. This fascinating question will be addressed in part here and further in Chapters 7 and 9.

Lipid molecules in the H_{II} phase pack inversely to the packing observed in the hexagonal I phase described above. The H_{II} phase puts the polar headgroups on the inside and the hydrocarbon tails on the outside. An extended array of molecules packed in this way will form tubes. Since the polar groups must contact water, they surround on aqueous channel at the center of the tube (Fig. 4.7).

This structure therefore leaves an exposed hydrophobic surface on the outside of the tubes. Because of this hydrophobic surface, the tubes tend to pack closely together to exclude water from their outside surfaces. They stack like pipes being readied for a pipeline, forming an hexagonal array in cross section. This may leave a finite hydrophobic surface in probable contact with water on the outside of the collection of tubes. However, the otherwise energetically favorable packing apparently stabilizes this phase as a whole. Or an outer monolayer of lipid coats the surface of the collection of tubes to protect that hydrophobic surface from interaction with the aqueous phase.

The hypothesis of a spontaneous radius of curvature for lipid assemblies was introduced to help understand the tendency for lipids like PE to form the H_{II} phase.[10] This hypothesis states, in brief, that surfaces made up of lipids in a liquid crystalline state have a spontaneous tendency to form a surface of defined radius of curvature, R_0, called the spontaneous radius of curvature. Within this hypothesis, phospholipids like PC have a large R_0 and spontaneously form surfaces with relatively large radii of curvature, as is observed. However, phospholipids like PE have a relatively small, negative R_0 and thus have a tendency to form curved surfaces, such as in the H_{II} phase. The coexistence of lipids with different R_0 values within the same bilayer would be expected to lead to packing stress within the bilayer.

How is R_0 determined? The formation of the H_{II} phase by PE and other lipids offers a means to measure R_0. Normally, the formation of the tubes of this phase leads to packing defects at the juncture of the tubes. Packing of cylinders in an hexagonal array leaves voids where each set of three tubes is in contact. Liquid crystalline hydrocarbon chains of the lipids can attempt to fill such voids, but this introduces stress into the packing

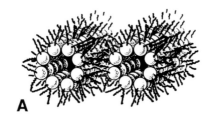

A

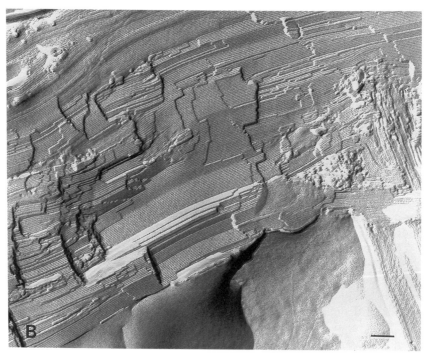

B

Fig. 4.7. (A) Representation of the structure of the hexagonal II phase. (B) Freeze-fracture electron micrograph of the hexagonal II phase of dilauryl PE. Bar: 100 nm. Micrograph courtesy of Dr. S. W. Hui.

of the lipid hydrocarbon chains. The unfavorable ΔG associated with this packing stress inhibits the formation, at equilibrium, of the H_{II} phase. Introduction of small amounts of liquid hydrocarbon would be expected to fill such voids, relieving the stress. Experimentally, dramatic stabilization of the H_{II} phase is observed by incorporation of small amounts of liquid hydrocarbon in the lipid matrix.[11] From the diameters of the water cores of the H_{II} tubes, whose formation is favored by the presence of liquid hydrocarbon, R_0 can be determined.

From such a hypothesis, one would conclude that stress introduced into a membrane by the coexistence of lipids with significantly different R_0 values would lead to an alteration of bulk membrane properties that could affect membrane function. To date, adequate means of testing this suggestion have not been reported.

The understanding of nonlamellar phases cultivated by the hypothesis of R_0 takes the understanding of the behavior of lipids like PE one step away from the (too) simple geometric arguments introduced above. Another significant step away from geometric arguments is forced by the observation that, in some cases, an increase in headgroup size leads to a *de*stabilization of the lamellar phase. For example, addition of an ethyl group to the methylene next to the nitrogen in the PE headgroup leads to a destabilization of the liquid crystalline lamellar phase and favors the formation of the H_{II} phase.[12]

Considering the thermodynamics of the interaction of water with a membrane surface defined by PE headgroups leads to an appreciation of entropy terms contributing to H_{II} phase formation. Phosphatidylethanolamino will be used as a example.

When in a bilayer, the headgroup of PE engages in intermolecular hydrogen bonds, N–H to phosphate.[13] This intermolecular hydrogen bond "neutralizes" the charges (positively charged amino and negatively charged phosphate) on the PE headgroup and competes effectively with water for binding to these charged groups. Fewer water molecules are bound by PE than by PC.[14] Therefore, more unbound water molecules would be needed to cover the surface of a PE bilayer than a PC bilayer. Such water molecules must be ordered by the surface, giving rise to an unfavorable entropy term. Since many of these water molecules are not bound, little compensating enthalpy is available to contribute to the overall ΔG. Therefore the interaction of the aqueous media with the PE surface is less favorable than the interaction of the aqueous media with the PC surface. This can be described in terms of the free energy of transfer, ΔG_t, of the phospholipid headgroups into water as (the contributions of the hydrocarbon chains to this ΔG are ignored because they are expected to be virtually the same for the two phospholipids)

$$\Delta G_{tPE} \approx \Delta G_{0PE} + \Delta H_{tPE} - T\Delta S_{tPE} \tag{4.2}$$

$$\Delta G_{tPC} \approx \Delta G_{0PC} + \Delta H_{tPC} - T\Delta S_{tPC} \tag{4.3}$$

The discussion above suggests $\Delta H_{tPE} < \Delta H_{tPC}$ and $\Delta S_{tPE} \gg \Delta S_{tPC}$. Therefore, $\Delta G_{tPE} > \Delta G_{tPC}$. In a sense, the surface of a PE bilayer is "hydrophobic" compared with the surface of a PC bilayer.

The structure of the phase formed by PE must therefore compensate for this hydrophobic effect. A structure that reduces the contact of the surface with water such as aggregation of bilayers will satisfy the need. Aggregation of PE bilayers, which excludes much of the water from between the bilayer surfaces, is commonly seen in aqueous dispersions of pure PE bilayers. An alternative means to compensate for the hydrophobic effect is formation of the H_{II} phase structure. This structure packs the headgroups of the PE molecules more closely together, on the inside surface at the hexagonal tubes. Tight packing of the headgroups reduces their contact with the aqueous phase, thereby reducing the amount of ordered, but unbound water at the surface. Furthermore, the small diameter of the tubes of the hexagonal phase (~20 Å) reduces the total amount of water encountering the PE surface. From this one can deduce that the spontaneous radius of curvature, R_0, would necessarily be smaller for PE than for PC as is observed.

Hexagonal II phase is favored by PE dispersions at elevated temperatures. The effect of increased temperature is to increase the surface area occupied by each PE headgroup, thereby magnifying the unfavorable contribution to ΔG of the structure of the surface of PE bilayers. For similar reasons, unsaturation in the hydrocarbon chains (which also increases the surface area the headgroup must cover) favors the hexagonal II phase.

There is direct evidence for this role of the hydrophobic effect in the structures formed by PE.[15] Chaotropic agents such as guanidine hydrochloride disrupt water structure, reducing the unfavorable nature of the interaction of a hydrophobic species with an aqueous phase. For example, guanidine hydrochloride can cause the unfolding or denaturation of proteins by stabilizing the interaction of hydrophobic amino acid side chains with the aqueous phase. With these considerations and the discussion above one would expect that chaotropic agents would stabilize the lamellar phase of PE. Such stabilization is observed experimentally (see Fig. 4.8). As mentioned above, addition of alkyl groups to the PE headgroup that do not interfere with the hydrogen-bonding capability of the amino group (such as the addition of an ethyl to C_2 of the ethanolamine) enhances the hydrophobic character of the headgroup. Such modified PE more readily forms hexagonal II phase than the corresponding unmodified PE.

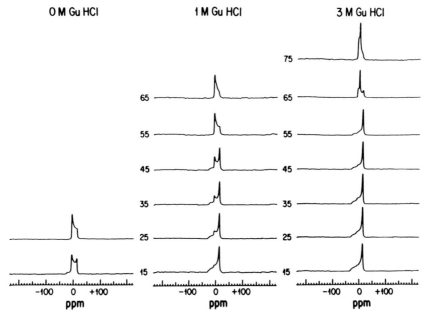

Fig. 4.8. ^{31}P NMR spectra of aqueous dispersions of soybean PE. The spectra in the lower right represent pure lamellar phase. The spectrum at the top center represents pure hexagonal II phase. In the absence of the chaotropic agent the lamellar phase is unstable even at 15°C. Guanidine hydrochloride (GuHCl) stabilizes the lamellar phase in proportion to concentration added, as can be seen reading across the figure. In 3 M GuHCl, the lamellar phase is stable up to 55°C. This is the same concentration range of GuHCl that is required to denature soluble proteins, due to the ability of the chaotropic agent to alter the structure and properties of the aqueous phase.

What lipids form H_{II} phase? Phosphatidylethanolamine with unsaturated hydrocarbon chains forms the hexagonal II phase readily. Diphosphatidylglycerol in the presence of calcium is also capable of forming the hexagonal II phase,[16] as is the glycolipid, monogalactosyldiglyceride, a component of the *Acholeplasma laidlawii* membrane.[17] In the presence of calcium, the lipids of the retinal rod outer segment disk membrane form hexagonal II phase.[18]

In the region where an approximate balance is achieved between hexagonal II and lamellar phase, a third type of structure is sometimes formed. In freeze-fracture electron micrographs spherical particles, called lipidic particles, appear in the membrane bilayer[19, 20] (Fig. 4.9). Although the structure of lipidic particles is the subject of controversy, the ^{31}P NMR spectra show isotropic motional averaging for some of the phospholipids when lipidic particles are present. This has led some investigators to

Fig. 4.9. Rows of lipidic particles in mixture of soybean PE and egg PC (9:1) at 30°C Bar: 100 nm. Micrograph courtesy of Dr. S. W. Hui.

postulate that lipidic particles are interlamellar attachments. Lipidic particles may be a nucleation point for formation of extended hexagonal II phase. Lipidic particles constitute discontinuities in the membrane that increase membrane permeability.

Figure 4.10 shows a phase diagram depicting the interplay of temperature and composition on the lamellar to hexagonal II phase transition of one mixed lipid system.[21] An increase in temperature at a given composition favors formation of the hexagonal II phase. An increase in the phosphatidylethanolamine content at a fixed temperature also drives the system toward the hexagonal II phase. The phase diagram was constructed using three physical techniques. One is ^{31}P NMR, where different powder patterns are observed for lamellar, hexagonal, and isotropic phases. The second is freeze-fracture electron microscopy, which is also sensitive to all three phases. The third is X-ray diffraction, which is sensitive to lamellar and hexagonal phase structures and provides definitive evidence of their existence. This study is an example of the use of several complementary techniques to obtain an adequate description of the lipid system.

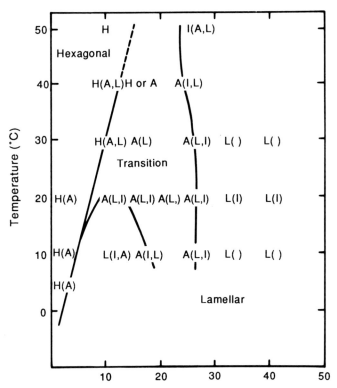

Fig. 4.10. "Phase diagram" of soybean PE and egg PC mixtures. L, lamellar; H, hexagonal; I, isotropic (most likely lipidic particles). Reprinted with permission from S. W. Hui, T. P. Stewart, P. L. Yeagle, and A. D. Albert, *Arch. Biochem. Biophys.* (1981): **207**, 227.

Mixing bilayer-forming lipids with lipids capable of forming a hexagonal II phase can produce some provocative effects. For example, mixing a modest amount of a bilayer-forming phospholipid, like phosphatidylcholine, into phosphatidylethanolamine can change the thermodynamically favored state of the phospholipid mixture from the hexagonal II to the bilayer structure.[22] The phase structure of the system then depends on the PC/PE mole ratio.

Other lipid mixtures can stabilize the bilayer. An example of bilayer stabilizing phase behavior can be seen in the mixing of lysophosphatidylcholine with fatty acids. If these two lipids are mixed in equimolar amounts, this mixture forms a bilayer.[23, 24] This structure results even though the individual components by themselves form micelles.

6. Cubic Phase

Another phase that can be exhibited by phospholipids is the cubic phase.[25-27]. This is nearly an isotropic phase, because phospholipids experience all possible orientations with respect to a laboratory reference frame on a relatively short time scale. However, the cubic phase is an extended structure and consists of short tubes connected in a hexagonal array. Cubic phase is sometimes found in mixtures of phospholipids that are changing from a lamellar phase to a hexagonal II phase. Its structure has been characterized primarily by X-ray diffraction and freeze-fracture electron microscopy. In ^{31}P NMR spectra, cubic phase behaves as an isotropic phase. A schematic representation of the structure of cubic phase appears in Fig. 4.15.

7. Subphase for Phospholipid Bilayers

At temperatures near 0°C, and on extended incubation, some phosphatidylcholines can enter a subphase. The kinetics of entering this phase are slow, much slower than for the main gel to liquid crystalline transition. This phase exhibits properties similar to that observed in a dehydrated solid state. Instead of slow axial rotation, apparently little or no phospholipid axial rotation takes place, nor is there any significant lateral diffusion. This phase, often an $L_{c'}$ phase, is dehydrated relative to the hydrated gel state, $L_{\beta'}$.

8. Solution Phase

In addition to forming the wide variety of phase structures described above, phospholipids can also exist in a solubilized form. Solubilization most readily occurs in organic solvent. For example, in methanol, phospholipids occur as monomers, dimers, and trimers in solution. However, in chloroform, particularly wet chloroform, the phospholipids inhabit inverted micelles. Any water in the system is trapped next to the headgroups in the center of the inverted micelle.

In addition, phospholipids can exist in solution in an aqueous phase. For more familiar phospholipids with long-chain fatty acids, the solubility in the aqueous phase is vanishingly small. However, dihexanoylphosphatidylcholine, because of its short chains, can exist as a monomer in the aqueous phase and has a CMC of about 10 mM. Apparently, the unfavorable entropy terms arising from contact between the solvent and the hydrocarbon chains are more than compensated by favorable terms due to the interaction of the polar headgroups with the water.

Cholesterol also is best solubilized in organic solvent, although it can self-associate even in nonaqueous solvents. Its solubility is small in the aqueous phase. However, the aqueous solubility of cholesterol is significant enough to play a role in the transfer of cholesterol from one membrane to another.

B. Phase Transitions

1. GEL TO LIQUID CRYSTALLINE PHASE TRANSITION

This discussion will now return to the structure of the lipid bilayer. The lipid bilayer represents the dominant form of assembly for lipids in cell membranes. Some of the unique properties of the phospholipid bilayer are bestowed on biological membranes.

There is some order to the structure of membranes; that is, there are limitations on the freedom of motion of molecules in the membranes. Because of the hydrophobic effect, the lipids of the bilayer cannot translate extensively in a direction normal to the bilayer surface. In other words, the lipids are not inclined to pop out of the membrane. However, the lipids can translate with considerable freedom and rapidity in the plane of the membrane, parallel to the membrane surface. Hence, to a first approximation the lipids in the membrane constitute a two-dimensional fluid. But the phospholipid bilayer is not a true liquid. Under most physiological conditions, the bilayer is in a liquid crystalline state (L_α) as opposed to a true solid or a true liquid. Consideration of the properties of a synthetic phosphatidylcholine will help illuminate this concept.

Dipalmitoylphosphatidylcholine (DPPC) is quite different in its fatty acid composition from most natural phosphatidylcholines (although it can be found in biological membranes as a rare species and in pulmonary surfactant as the major species). Instead of having a saturated hydrocarbon chain at position 1 and an unsaturated hydrocarbon chain at position 2, it contains palmitic acid (16:0) at both positions. When this phospholipid is hydrated it forms multilamellar liposomes, just like other phosphatidyl-cholines and several other classes of phospholipids.

Above 42°C, these model membranes are in a liquid crystalline state, L_α. Laterally, in the plane of the membrane, the phospholipid bilayer is disordered, because of two-dimensional lateral diffusion of the phospholipids. Nevertheless, phospholipid movement is inhibited by some organizational constraints as described above. The phospholipids are therefore not in a true liquid state in which their movement would be isotropic (in other words movement in any direction would be equally likely). Neither are these phospholipids in a solid state. This liquid crystal state of the phospholipids is therefore distinguished from the solid and the liquid state. A liquid crystal retains at least one dimension of order relative to the solid state.

Phospholipids are not the only species capable of existing in a liquid crystal state. Another class of molecules common in biochemistry that can exist in a liquid crystalline state is cholesterol ester, which forms liquid crystals in the interior of serum lipoproteins and in atherosclerotic plaques.

When the temperature of a suspension of dipalmitoylphosphatidylcholine is lowered below 42°C, the bilayer undergoes a liquid crystalline to gel phase transition. The gel state is a solid state in which little movement occurs. The rate of lateral diffusion of the lipids decreases by at least two orders of magnitude. The phospholipids lose much of their internal freedom of motion and adopt conformations more analogous to stiff rods. Raman spectroscopic studies indicate that the hydrocarbon chains loose their tendency to isomerize to alternate conformations about the carbon–carbon bonds. Therefore the carbon–carbon bonds tend to adopt an all-trans configuration in the gel state.

The phase transition from a gel state to a liquid crystalline state is endothermic and can be detected in a differential scanning calorimeter.[28] A scan for an aqueous dispersion (in excess water) of dipalmitoylphosphatidylcholine is shown in Fig. 4.11. The midpoint of the main transition is at about 42°C. Below 42°C, the phospholipid is in a gel state and above 42°C, the phospholipid is in a liquid crystalline state. The endothermic

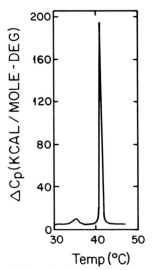

Fig. 4.11. Phase behavior of dipalmitoylphosphatidylcholine as detected by differential scanning calorimetry. The main transition is about 42°C: the lower temperature transition is referred to as the pretransition, which may reflect a change in tilt of the hydrocarbon chains, prior to the highly cooperative main transition.

transition is a fact of physical chemistry. In the transition region, energy can be added to the system with little change in temperature. The narrow peak is further representative of a highly cooperative transition, that is, a transition that involves many molecules as a unit. Broader peaks are indicative of a smaller cooperative unit undergoing the phase transition.

The phase transition to the gel state of DPPC is a bit more complicated than represented above. A second transition also appears in the calorimetric scan about 8°C lower than the main transition. This smaller transition is often referred to as a pretransition and generally only appears in pure phospholipids, and not in all of those. For example, it is largely absent in PE dispersions. There is a change in tilt of the hydrocarbon chains of the DPPC and a "melting" of the headgroups of DPPC at the pretransition.

Between the pretransition and the main transition temperatures, a rippled phase is observed in freeze-fracture electron microscopy. This is called the $P_{\beta'}$ phase. There is evidence that the hydrocarbon chains adopt an orientation perpendicular to the membrane surface in the $P_{\beta'}$ phase. The pretransition also seems to have something to do with the onset of rapid axial rotation of the phospholipid about an axis perpendicular to the membrane surface. Such axial rotation is found uniformly in the liquid crystal state. Axial rotation is probably hindered below the pretransition due to the tilt of the hydrocarbon chains. Apparently, the phospholipid headgroups adopt the properties of the liquid crystal state at the pretransition of DPPC, whereas the hydrocarbon chains do not gain the properties of the liquid crystal state until the main calorimetric transition.

Below the pretransition, the $L_{\beta'}$ phase is found. Thus the observable phase transitions for aqueous dispersions of dipalmitoylphosphatidylcholine are, starting from the lowest temperature phase,

$$18°C \quad 35°C \quad 41°C$$
$$L_{c'} \leftrightarrow L_{\beta'} \leftrightarrow P_{\beta'} \leftrightarrow L_{\alpha}$$

Because of the orderliness of the bilayer when it is in the gel state, one might expect X-ray diffraction to be a powerful tool in describing the structure of phospholipid phases. Wide angle X-ray diffraction is uniquely sensitive to the gel state of the phospholipids. A sharp 4.2-Å line is generally seen for gel state lipids, whereas a diffuse 4.6-Å reflection is usually observed for lipids in the liquid crystalline state. The change reflects the increase in packing density and order of the acyl chains in the gel state.

Using techniques such as calorimetry and X-ray diffraction, phase transition temperatures for the main transition have been accurately measured for several different phospholipids with different hydrocarbon chains. Within a particular class of phospholipids, chain length is one of the determinants of the transition temperature from the gel state to the liquid crystalline state. The longer the chain, the higher the temperature of transition to the liquid crystalline state, presumably because of favorable chain–chain interactions between the longer hydrocarbon chains when they pack into the gel state. The chain length dependence is represented graphically in Fig. 4.12 for a series of saturated phosphatidylcholines with homogeneous acyl chain composition.[29]

The relationship between the acyl chain composition of saturated phosphatidylcholines and their phase transition temperatures has been described empirically by [29]

$$T_m = 154.2 + 2\Delta C - 142.8\frac{\Delta C}{CL} - \frac{1512.5}{CL}, \tag{4.4}$$

where ΔC is the effective difference in chain length in carbon–carbon bonds in the all-trans configuration, and CL is the effective length of the longer of the two acyl chains in carbon–carbon bonds in the all-trans configuration. Because of the conformation of the phospholipid, the effective chain length of the sn-2 chain is inherently 1.5 carbon–carbon bond lengths shorter than the sn-1 chain (see Section II,A). Thus for 16:0, 16:0 PC, $\Delta C = 1.5$ and $CL = 15$ and $\Delta C/CL = 0.1$. Using Eq. (4.4),

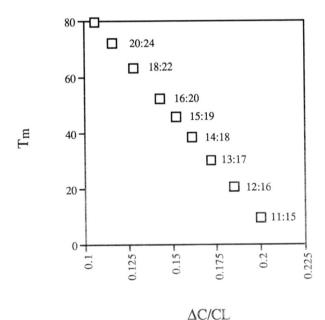

Fig. 4.12. Relationship described by Eq. (4.4). The hydrocarbon chain composition is indicated next to the data points. In general, an increase in chain length is associated with an increase in T_m. Data derived from C. Huang, *Biochemistry* **30** (1991): 26–30.

$T_m = 42.1°C$, compared to the experimentally reported value of 41.6°C. This equation can be used to accurately predict the T_m for virtually all possible disaturated phospholipids with acyl chain lengths between 8 and 26.

An effect on T_m more powerful than the chain length is the presence of double bonds in the phospholipid hydrocarbon chains. This effect can be seen in the difference between the gel to liquid crystalline phase transition temperatures of phosphatidylcholine with stearic acid or oleic acid hydrocarbon chains. The dramatic decrease in that phase transition temperature is caused by the disorganizing (not disordering) effect of the cis double bond. A permanent kink is formed in the hydrocarbon chain by the double bond that inhibits packing of the hydrocarbon chains of the lipid in a gel state. The unsaturation destabilizes the gel state and the phase transition therefore occurs at a lower temperature.

Common monounsaturated fatty acids found in phospholipids usually contain the carbon–carbon double bond between carbons 9 and 10. The placement of the double bond arises through the biosynthetic machinery in the cell and that placement produces a considerable effect on the phase behavior of the lipid. In fact, the double bond is located precisely where the maximal effect on the temperature of the gel to liquid crystal phase transition is achieved. The phase transition temperature of the monounsaturated fatty acid increases in isomers where the double bond is placed nearer to the ends of the fatty acid. When the double bond is placed as close to the end of the hydrocarbon chain as possible, the phase transition temperature of the phospholipid is nearly the same as that exhibited by the fully saturated phospholipid derivative (Fig. 4.13).[30]

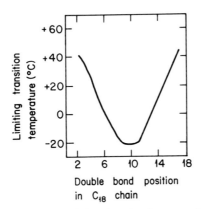

Fig. 4.13. Effect of position of double bond on the phase transition temperature of PC. Reprinted with permission from P. G. Barton and F. D. Gunstone, *J. Biol. Chem.* **250**, (1975): 4470.

Also playing an important role in the phase transition behavior of the phospholipids is the structure of the polar headgroup. Whereas dipalmitoylphosphatidylglycerol has a transition temperature similar to that of dipalmitoylphosphatidylcholine, dipalmitoylphosphatidylethanolamine converts to the liquid crystal about 20°C higher than the former two species. This is probably because the ethanolamine headgroups can intermolecularly hydrogen bond from the amino of one headgroup to phosphates of neighbor phospholipids. Such hydrogen bonds stabilize the gel state relative to the liquid crystal state. This stabilization may also result from the relatively poor hydration of the PE headgroup.

Other effects on the phase transition temperatures are mediated by the phospholipid headgroups. For example, dehydration of the phospholipid, which primarily means dehydration of the headgroup where the water is bound, can convert a phospholipid bilayer into a gel state. As an example, dehydrated dipalmitoylphosphatidylcholine (or the experimentally accessible dihydrate) has a gel to liquid crystalline phase transition temperature 60°C or more higher than the fully hydrated lipid. The $L_{c'}$ phase is a largely dehydrated phase, similar to that found in the crystals of the phospholipids. Forced dehydration of a phospholipid dispersion therefore favors the formation of the $L_{c'}$ phase. Dehydration to this extent appears to require a fair amount of energy. It becomes sequentially more difficult as more waters are removed.

Another membrane phenomenon mediated by the phospholipid headgroups is the binding of cations to the membrane surface. Such binding dehydrates the lipids, increasing the temperature of the gel to liquid crystal phase transition. For example, the gel to liquid crystalline phase transition temperature of PS increases upon calcium binding. Many cations such as calcium mediate important biological processes in the cell. Therefore these phenomena may be important to cell biology, possibly including the fusion of two membranes. Monovalent cations display much less of an effect, presumably because they bind less tightly to the membrane surface.

Whereas pure phospholipids, particularly those with long saturated fatty acids esterified to them, show prominent gel to liquid crystalline phase transitions, it is important to ask what relevance this might have to biological membrane systems. Although there are not many well-documented examples of such phase transitions in native biological membranes near physiological temperature, calorimetric studies of fatty acid-modified cell membranes have provided some of the evidence for the presence of a phospholipid bilayer in those membranes. *Acholeplasma laidlawii* can be supplemented with exogenous fatty acids, thereby changing the fatty acid composition of the membrane. For example, if the fatty acid is palmitic acid, a highly saturated membrane results and it exhibits a broad gel to liquid crystal phase transition centered at 38°C. These calorimetric data appear in Fig. 4.14.[31] The phase transition temperature reflects the nature of the exogenous fatty acid; i.e., the longer the fatty acid and the more saturated the fatty acid, the higher the phase transition temperature. The growth of *A. laidlawii* was also measured as a function of the exogenous fatty acids. Table 4.2 shows the results. The minimum growth temperature shows considerable sensitivity to the fatty acid supplementation, whereas the optimum growth temperature and the maximum growth temperature appear to be determined by other parameters. Note that at the minimum

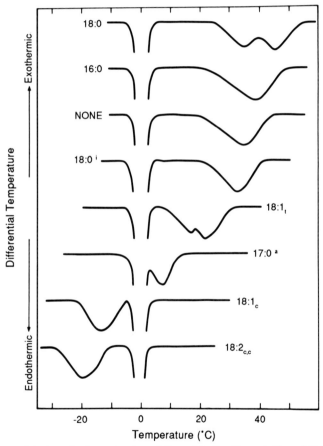

Fig. 4.14. Calorimetric determination of the gel to liquid crystalline phase transition temperature of *Acholeplasma laidlawii* membranes as a function of exogenous fatty acid supplementation. Reproduced with permission from R. McElhaney, *J. Mol. Biol.* **84** (1974): 145.

TABLE 4.2

The Minimum, Optimum, and Maximum Growth Temperatures and the Membrane Lipid Phase Transition Parameters of *Acholeplasma laidlawii* B Cells Grown in Various Fatty Acids

Fatty acid added	Growth temperatures (°C)			Transition midpoint (°C)	Transition range (°C)
	Minimum	Optimum	Maximum		
18:0	28	38	44	41	25–55
16:0	22	36	44	38	20–50
None	20	36	44	34	18–45
18:0^i	18	36	44	32	18–42
18:1$_t$	10	36	44	21	5–32
17:0^a	8	36	44	7	0^a–15
18:1$_c$	8	34	40	−13	−22--4^a
18:2$_{c,c}$	8	32	38	−19	−30--−10

^{a} These temperatures are estimates because a portion of the lipid phase transition endotherms was obscured by the melting of the ice from the excess water associated with the membrane preparations.

growth temperature, a large fraction of the membrane lipids are in the gel state for many of the cases measured. However, in no case when the membrane is fully gel is growth supported. Apparently some liquid crystalline domains are required in the membrane for proper biological function.

Some natural membranes, including microsomal membranes, may also exhibit a gel to liquid crystal phase transition, though not normally in the region of physiological temperature. Furthermore, there may be microscopic domains of lipid in biological membranes that undergo phase transitions near physiological temperatures. Sphingomyelin is a commonly occurring lipid in biological membranes (for example, plasma membranes) that, when isolated in pure form, undergoes a gel to liquid crystalline phase transition near physiological temperature. Thus speculation exists on the presence of sphingomyelin-rich domains in plasma membranes that may be in the gel state under certain conditions. Much remains to be elucidated.[32]

2. LAMELLAR TO HEXAGONAL II PHASE TRANSITION

The transition from the lamellar phase to the hexagonal phase can be detected by a variety of methods. X-ray diffraction shows a distinct change in the spacing of the reflections: $1:2:3$ (lamellar) to $1:\sqrt{3}:2:\sqrt{7}$ (H_{II}). Freeze-fracture electron microscopy reveals the development of tube-like structures characteristic of that phase. ^{31}P nuclear magnetic resonance (NMR) shows pronounced changes in the spectral shape (^{31}P powder pattern) arising from the phospholipids due to the tubular shape of the H_{II} phase (see Fig. 4.7). Calorimetry does not sense as distinctive a change as it does in the case of a gel to liquid crystalline phase transition. An endotherm is observed, but of less magnitude than observed in the gel to liquid crystalline phase transition. Therefore the lamellar to hexagonal phase transition is different in its properties from the gel to liquid crystalline phase transition.

Table 4.3 offers some examples of midpoint temperatures, T_h, for L_α to H_{II} phase transitions of some lipids. Some of the influences on this phase transition temperature, such as unsaturation, are evident in Table 4.3.

Another important observation regarding the L_α to H_{II} phase transition is that the kinetics of the phase change can be much slower than the gel to liquid crystalline phase transition. Particularly notable is that the reversibility of this phase change (H_{II} to L_α) is often not demonstrable on a time scale of seconds or minutes. The kinetics of the latter transition are in fact even slower than the L_α to H_{II} phase transition, creating considerable hysteresis in measurements of this phase transition. Commonly, the L_α to H_{II} phase transition occurs rapidly on a time scale of minutes or faster during the heating of the sample. In contrast the H_{II} to L_α phase transition often requires prolonged incubation (hours or days) at temperatures below the transition temperature. Repeated cycling of the sample through T_h can convert the system to the cubic phase.[26]

These observations suggest unusual kinetics and pathways for this transition. In fact, it is not clear what pathway this transition follows. Some investigators suggest that the transition involves an interlamellar attachment that has geometric similarities with cubic phase.[27] Figure 4.15 shows one model of an intermediate between the L_α and the H_{II} phase. The kinetics of the pathway followed appears to depend on the lipid and ion

TABLE 4.3
Transition Temperatures for the
L_a − H_{II} Phase Transition[a]

Species of PE	T_h (°C)
16:0, 16:0	123
16:0, 18:1	71
18:1, 18:1	10
18:1, 18:1 monomethyl	73

[a] Data from D. Marsh, "CRC Handbook of Lipid Bilayers," CRC Press, Boca Raton, 1990.

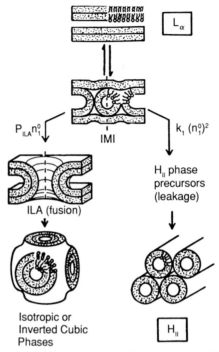

Fig. 4.15. Putative intermediates in the lamellar to hexagonal II phase transition. A representation of the cubic phase is also given. Reprinted with permission from H. Ellens, D. P. Siegel, D. Alford, P. L. Yeagle, L. Boni, L. J. Lis, P. J. Quinn, and J. Bentz, *Biochemistry* **28**, (1989): 3692–3703. Copyright 1989 American Chemical Society.

composition of the system under study. Contact between membranes may be important to facilitate the formation of the tubes of the H_{II} phase.

No structures like those found in the H_{II} phase have been observed (in abundance) in biological membranes. The ramifications of the L_α to H_{II} phase transition for biological membranes would be considerable. The permeability barrier of the cell membrane would be seriously compromised by such discontinuities in the bilayer structure. However, transient formation of nonlamellar (though not H_{II}) structures could occur that would significantly, but not radically, alter the permeability of the membrane. An intriguing possibility is that such transient defects could be controlled by calcium or the levels of diacylglycerol in the membrane. Such transient defects could also be involved in events such as membrane fusion. In the case of L_α to H_{II} transitions that are out of control, degenerative disease states could result.

II. LIPID CONFORMATION IN MEMBRANES

Several aspects of the conformation of lipids in membranes have been established in recent years, mostly from studies on phospholipid model membrane systems. One fundamental aspect has already been alluded to: the lipid molecule, on the average is perpendicular to the membrane surface. One of the possible motions readily available to the lipid is then rotation about an axis perpendicular to the membrane surface. Other motions available to the molecule will be discussed.

The discussion will start by considering the conformation of phospholipids in membranes, with separate examinations of the three structural regions of the lipid: headgroup, glycerol backbone, and hydrocarbon chains. The examples will be primarily derived from phospholipids, because they have been most extensively studied. However, the conformation of the hydrophobic portion of some other lipids, such as simple glycolipids, is well modeled by the phospholipids.

A. Hydrocarbon Chains of the Phospholipids

The hydrophobic region of the lipid molecule usually consists of a linear series of methylene segments. One might expect, therefore, that the two hydrocarbon chains would behave identically. However, several lines of evidence have indicated that the two hydrocarbon chains of the phospholipid are not equivalent in structure. The chemical structure is somewhat different to begin with. One chain is esterified to position 1' of the glycerol and another to position 2' of the glycerol, the latter of which contains an asymmetric carbon. ^{2}H NMR experiments have shown experimentally the nonequivalence at carbon 2 of the hydrocarbon chains by a difference in quadrupole splittings for deuterium attached to that carbon.[33] These data have been interpreted with a model in which the hydrocarbon chain attached to position 2' of the glycerol runs parallel to the surface up to carbon 3, and beyond carbon 3 changes direction to run parallel to chain 1 and perpendicular to the membrane surface. The hydrocarbon chain at position 1' of the glycerol remains perpendicular to the membrane surface along its whole length. Such a conformation has been observed in the X-ray crystal structures available for phospholipids:[34] for the crystal structure of dimyristoylphosphatidylcholine, see Fig. 4.16.[35] The hydrocarbon chain at position 2' of the glycerol does not extend as far toward the bilayer center as does the hydrocarbon chain at position 1' when the two chains have the same number of carbon atoms. The same conformation with respect to the glycerol is observed both in the crystal and in the liquid crystal of a phospholipid membrane.

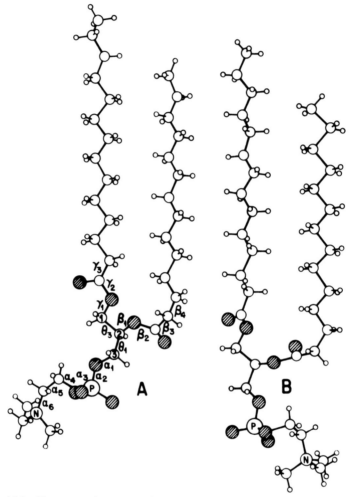

Fig. 4.16. X-ray crystal structure of dimyristoylphosphatidylcholine. (A, B) Two different structures in the unit cell. Reprinted with the permission from R. H. Pearson and I. Pascher, *Nature (London)* **281** (1979): 499.

The inequivalence of the two hydrocarbon chains extends much deeper into the bilayer than the region just described. At position 10 of an oleic acid attached to a dioleoylphosphatidylcholine, the inequivalence is again apparent. ^{2}H NMR spectra of this phospholipid show distinguishable resonances for the two chains.[36] This unambiguously demonstrates that the two chains are not identical even at this position.

What conformation do unsaturated chains adopt when in a phospholipid bilayer? The crystal structure of $18:2 \, \Delta_{9, \, 12}$ has been determined and is presented in Fig. 4.17.[37] Recent elegant ^{2}H NMR studies have examined the conformation of this fatty acid when part of PC in a hydrated bilayer.[38] The results showed that the linoleoyl chain adopted the same conformation in the bilayer that it showed in the crystal structure. More detailed analysis revealed that there were two equivalent conformers and that the chain rapidly jumps between these two conformers.

Such a conformation clearly suggests that unsaturation in lipid hydrocarbon chains leads to the formation of packing defects in the interior of the lipid bilayer. The packing defects enhanced by the presence of cis unsaturated fatty acids are transient voids in the bilayer structure and collectively lead to a "free volume" that can be utilized by membrane proteins to enhance membrane function. The concept of "free volume" will be examined in some detail in Chapter 7 in relationship to membrane protein function.

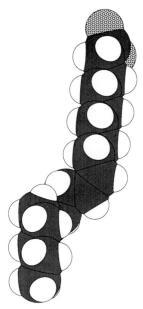

Fig. 4.17. Crystal structure of linoleic acid. Data derived from J. Ernst, W. S. Sheldrick, and J.-H. Fuhrhop, *Z. Naturforsch. B* **34** (1979):706–711.

B. Conformation of the Glycerol Region of the Glycerolipids

The conformation described above requires that the two carbonyls of the ester bonds to the glycerol of the two acyl chains be inequivalent. This inequivalence has been verified experimentally.[39] One of the carbonyls appears to be closer to the membrane surface than the other. The carbonyl closer to the surface is consequently more fully hydrated than the other carbonyl.[40] The carbon backbone of the glycerol is oriented nearly perpendicular to the membrane surface, as shown by the X-ray crystal structures. These structural features are found in glycolipids as well as phospholipids, and in biological membranes as well as model membranes.

C. Phospholipid Headgroup Conformation

Nuclear magnetic resonance studies led to the conclusion that the polar headgroup of phosphatidylcholine is oriented approximately parallel to the membrane surface.[41] This suggestion was strongly supported by some elegant neutron diffraction studies.[42] Electron density profiles of phospholipid membranes are sensitive to the position of deuterium because of its different scattering factors for neutrons. A phospholipid specifically deuterated in a single position will create a characteristic deformation of the profile of the membrane, thereby revealing its position. Using this approach it was possible to measure the distance between each deuterated position on the phosphatidylcholine headgroup and the center of the membrane. Those distances were approximately the same for each position in the phospholipid headgroup, thereby supporting a headgroup orientation parallel to the membrane surface (see Table 4.4).

In addition, the phosphatidylcholine headgroups appear to be interacting intermolecularly, with the positively charged N-methyl group interacting

TABLE 4.4

Distance of Labels in DPPC Bilayer from Bilayer Center[a]

Label position	Distance
$C\alpha$	21.0
$C\beta$	21.2
$C\gamma$	21.8

[a] Data at 50°C in L_α phase. From G. Büldt, H. U. Gally, A. Seelig, J. Seelig, and G. Zaccai, *Nature (London)* **271** (1978): 182–184.

with a phosphate on a neighboring phospholipid. Such intermolecular interactions may contribute to the organization of a membrane, in particular, the organization of the membrane surface.

One method of detecting this interaction involved the measurement of dipolar interaction between the N-methyl hydrogens of PC and the phosphate of a neighboring PC.[43] A nuclear Overhauser effect was observed from the N-methyl hydrogen to the ^{31}P nucleus in the PC phosphate (see Fig. 4.18), indicative of close association of these two portions of the PC molecule. Observation of a similar interaction between the N-methyl hydrogens of PC and the phosphate of PE led to the conclusion that the interaction must be intermolecular. Phosphatidylethanolamine appears to have even a stronger propensity for such intermolecular interactions. Evidence points to a hydrogen bond between the amino group and the phosphate.

Similar conclusions have been reached concerning the conformation of phosphatidylserine, phosphatidylinositol, and phosphatidylglycerol. However, there are some differences. ^2H NMR data suggest that the glycerol portion of PS is similar to that of PC.[44] However, the headgroup behavior is different between PS and PC. These data indicate that the serine headgroup is relatively rigid, compared to that of the PC headgroup.

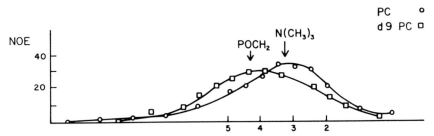

Fig. 4.18. Use of the ^{31}P [^1H] NOE to study the headgroup conformation of PC. The NOE arises from internuclear dipolar interactions that depend in part on the internuclear distance between the protons and the phosphorus. These interactions are revealed by irradiating particular regions of the ^1H NMR spectrum (horizontal axis represents the point of irradiation in the ^1H NMR spectrum employing the normal ^1H NMR chemical-shift scale) and looking for a change in resonance intensity of the ^{31}P NMR resonance of the phosphate (vertical scale in figure). This work demonstrated a close interaction between the N-methyl protons of PC and the phosphate of PC because of the maximum in the NOE at the position of the N-methyl protons. This was confirmed by the removal of the N-methyl protons (by deuteration of the N-methyl protons) and observing a shift in the source of NOE to the other protons in the lipid headgroup. The intermolecular nature of the interactions was discovered by observing NOE in the ^{31}P resonance from PE while irradiating the N-methyls of PC. Reprinted with permission from P. L. Yeagle, W. C. Hutton, C.-H. Huang, and R. B. Martin, *Biochemistry* **16** (1977): 4344, American Chemical Society.

Apparently more than one conformation is possible, and exchange between the available conformations is slow. In contrast to PC where more than one conformation is also possible, exchange between the PS conformations is slow.

The headgroup of sphingomyelin also appears to be parallel to the membrane surface, as in the case of phosphatidylcholine and phosphatidylethanolamine. The summation of available data suggests that these results may be applicable to a wide variety of biological membranes. A model embracing these ideas of phospholipid conformation ideas appears in Fig. 4.19.

The conformation of the polar headgroups of glycolipids is not generalizable. What is clear is that these headgroups probably describe relatively specific structures, rather than randomly orienting themselves in a time-dependent manner at the membrane surface.[45, 46] Thus recognizable structures may be provided at the membrane surface for binding of exogenous molecules including antibodies. The investigation of such structures is a

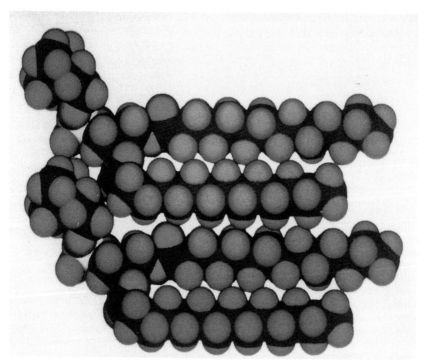

Fig. 4.19. Space-filling model representation of the conformation of dimyristoyl-PC. Figure courtesy of Dr. W. Daux and Dr. J. Griffin.

relatively new field and promises to bring considerable excitement to the study of membranes in the future.

To date, limited measures of phospholipid conformation are available for biological membranes. To the limited extent to which this question has been investigated, the conformations of the phospholipids appear similar in model systems and biological systems.

D. Water

The question of the location of water in the membrane is well studied. Of course, water is largely excluded from the hydrophobic region of the membrane for thermodynamic reasons, arising from the hydrophobic effect. However, this does not mean that water cannot transit the membrane. In fact, water does rapidly transit both model and biological membranes relatively freely. This phenomenon is the basis for the hemolysis of red blood cells, a common experimental procedure in the isolation of red cell components. Decreasing the osmotic strength of the media causes water to flow inside the cell, across the plasma membrane. As a result the cell will burst when the cell's water content swells the erythrocyte membrane beyond its elastic capability.

However, water is not resident inside the membrane for long periods of time. Three sites are known to bind water molecules: the ester carbonyls, the phosphate, and, in the case of PC, the N-methyls.[47] The extent of hydration is a somewhat debatable point, depending upon the method of measurement. However, 16 and possibly more waters are bound per phosphatidylcholine, at full hydration. Fewer are bound to PE. In the crystal structure of DMPC, two water molecules are bound in a hydrogen-bonded structure involving the phosphates. This represents the nearly dehydrated state. The two water molecules are located at the sites of strongest interaction with the lipid.[35]

III. PHOSPHOLIPID MOTIONAL PROPERTIES

Conformational flexibility is the rule, not the exception, in structure. Some of the unique properties of membranes arise from conformational flexibility (and the limits thereof) of the membrane lipids. Conformational flexibility implies motion. The challenge in this section, then, is to get a good picture of the motional characteristics of lipids in membranes. Modern magnetic resonance techniques, including electron spin resonance and nuclear magnetic resonance, have contributed greatly to an understanding of the motional characteristics of phospholipids in membrane bilayers.

Both electron spin resonance and nuclear magnetic resonance are observable because of transitions of electrons or nuclei between energy levels induced by the presence of a large external magnetic field. The structure of these energy levels is described by quantum mechanics and arises from an interaction between the magnetic moment of the electron, or nucleus, and the external magnetic field. Transitions between these energy levels are induced by a time-dependent mixing of the quantum states by an external energy source. Input of the appropriate energy, corresponding to the energy differences between the energy levels, is required for a transition to occur. The energy required in the electron spin resonance experiment is in the microwave region. The energy required for the nuclear magnetic resonance experiment is in the radiofrequency region (FM region).

A. Electron Spin Resonance

Electron spin resonance (ESR) in membranes normally involves the use of spin labels.[48] In order to observe transitions involving electrons, free, unpaired electrons must be available. Spin labels are free radicals containing an unpaired electron whose existence as such is stabilized by the chemical structure of the spin label. An example of this structure is given in Fig. 4.20. This is the spin label Tempo, which was widely used in early ESR studies of membranes. The extra methyl groups around the nitroxide moiety stabilize the free radical. Spin labels of other structures have been more useful in probing the behavior of lipid hydrocarbon chains in the membrane (see Fig. 4.21). Here, the same nitroxide structure is incorporated at a particular position in the hydrocarbon chain of the lipid, in this case a fatty acid. These fatty acids can be used by themselves, or attached to a phospholipid.

One might expect the ESR resonance from such a structure to consist of a single resonance line, because only one isolated electron is available. In fact, as shown in Fig. 4.22, the resonance consists of three lines. This is because the unpaired electron in the nitroxide does not spend all its time on the oxygen, but also resides some of the time on the nitrogen. Therefore the spin system of the electron is perturbed by the nuclear magnetic moment of the nitrogen. Since the spin of the nitrogen nucleus (^{14}N) is 1, the electron spin system is modulated so that transitions of three different energies are possible. This is because the nitrogen nucleus can exist in three spin states designated $+1$, 0, -1. Hence three resonances are observed when these spin labels are incorporated in membranes.

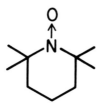

Fig. 4.20. Chemical structure of the spin label Tempo.

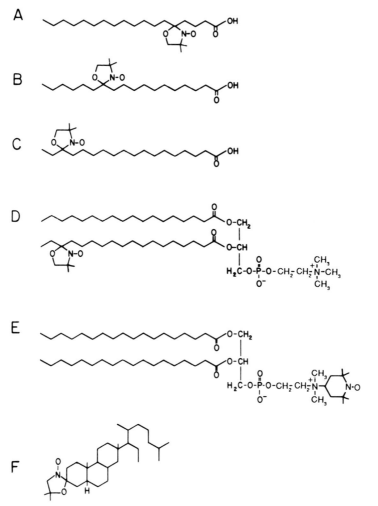

Fig. 4.21. Chemical structures of some spin labels used in membrane studies. (A–C) Fatty acid spin labels; (D, E) phospholipid spin labels; (F) steroid spin label.

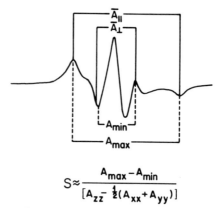

$$S \approx \frac{A_{max} - A_{min}}{[A_{zz} - \frac{1}{2}(A_{xx} + A_{yy})]}$$

Fig. 4.22. Representation of the derivation of an order parameter, S, from the ESR spectra of nitroxide spin labels like (A)–(D) of Fig. 4.21.

B. Nuclear Magnetic Resonance

A complementary technique for studying the behavior of the lipid hydrocarbon chains in membranes, deuterium nuclear magnetic resonance (^{2}H NMR), exploits the use of specifically deuterated lipids to explore lipid behavior in membranes.[49] This approach has advantages over the ESR studies, in that the incorporation of a deuterium at a specific position in a hydrocarbon chain of a lipid is less of a structural perturbation than incorporation of a spin label at the same position. This is readily apparent when comparing the size of a spin label with a deuterium. Also apparent is that the spin label is not as hydrophobic as the unmodified chain, whereas the deuterated lipid is virtually indistinguishable from the unmodified chain in that regard. For both these reasons, ^{2}H NMR of specifically labeled phospholipids in membranes has provided what is probably more accurate information concerning the membrane structure. It is only fair to point out, however, that the ESR experiments, which for the most part came earlier than the ^{2}H NMR experiments, did not lead the field far astray in developing a picture of membrane structure.

^{2}H NMR of specifically deuterated phospholipids is not without problems, either. The sensitivity of the ^{2}H NMR experiment is much less than that of the ESR experiment. Therefore any potential perturbing properties become magnified in the ^{2}H NMR experiment because all or nearly all the lipids in the membrane under study must be deuterated to obtain sufficient sensitivity. In the ESR experiment, one usually uses spin labels at about the 1 mol% level with respect to the normal phospholipids.

Furthermore, fully deuterated phospholipids exhibit a phase transition temperature 2 to 4°C higher than the fully protonated species. Finally, note that the molecular order parameter, S_{mol}, is indirectly obtained (calculated, using a model) in the ^{2}H NMR experiment; S_{CD}, the motional order of the C–^{2}H bond, is directly obtained from the quadrupole splitting observed in the ^{2}H NMR experiment.

With these precautions in mind, ^{2}H NMR has proven to be a powerful technique for studying membrane structure. As in the case of ESR, the technique involves transitions between energy levels, but in this case, nuclei of deuteriums provide the spin system. In the case of an isolated deuterium on a lipid, one might expect a single resonance. However, there is a complication. The nuclear magnetic moment interacts with an electric field gradient controlled by the electrons surrounding the deuterium nucleus. This interaction creates a quadrupole coupling that produces two overlapping powder patterns that are mirror images of each other, rather than one resonance. So in this case rather than the nuclear magnetic moment perturbing the spin system of the electrons (ESR), the electric field gradient (of the electrons) perturbs the nuclear spin system.

Consequently, in the ^{2}H NMR spectra of specifically deuterated lipids in membranes, a doublet is observed. The frequency difference between the maxima of the two overlapping powder patterns corresponds to the observed quadrupole splitting. An example is given in Fig. 4.23.

Nuclear magnetic resonance and ESR measurements, as described, provide information on the motional rates and motional order for lipids in membranes. It is important to have a clear understanding of the differences between these two concepts, because the literature is somewhat confusing (and confused) on these subjects. The result is that one can easily be trapped into incorrect interpretations of experimental data.

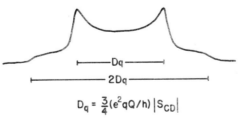

$$D_q = \tfrac{3}{4}(e^2qQ/h)\,|S_{CD}|$$

Fig. 4.23. Representation of the derivation of an order parameter for the C–D bond segment from the ^{2}H NMR spectrum of a specifically deuterated phospholipid.

C. Motional Order

1. METHODS OF MEASUREMENT

The motional order (orientational order) of a lipid molecule or a portion of a lipid molecule is a concept that is theoretically independent of the rates of motion of the same molecule or portions of molecule. Motional order refers to the number of degrees of motional freedom expressed by the particular motion of concern. The order of a motion is normally described by an order parameter. In general, the smaller the value of an order parameter, the greater the number of degrees of freedom experienced by the molecule or molecular segment. Thus in solution, where a molecule experiences extended freedom of motion of all modes of diffusion, often described as isotropic motion, the order parameter takes on its smallest value. In contrast, in a crystal where little motion occurs, the molecule is in its most highly ordered state and the order parameter takes on its highest value. The limits of the order parameter depend on its definition, which can change according to the method of measurement. However, the general features of all the order parameters are the same.

Order parameters for the labeled portion of the lipid hydrocarbon chain can be readily obtained from ESR and ^{2}H NMR measurements. In ESR measurements the order parameter can be derived from the splittings of the spectrum arising from the ESR spin label. One form of that definition is expressed in Fig. 4.22. In ^{2}H NMR measurements the order parameter can be derived from the value of the quadrupole splitting. An example of this is shown in Fig. 4.23. From this measurement, S_{mol} can be calculated, taking into account the orientation of the C–^{2}H bond with respect to the major axis of rotational diffusion for the molecule. In each case, the values of the spectral splittings observed in a membrane are compared to a reference state in which no motion is taking place, such as in a dry powder or in a crystal, to obtain the order parameter.

Fourier transform infrared (FTIR) and Raman spectroscopy have been shown to be sensitive to the conformations of the lipid hydrocarbon chains. In particular, the gauche and trans conformers that populate the hydrocarbon chains can be quantitatively measured. From such analyses, motional order can be assessed.[50-53]

The evaluation of these order parameters is not quite as simple as just described. The rate of motion is involved to a limited extent. If a motion is not fast relative to the time scale of the method of measurement (10^{-8}sec for ESR and 10^{-5}sec for ^{2}H NMR) then that motion will not have any effect on the splittings observed. It will not be counted in the degrees of freedom available to a particular molecular segment, as measured by that technique. Therefore it is possible to get different order parameters by different techniques.

Another important point is that the average orientation of a segment of the hydrocarbon chains relative to the major axis of phospholipid rotation can be important to the value of the splittings obtained. For example, if the orientation is close to 54° with respect to the molecular rotation axis, the splitting can decrease to zero, regardless of the motional order describing that molecular segment. This is because there is a scaling factor of $\langle 3 \cos^2 \theta - 1 \rangle$, where θ is the angle between the labeled segment and the major axis of rotational diffusion. When θ approaches 54.7° (the "magic angle") then the scaling factor is reduced toward zero and the observed splitting vanishes.

2. ORDER PARAMETERS IN MEMBRANES

With these cautions in mind, examine the results for the order parameters as a function of position on the phospholipid hydrocarbon chain. Figure 4.24 shows the data for a membrane consisting of dipalmitoylphosphatidylcholine, in which specific positions have been deuterated.[54] Also shown in this figure are order parameter results obtained from ESR spin labels at various positions of the chains for comparison with the ^{2}H NMR results.

Similar results are obtained from model membranes and from biological membranes. The order parameter profile is characterized by a plateau of constant motional order in the region between the carbonyl and the middle of the chain. In the region near the chain terminus, the motional order decreases significantly. Thus at the ends of the hydrocarbon chains, the molecule experiences considerable motional freedom, compared to the region of the hydrocarbon chain near the glycerol.

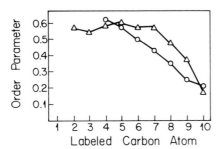

Fig. 4.24. Results from ESR (O) and ^{2}H NMR ($\triangle$) for the determination in a bilayer. Reprinted with permission from J. Seelig, *Q. Rev. Biophys.* **10** (1977): 353–418, Cambridge University Press.

One exception from this general model is found for membranes made of phospholipids with very different chain lengths on positions 1' and 2' of the glycerol. In such cases, it is possible for the hydrocarbon chains to interdigitate. In extreme cases, the terminal methyl of the long chain may reach nearly to the opposite surface of the bilayer. The motional order across this membrane would be different than that for the membranes just discussed.

3. ANALYTICAL DESCRIPTION OF ORDER PARAMETER

Seelig and Seelig[55] have elegantly described the order parameters in terms of molecular parameters. The segmental order parameter for a particular segment of the phospholipid acyl chain can be represented by

$$S = \frac{1}{2}(3 <\cos^2 \theta> - 1), \tag{4.5}$$

where θ is the angle the particular chain segment makes at a time, t, with respect to the bilayer normal (which is the average orientation of the acyl chain as a whole in the liquid crystal state). The segmental order parameter is then obtained as a time average of θ. This is the function that is plotted in Fig. 4.24.

To understand better the behavior of the acyl chains, examine the molecular details of their conformations, as a consequence of their structure and the order imposed upon them by the membrane. For molecules of this kind, the carbon–carbon bond segments can exist in a trans configuration or in one of two gauche configurations. Figure 4.25 shows that the average orientation of a fully saturated lipid acyl chain in the all-trans configuration is perpendicular to the bilayer surface. One can then determine the consequences of the various possible conformations. In the most simple approach, the individual carbon–carbon single bonds can take on two orientations. The direction of the first segment of the chain is parallel to the bilayer normal. The chain segments can be aligned either parallel to the bilayer normal, or at an angle of 60° with respect to the bilayer normal. Figure 4.25 shows some structural possibilities based on combinations of gauche–trans isomerizations about the carbon–carbon bonds.

Designate the probability that a segment is oriented parallel to the bilayer normal by P_A. The probability that a segment is oriented at an angle of 60° with respect to the bilayer normal is designated by P_B. In this simple model in which these are the only two possible orientations,

$$P_A + P_B = 1. \tag{4.6}$$

The order parameter for a particular segment, S_{mol}, is then given by

Structural defect	Conformation	Chain skeleton
All-trans	t t t t t t t t t t t	
2g1 kink	t t t t g^+ t g^- t t t t t t t t g^- t g^+ t t t t	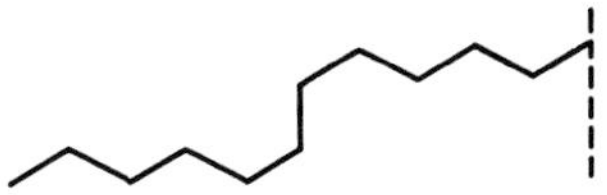
2g2 jog	t t t g^+ t t t g^- t t t t t t g^- t t t g^+ t t t	
3g2	t t t g^+ t g^+ t g^+ t t t t t t g^- t g^- t g^- t t t	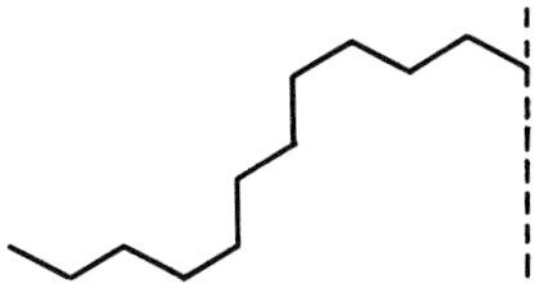
2g3 jog	t t g^+ t t t t t g^- t t t t g^- t t t t t g^+ t t	
3g3	t t g^+ t t t g^+ t g^+ t t t t g^- t t t g^- t g^- t t t t g^+ t g^+ t t t g^+ t t t t g^- t g^- t t t g^- t t	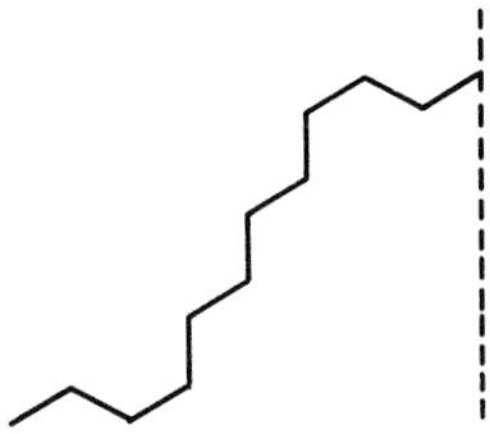
4g3	t t g^+ t g^+ t g^- t g^- t t t t g^- t g^- t g^+ t g^+ t t	

Fig. 4.25. Chain conformations in a bilayer. The first number in the structural defect column gives the total of gauche conformations. The second number indicates the shortening of the chain in units of $l = 1.25$ Å compared to all-trans conformations.

$$S_{mol} = P_A(\tfrac{1}{2})(3\cos^2 0° - 1) + P_B(\tfrac{1}{2})(3\cos^2 60° - 1). \quad (4.7)$$

Solving for P_B then provides

$$P_B = \frac{(1 - S_{mol})}{1.125}. \quad (4.8)$$

In this approximation of the acyl chain behavior, it is possible to calculate the thickness of a bilayer, based on the order parameters, again as shown by Seelig and Seelig. The effective length of a carbon–carbon bond segment is 1.25 Å, projected along the bilayer normal. This will be referred to as l. Therefore if the segment is oriented parallel to the bilayer normal, the effective length is $d = l$. If the segment is oriented at the angle of 60°, then the length, d, is

$$d = l \cos 60°. \tag{4.9}$$

With these definitions, the average effective length of a hydrocarbon chain can be expressed as

$$<L> = \sum_{i=1}^{15} <l_i> = \sum_{i=1}^{15} l(1 - 0.5P_{iB}), \tag{4.10}$$

where the time average is actually being calculated. One can now substitute the results from Fig. 4.24 into Eq. (4.10) using Eq. (4.8). This predicts a chain length of 13.15 Å at 50°. The same chain in the all-trans configuration would be $L = 15(l) = 18.75$ Å. Therefore, the difference in chain length between the liquid crystal state and the gel state (all-trans) is about 5.6 Å. Since two chains make up the phospholipid bilayer this analysis predicts that the average bilayer thickness will be reduced by 11.2 Å when the bilayer goes from the gel to the liquid crystal state. Experimental data from X-ray diffraction indicate that a reduction of bilayer thickness, as measured between the two phosphate regions, of 11.6 Å actually occurs upon going from the gel to the liquid crystal state. Thus the model of Seelig and Seelig is useful in understanding the order parameters describing the hydrocarbon chains of the phospholipids.

4. EFFECTS OF PHASE TRANSITION ON ORDER PARAMETERS

The effects of a gel to liquid crystalline phase transition on the motional behavior of a phospholipid can now be more fully described. In the gel state, in contrast to the liquid crystal state, the motional order of the hydrocarbon chains increases in all chain segments. This increase in order reflects a decrease in the extent of gauche–trans isomerization for the lipid hydrocarbon chains in the gel state. In fact, the chains tend to adopt the all-trans configuration in the gel state, as just described. In addition, the rate of phospholipid motion and the rate of lateral diffusion of the phospholipids in the plane of the membrane decrease in the gel state (see later in this chapter). Therefore, the gel state corresponds to a highly motionally restricted state for the hydrocarbon chains, both in motional freedom and in rates of motion.

The effects of an L_α to H_{II} phase transition on the motional order profile of the lipid hydrocarbon chains are not as dramatic as for the transition to the gel state. A gradient in the order parameters along the length of the lipid hydrocarbon chain, similar to that seen in the L_α phase, is preserved in the H_{II} phase. In general, most positions in the hydrocarbon chain in the H_{II} phase appear to be slightly disordered compared to the L_α phase. However, the observable, for example, the ^2H quadrupole coupling, decreases by a factor of 2, due to the rapid diffusion about the cylinders of the H_{II} phase.

5. EFFECTS OF DOUBLE BONDS ON ORDER PARAMETERS

The effects of double bonds on motional order are not always what one might expect. When the order parameters are properly corrected for orientation effects (of great importance, for example, with the 10 position of oleic acid) the profile of motional order is not much different in the presence of one double bond than in the absence of that double bond. (In fact double bonds may actually increase slightly the order of a membrane lipid hydrocarbon chain due to the inhibition of rotation about the double bond.) This is in spite of the fact that the presence of a double bond lowers the gel to liquid crystalline phase transition temperature by 60°C or more. Therefore the energy required to pack phospholipids in a uniform array (to be able to enter the gel state) is not reflected in the order parameter profile. Perhaps for this reason, the order parameter profiles of biological membranes are the same as for model membranes. It also indicates that intuitive feelings about lipid behavior are not always borne out, and one must be very careful of definitions.

Incorporation of cyclopropane into a hydrocarbon chain of a phospholipid produces an effect similar to a carbon–carbon double bond. The gel to liquid crystal phase transition temperature is dramatically lowered. Hence microorganisms that produce large amounts of such derivatized lipids can maintain their membranes in the liquid crystal state without carbon–carbon double bonds.

The effect of the cyclopropane ring on motional order is more rigorously described than the carbon–carbon double bond. This is because the geometric factors governing orientation are more clearly defined. The cyclopropane ring actually increases the motional order of the hydrocarbon chain in the region of the ring (see Fig. 4.26).[56] This is because the cyclopropane structure reduces the number of rotational isomers possible. Hence motional order is essentially an entropic parameter.

The divergence between intuition and motional order is now even more clear. Both carbon–carbon double bonds and cyclopropane rings inhibit packing of hydrocarbon chains in a gel state. Yet neither structure de-

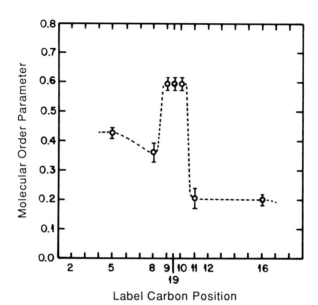

Fig. 4.26. Effect of a cyclopropane substituent on the lipid hydrocarbon chain on the outer parameter profile of a bilayer. Reproduced with permission from E. J. Dufourc, I. C. P. Smith, and H. C. Jarrell, *Chem. Phys. Lipids* **33** (1983): 153–177.

creases the motional order of the lipid in the membrane. In fact one, and perhaps both, actually increases motional order of the individual lipid chains. Therefore, the role of multiple (i.e., more than one per lipid) double bonds in membrane lipids is not simply to make membranes less motionally ordered. The role of unsaturation of lipid hydrocarbon chains in membrane structure and function will be considered in more depth in Chapter 7.

6. PERMEABILITY AND MEMBRANE ORDER

The behavior of the hydrocarbon chains in the membrane suggests an interesting mechanism for rapid permeation of a membrane by small molecules like water. The formation of trans–gauche isomers can lead to the formation of kinks in the chain, as described above. Because coupled trans–gauche isomerizations can occur, it is feasible for these kinks to diffuse up and down the chain. More than one such kink will be in each chain, on average, in the liquid crystal state. These kinks provide structural discontinuities in the bilayer in which small molecules can reside, even if only transiently. Therefore, small molecules can move through the membrane following kink diffusion, or by jumping from one discontinuity to another. For the smallest of molecules, such as H_2O, diffusion across model and biological membranes is therefore fast, even though the partition coefficient for water between the aqueous phase and the liquid crystal of the membrane heavily favors the aqueous phase.

7. EFFECTS OF STEROLS

The emphasis so far has been on the behavior of phospholipids in bilayers. The question of the effects of cholesterol will now be considered briefly (see Chapter 5 for more detail). ^{2}H NMR measurements indicated that this sterol experiences axial diffusion about the long molecular axis of the molecule.[57] The motion of this rotational axis experiences a high degree of order and is oriented perpendicular to the membrane surface. Cholesterol is located primarily in the region of the membrane that is described by the plateau of order parameters (for the lipid hydrocarbon chains). The order parameter of the steroid nucleus of cholesterol therefore mimics the order of the plateau region associated with the upper portion of the phospholipid hydrocarbon chains.

The motional order characterizing the dynamics of the cholesterol molecule has been determined. $S_{\perp c}$, characterizing the "wobble" or the order of the axis for rotational diffusion for cholesterol, is in the range 0.45–0.5.[58] This order parameter appears to be relatively insensitive to temperature and cholesterol concentration in the liquid crystal state. These results were in marked contrast to the increase in ordering of membrane lipids observed on introduction of cholesterol into the lipid bilayer.

The effects of cholesterol on the order parameter profile of the membrane lipid hydrocarbon chains are more predictable. Because cholesterol has a rigid steroid nucleus, one might expect that the presence of cholesterol in the membrane would decrease the ability of hydrocarbon chains to undergo gauche–trans isomerizations. This would be reflected in an increase in order parameter for the lipid hydrocarbon chains, as shown in Fig. 4.27.[59]

D. Motional Rates

Rates of motion refer to rates of axial diffusion, translational diffusion, etc. The rates of motion are characterized by correlation times, which in turn can be inferred to represent an exponential correlation function describing a motion, in the simplest of cases. In principle, the rates of motion are independent of the motional order. This is sometimes confusing, since, in practice, changes in motional rates often parallel changes in motional order. However, such apparent coordinate behavior of motional order and motional rates is not required. Some cases exist that display the mutual independence of motional order and motional rates.

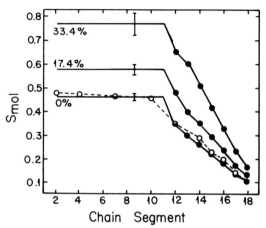

Fig. 4.27. Effect of cholesterol on the molecular order parameter determined from the ^{2}H NMR of deuterated fatty acids in egg PC, at 0, 17.4, and 33.4 mol% cholesterol. Reprinted with the permission from B. W. Stockton and I. C. P. Smith, *Chem. Phys. Lipids* **17** (1976): 251.

Three types of motion have been identified as important to understanding lipid dynamics: axial diffusion, time-dependent orientation of the director for axial diffusion or wobble, and rotation about single bonds. The latter manifests itself as trans–gauche isomerization about carbon–carbon single bonds in the lipid hydrocarbon chains. Figure 4.28 presents a schematic representation of these motions for both the phospholipid headgroups and the hydrocarbon chains.

In this section the phospholipid hydrocarbon chains and headgroups will be considered separately. As will be seen, there is evidence that the headgroups and hydrocarbon chains have some independence of motion and thus a separation of their behavior in this discussion is appropriate.

The correlation times for axial diffusion, wobble, and single bond rotation are on different time scales. The most rapid of these motions is rotation about single bonds, although not all single bonds may be subject to rapid rotation. This motion is characterized by the correlation time τ_J and often $\tau_J < 10^{-10}$ sec. In the case of lipid hydrocarbon chains, there are some steric hindrances to bond rotation such that gauche $(+,-)$ and trans conformers are usually the most populated.

The next most rapid motion is generally considered to be axial diffusion. This motion is characterized by the correlation time $\tau_\parallel$. The director for axial diffusion in the liquid crystal state is normally perpendicular to the bilayer surface. In some gel phases, the director for axial diffusion of the lipid hydrocarbon chains is tilted away from the perpendicular to the bilayer surface.

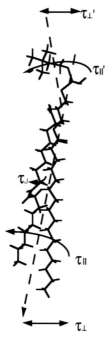

Fig. 4.28. Schematic representation of the dominant motions described for a lipid molecule in a bilayer. The individual motions are defined in the text.

The third motion important to lipid dynamics as they can currently be measured is time dependence of the orientation of the director for axial diffusion, or wobble. This is characterized by the correlation time $\tau_\perp$. Another view on this motion is to picture the axis for axial diffusion randomly sampling a cone of finite dimensions. Thus this is a restricted motion, in that in a lipid bilayer, not all orientations of this axis (with respect to the normal to the bilayer surface) are equally probable. In fact, only limited deviations from the bilayer normal are typically probable.

Magnetic resonance and time-resolved fluorescence provide the primary means to probe lipid dynamics. The nuclei most commonly employed in NMR studies for probing lipid dynamics are ^{2}H NMR and ^{31}P NMR. Measurement of ^{2}H relaxation provides information on the correlation times characterizing the relevant motional rates. ^{31}P relaxation provides information on the correlation times characterizing the motions relevant to phospholipid headgroups. Electron spin resonance of spin labels provides an alternative view of the lipid bilayer, complementary to the view from ^{2}H NMR.[60] In this case lineshape analysis and relaxation studies offer information on motional rates.

Utilizing different technology, the decay from the excited state of fluorophores in a membrane can be exploited to obtain information on lipid dynamics. In particular the decay of the anisotropy of emission can provide correlation times for important lipid motions if sufficient information about the fluorophore is available. The residual anisotropy at long decay times may reflect the motional order of the system. Since the lifetime of the excited state of common membrane fluorescent probes is in the range of 1 to 10 nsec, this technique usually provides information on correlation times near that time scale. Alternative measurements with phosphorescence of membrane probes can identify correlation times several orders of magnitude longer. For example, rotational diffusion of membrane proteins with correlation times in the neighborhood of 10–100 μsec can be measured.

1. MOTIONAL RATES FOR THE DOMINANT HYDROCARBON CHAIN MOTIONS IN THE LIPID BILAYER

The trans–gauche ($\pm$) isomerizations in the liquid crystalline L_α phase are rapid; $\tau_J \approx 10^{-11} - 10^{-12}$ sec. The activation energy for this process in the L_α phase as determined from Arrhenius plots is 8–15 kJ/mol. The correlation time for rotation of the terminal methyl group of the myristic acid was nearly an order of magnitude shorter than τ_J.

Nuclear magnetic resonance relaxation data show a mobility gradient along the hydrocarbon chains.[61] In particular, the values for τ_J are relatively long near the glycerol backbone. The values for τ_J are much shorter for the rest of the hydrocarbon chain and vary little from the middle of the chain to its terminus, in contrast to the profile across the bilayer of order parameter. Both $\tau_\parallel$ and $\tau_\perp$ have been determined from ^{2}H NMR of specifically deuterated DMPC in the L_α phase. The values for $\tau_\perp$ decrease from 5×10^{-8} s to 1×10^{-8} s in the temperature range of 23 to 60°C. This is a much stronger temperature dependence than that seen for τ_J. The activation energy for this process as derived from an Arrhenius plot is in the range 60–90 kJ/mol. This represents an energetically more difficult process than that represented by τ_J. The motion characterized by $\tau_\perp$ involves intermolecular interactions in the bilayer since it represents motion of the chain as a whole. Wobble of one phospholipid necessarily influences (and is influenced by) a neighboring phospholipid. In other words, one phospholipid cannot "wobble" into space occupied by another phospholipid unless the other phospholipid leaves that same space on the same time scale.

In contrast, axial diffusion was characterized by a correlation time about 10 times shorter than $\tau_\perp$, but with an activation energy similar to that for wobble.

The values for $\tau_\perp$ and $\tau_\parallel$ were dramatically longer in the $P_{\beta'}$ phase of DMPC. The temperature dependence of $\tau_\perp$ and $\tau_\parallel$ was characterized by the same activation energy as found in the L_α state. In the $P_{\beta'}$ phase, $\tau_\perp$ was in the range 5×10^{-7} to 5×10^{-6} sec and $\tau_\parallel$ was again an order of magnitude longer. The values for τ_J for th $P_{\beta'}$ phase are significantly longer than in the L_α phase.

In the $L_{\beta'}$ phase, $\tau_\perp$ and $\tau_\parallel$ become significantly longer with decreasing temperature, but with no discontinuity characterizing the transition from the $P_{\beta'}$ phase to the $L_{\beta'}$ phase. τ_J does not change from the $P_{\beta'}$ phase to the $L_{\beta'}$ phase, and τ_J is largely independent of temperature in the $L_{\beta'}$ phase.

The rates of motion in the liquid crystalline state and in the gel state as determined from ^{2}H NMR can now be summarized. The rates of trans–gauche isomerizations display low apparent activation energies in the gel state and in the liquid crystalline state since they only involve single bond rotations and thus are strictly intramolecular, but show a large change in magnitude at the L_α to $P_{\beta'}$ phase transition. The rates of wobble and axial diffusion appear to have a large activation energy, presumably because they involve effects from more than one phospholipid molecule and thus in a sense are intermolecular. The rates of wobble and axial diffusion change at the L_α to $P_{\beta'}$ phase boundary, decreasing nearly two orders of magnitude from the L_α phase to the $P_{\beta'}$ phase.

2. PHOSPHOLIPID HEADGROUP MOTIONS

The data on phospholipid headgroup dynamics are not yet as clear as those for the phospholipid hydrocarbon chains. However, it is possible to identify the dominant motions of the lipid headgroups in a manner analogous to the treatment of the phospholipid hydrocarbon chains.

Three kinds of motions can be defined for the phospholipid headgroups. Two of these are represented schematically in Fig. 4.28. One, characterized by $\tau_{\parallel'}$, reflects axial diffusion of the headgroup. Another, characterized by $\tau_{\perp'}$, reflects time dependence of the orientation of the director for headgroup axial diffusion, or wobble of the headgroup. The last kind of motion, characterized by $\tau_{J'}$, reflects internal motions of the headgroup dominated by isomerizations around the chemical bonds of the headgroup structure.

In the liquid crystal state, $\tau_{\parallel'}$ is in the nanosecond range.[62, 63] Unfortunately, values for the other correlation times mentioned above have not been well defined.

The dynamics of the phospholipid headgroups and the hydrocarbon chains are not the same. $\tau_{\parallel'}$ for the headgroups is an order of magnitude shorter than $\tau_\parallel$ for the hydrocarbon chains. Thus the axial diffusion of the phospholipid headgroups is more rapid than the axial diffusion characterizing the phospholipid hydrocarbon chains. This indicates some independence in the dynamics of the phospholipid headgroups from the phospholipid hydrocarbon chains. That independence was further defined as an observation of "melting" of the phospholipid headgroups of dipalmitoylphosphatidylcholine at the $P_{\beta'}/L_{\beta'}$ phase boundary, whereas the main chain melt is at the $L_\alpha/P_{\beta'}$ phase boundary.[64]

3. STEROL DYNAMICS

Sterol dynamics can be described by $\tau_{\perp c}$, a correlation time for "wobble" of the sterol about its long axis, and $\tau_{\parallel c}$, characterizing axial diffusion. $\tau_\perp$ in phosphatidylcholine bilayers is about 1–3 nsec, according to time resolved fluorescence of cholestatrienol, a fluorescent analog of cholesterol. There is little dependence of $\tau_\perp$ on cholesterol content or on temperature.[65] Spin label studies with a sterol analog suggested that $\tau_{\parallel c} = 0.1$ nsec, the correlation time for axial diffusion.[66]

E. Motional Order and Dynamics: A Combined View

Through a combination of the information on motional order and motional dynamics of lipids in bilayers, an interesting picture of the profile of a membrane emerges. In the midplane of the membrane bilayer (in noninterdigitated phases), the terminal methyls, and nearby methylenes, of the lipid hydrocarbon chains experience considerable motional freedom. The rates of motion in this region of the hydrocarbon are not exceeded by any other regions of the hydrocarbon chain. The bilayer midplane is thus the region of the membrane that most nearly approximates an isotropic organic hydrocarbon phase and is where some hydrophobic molecules will dissolve when introduced to the membrane.

In contrast, the region of the membrane defined by carbons 2–9 of the hydrocarbon chains is much more ordered than the terminal regions. This characterizes a highly anisotropic region of the lipid bilayer, distinctly different from any liquid hydrocarbon. The glycerol region of the phospholipid is highly ordered. The glycerol is therefore referred to as the backbone of the lipid in the bilayer. These conclusions suggest that a foreign molecule entering a membrane or traversing a membrane would encounter a highly anisotropic medium.

F. Fluidity

With this understanding of phospholipid properties in membranes, it is now possible to address a widely used, but poorly defined, concept in membrane biology: fluidity. Some of the connotations of the word fluid seem to apply to the nature of a lipid bilayer. Molecules can move laterally in the membrane, diffuse about their long axes, and oscillate about the normal to the membrane. Intuitively it seems feasible to define some parameter that would quantitatively describe how fluid is the interior of a membrane. One might use the viscosity experienced by a molecule in the membrane or a measure of the freedom of lipids and other molecules to move in the bilayer. Because of the strong intuition surrounding this concept, fluidity has come into general use by the community of investigators concerned with membrane properties and functions.

The problem with its usage is that the term fluidity has no precise meaning in the context of cell membrane structure. In physics, fluidity is the inverse of viscosity and is defined in an isotropic fluid. However, biological membranes and lipid bilayers are highly anisotropic. The membrane and its components are largely confined to two dimensions, with only a limited third dimension available. The hydrophobic membrane interior cannot be mimicked by any isotropic organic phases, as has been seen earlier in this chapter. The order parameter analysis shows that the center of the bilayer is nearly isotropic, but that only a few angstroms away toward the membrane surface, the system is highly ordered. Thus not only is the phospholipid behavior anisotropic, but the environment for a foreign molecule is dramatically dependent on the location of that foreign molecule in the membrane. To describe this behavior, a variety of position-dependent order parameters are required. Therefore, the problem with attempting a description of the interior of a membrane with a single fluidity parameter should be evident.

However, experiments that intend to measure membrane fluidity have been performed. The data obtained in such experiments have shown sensitivity to changes in the condition of the membrane being probed. An example of a measurement designed to assess membrane fluidity is the steady-state anisotropy of the fluorescence of the hydrophobic probe molecule diphenylhexatriene (DPH).[67] The fluorescent properties of this molecule are sensitive to a gel to liquid crystalline phase transition. Furthermore, the same measurement shows sensitivity to the presence of membrane proteins. Therefore, it is important to determine what is being measured or to redefine the concept of fluidity.

The physical measurement of fluorescence anisotropy is an optical measurement that refers to a short-term scale. The fluorescence anisotropy is a measure of how effectively a probe molecule can depolarize plane polarized excitation light during the interval between absorption of the polarized light and subsequent emission of fluorescence. In many systems this depolarization occurs via a randomization of molecular orientation on a time scale comparable to the fluorescence lifetime. The experimental result is an emission with a degree of polarization different from that exhibited by the excitation beam. To be effective, the motion must occur during the lifetime of the excited state, a period of nanoseconds or tens of nanoseconds. This measurement therefore contains information about molecular rotation. It should be noted that the fluorescent probes used for probing membrane properties in this manner exhibit anisotropic motion in the membrane. Therefore probe dynamics must be described by several different motions, each with potentially different rotational correlation times. This should come as no surprise, considering the anisotropic structure of a membrane and the properties of the constituent lipid molecules.

However, since this fluorescence depolarization is sensitive to molecular motion, studies of this sort do provide information concerning the environment in which the probe resides. Therefore, one could expect that anything that changes the environment of the membrane may also change the motional characteristics of a probe molecule in that membrane.

However, the question of membrane fluidity still remains. Fluidity is clearly not describable by a single parameter. Perhaps the concept can best be appreciated from the point of a view of a foreign molecule in the membrane: fluidity might then represent the ability of a foreign molecule to move in the membrane and to experience internal conformational flexibility.

From this viewpoint fluidity must be dependent on size. A molecule that is small relative to the size of a phospholipid will be sensitive to the order parameter gradient across a membrane. In contrast, a large molecule, like an intrinsic membrane protein, will likely be much less sensitive to such a variation in properties across the membrane. For a small molecule, lateral movement will be different in the center of the bilayer than in the phospholipid ester carbonyl region of the membrane. The rotational characteristics of the small molecule will depend on whether it induces its own microenvironment, or whether it seeks its own microenvironment in the membrane. It may, for example, either create or seek out pockets where it can tumble relatively freely, more freely than one might otherwise expect. The state of the membrane as a whole will also affect the behavior of this molecule. For example, if the membrane is in the gel state, all forms of translational diffusion will be severely inhibited.

For a small molecule as defined here, fluidity is closely related to the rate of motion leading to conformational changes among the lipids constituting the bilayer in the membrane. The reason for this can be seen in the following model. In order for a small molecule to move through a membrane, space must be made for it by the phospholipids where no space existed previously (one cannot have large areas of vacuum on the molecular level existing in the membrane). The space for the foreign molecule must be created by changing the conformation of the lipid hydrocarbon chains, as in the case of kink formation. The rate at which such spaces form is dependent upon the rate of conformational change, which is in turn dependent on the rate of rotation of carbon–carbon bond segments on the lipids. Therefore one might expect a relationship between the rate of motion as just described and the fluidity of a membrane. The molecular order of the membrane is a largely independent parameter. Since fluidity, as it has been defined here, is highly location dependent, the location of the small foreign molecule must be known before meaningful information can be extracted concerning membrane properties.

For an intermediate-size molecule, that is, one similar in size to a lipid, rotation will be affected by the same parameters. Rotation of the molecule will require a change in the shape of the defect in which the foreign molecule resides. The change in shape can be achieved only by a change in lipid conformation. Translation of this molecule is comparable to translation of the lipid since the probe and the lipid are of similar size. Therefore, these probes can be good reporters of lipid translational diffusion and have been so used. For molecules of this size, conformational flexibility is likely not affected by the environmental variations found in the structure of the membrane.

For the large molecule, such as a transmembrane protein, the concept of fluidity has little meaning. Translational diffusion is not very dependent on size, at this level. Since it encounters all portions of the bilayer cross section, the gradient of order parameters and motional rates is not likely to be important to the movement of protein in the membrane. Only dramatic changes in the membrane state have effects on protein activity: a change from a liquid crystalline to a gel state of the lipid can strongly affect enzymatic activity, which might reflect a perturbation of conformation or conformational flexibility. Reactions that involve proteins colliding through lateral diffusion processes will be inhibited by gel state liquid, due to the reduction in lateral diffusion rates of the protein.

What about the case of a long, rod-like molecule, like the fluorescent probe parinaric acid? The rotational diffusion of these molecules will be strongly affected by their environment. In solution, rotational diffusion about the long molecular axis is favored over rotation about other axes, i.e., tumbling of the long axis. In the ordered environment of the membrane, this favoritism is enhanced. Rotation about the long axis is not greatly affected. Tumbling of the axis is severely limited by the organization of the bilayer. Measurements of fluidity from these probes largely reflect the extent to which the long axis of the molecule can experience deviations from an orientation perpendicular to the membrane surface.

Careful time-resolved analysis of such data has now provided true order parameters. The fluorescent probes have therefore proven very useful in membrane studies.

In conclusion, the term fluidity as it is usually used is not a useful term to describe membrane properties. Fluidity suggests that order and dynamics should be linked, but this is not necessarily the case. For these reasons, it would seem wise to avoid the term fluidity, and use only the well-defined concepts of motional order and motional rates. A third concept, free volume, will be discussed in more detail in Chapter 7.

G. Lateral Diffusion

Translational diffusion of membrane components in the plane of the membrane is called lateral diffusion. Both lipids and proteins can diffuse laterally, and measurements have been made of diffusion rates for a number of representatives of each of those classes of membrane components.

One of the early graphic demonstrations of such phenomena in biological membranes was reported by Frye and Edidin and described in Chapter 2. These experiments demonstrated that the labeled proteins were diffusing over the surface of the hybrid cell. Therefore, at least with respect to these labeled components, the cell plasma membrane allowed lateral diffusion.

Although the picture today has become more complicated, the initial impression left by this work is largely correct. Both the lipid components and the protein components of biological membranes, and model membranes, are capable of lateral diffusion (although all do not show it). A variety of techniques have been developed to measure such phenomena. Many involve fluorescence of probe molecules in the membrane or antibodies bound to the membrane components. Generally, these techniques involve the bleaching of some of the probe chromophores in the membrane, in particular those residing at a particular time in a defined area of the membrane surface. The rate at which the fluorescent probes then regenerate a random pattern in the membrane surface is measured. From this a diffusion coefficient can be obtained.

On obtaining diffusion coefficients for a variety of membrane components, a pattern emerges. Examples are given in Table 4.5. Nearly all the lipids measured, whether in biological membranes or in model membranes, exhibit similar diffusion coefficients. This is another piece of evidence that the phospholipid bilayer is an important element of biological membrane structure. The major perturbant on this lateral diffusion is a liquid crystal to gel phase transition. As expected, gel phase dramatically reduces the

TABLE 4.5
Lateral Diffusion in Membranes

Component	System	T (°C)	D (cm²/sec)
Lipid	DMPC	12	2×10^{-10}
Lipid	DMPC	21	3×10^{-8}
Lipid	Egg PC	20	2×10^{-8}
Lipid	Fibroblasts	—	10^{-8}
Proteins			
Glycophorin	DMPC	15	10^{-8}
Glycophorin	DMPC	25	10^{-10}
M-13 coat protein	DMPC	25	3×10^{-8}
Rhodopsin	DMPC	25	2×10^{-8}
Rhodopsin	ROS disk membrane	—	4×10^{-9}
Band 3	DMPC	—	10^{-8}
Band 3	Erythrocyte	—	5×10^{-11}
Acetylcholine receptor	DMPC	—	$1\text{-}3 \times 10^{-8}$
Acetylcholine receptor	End plates	—	Immobile

rate of lateral diffusion. In fact, the lateral diffusion coefficient decreases by about three orders of magnitude.

The results for membrane proteins are more complicated. As predicted there is not much size dependence on the lateral diffusion coefficient, so that even molecules as large as proteins can diffuse laterally in the plane of the membrane. However, in some membranes not all the membrane proteins are freely diffusable. As an example, in the erythrocyte membrane, some of the band 3 proteins are mobile and some are immobile (the latter condition due to links of band 3 to the membrane skeleton).

H. Lateral Phase Separation

If lipids can diffuse freely and rapidly in the plane of the membrane in both model and biological systems, then one might expect that the surface of a membrane will be homogeneous with respect to lipid composition. However, surprises are in store. It appears that it is possible to have domains of different lipid composition coexisting with other domains in the same membrane. The best known are those that result from some environmental perturbation of the membrane. However, many hints are available that in biological membranes, with their complex lipid composition, microdomains exist and may play functional roles. These suggestions are largely still in the realm of speculation for biological membranes because of a lack of adequate measuring techniques. Therefore this discussion will emphasize examples better known and studied in model membrane systems.

One such example is given by lamellar phases undergoing a gel to liquid crystalline phase transition. For a single species of lipid, such as dimyristoyl phosphatidylcholine, the phase transition is nearly a two-state process. The transition occurs over a narrow temperature range, and therefore, at any given temperature, the system is essentially all in the gel state, or all in the liquid crystal state.

However, in mixtures of lipids, the systems become more complex. As an example, consider the dimyristoylphosphatidylcholine (DMPC)/distearoylphosphatidylcholine system (DSPC). The phase diagram of this system is shown in Fig. 4.29.[68] Such phase diagrams can be constructed from, for example, the ESR spectra of a membrane probe like Tempo. Tempo shows a differential solubility in liquid crystal and gel state bilayers. Therefore the temperature dependence of the ESR spectra shows sensitivity to phase boundaries.

This diagram can be read in the following way. Start at a given total membrane composition. At a 50:50 overall mixture and at 20°C, the system is totally in a solid state, with a composition equal to the overall composition. At higher temperatures and at the same overall composition, one approaches a phase boundary. At 30°C, an interesting result obtains: there are two kinds of domains of phospholipids in the membrane at this temperature, each having a different lipid composition. The two domains are solid (S) and fluid (F), or gel and liquid crystal. The fluid domain has a composition that is dominated by DMPC; the solid domain, on the other hand, consists largely of DSPC. When the temperature is increased above 50°C at this same composition, all the phospholipid is found in the liquid

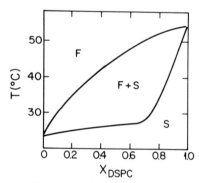

Fig. 4.29. Phase diagram of the system DSPC/DMPC. Reprinted with permission from E. J. Shimshick, W. Kleeman, W. L. Hubbell, and H. M. McConnell, *J. Supramol. Struct.* **1** (1973): 285, A. R. Liss.

crystal state and is apparently homogeneous in lateral distribution. One should note, however, that just because other phase boundaries are not detected does not necessarily mean that they do not exist. Therefore, although the liquid crystal state may be homogeneous, one should be careful not to make too many assumptions. Finally lateral diffusion across the phase boundary, from one microdomain to another, is slow for the phospholipids.

Another example of the coexistence of microdomains of differing composition concerns membranes containing cholesterol. A phase diagram of the system DPPC/cholesterol shows that this system exhibits immiscibility of domains, even in the liquid crystal state (see Fig. 4.30).[69] Considerable regions exist on the phase diagram in which gel and liquid crystal microdomains coexist and other regions where two liquid domains exist. The liquid crystal domains are enriched in the sterol, which tends to inhibit

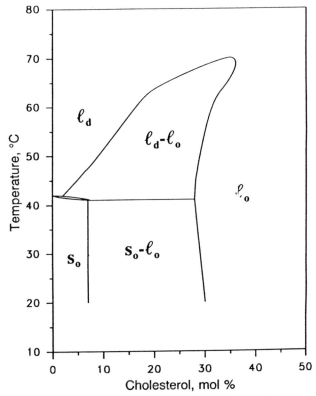

Fig. 4.30. Phase diagram of the DPPC/cholesterol system. Reprinted with permission from M. B. Sankaram and T. E. Thompson, *Proc. Natl. Acad. Sci. U.S.A.* **88** (1991): 8686–8690.

the formation of the gel phase by phospholipids. In fact, at concentrations as low as 5 mol% cholesterol, lateral inhomogeneities in the plane of the membrane exist. The two liquid crystal states in this phase diagram are referred to as liquid-ordered and liquid-disordered, which reflects differences in the dynamics of the lipids in these phases as detected by physical methods.

Other factors can influence the lateral distribution of membrane lipids. For example, calcium can influence the condition of membranes containing phosphatidylserine. In membranes containing phosphatidylcholine and phosphatidylserine, in which the phosphatidylserine is a major component, addition of millimolar calcium can cause a phase separation of a phosphatidylserine-rich microdomain, from phosphatidylcholine-rich domains. This is caused by a complexation of calcium with the negatively charged phosphatidylserine. In fact, this binding of calcium to the negatively charged PS actually causes the PS to enter a gel state apparently due to dehydration of the PS headgroup. The phase separation caused by calcium is related to the phase separation noted above in which temperature induced the formation of microdomains of gel state lipid of different composition from the remainder of the membrane.

The final example to be considered is the phase separation of glycolipids in a phospholipid bilayer.[70] At low levels, glycolipids are miscible in phospholipid bilayers, but as concentrations are increased, a point is reached at which a phase separation can be observed. Under certain conditions, the phase separation can be observed as linear arrays or ribbons of lipids alternating in composition between rich in glycolipid and poor in glycolipid. Interestingly, some models of phase separations of cholesterol-rich phases in phospholipid bilayers also invoke a ribbon-like structure.

An interesting consequence of phase separations such as those described above is the connectivity of the two phases in equilibrium with each other. One can imagine one phase as a fully connected reticulum and the other as disconnected patches. A point in the two-phase region of the phase diagram can be identified where one of the phases goes from a disconnected to connected structure. This point has been called the percolation point and provides an interesting base for speculation about the role of phase separations in membrane structure and function.[71]

I. Transmembrane Movement of Phospholipids

Can phospholipids translocate from one side of the membrane to the other on a finite time scale? The hydrophobic effect predicts a negative response to this question. Translocation of a highly polar phospholipid headgroup through a hydrophobic membrane interior is energetically unfavorable. However, such a thermodynamic argument relates only to the partition coefficient of a polar headgroup into the hydrophobic phase. The kinetics of possible movement through such a phase constitutes another subject. Thus transmembrane movement of phospholipids, or "flip-flop," has been the subject of considerable experimentation. The results are in some cases surprising.

The first direct experimentation on this question involved spin labels and ESR. The experiment examined whether any spin labels moved from the inside to the outside of a small phospholipid vesicle as a function of time. The population on the exterior was ascertained by reducing the exterior-facing spin labels with ascorbic acid. The main difficulty with the experiment was controlling the transmembrane movement of the reducing agent.[72]

Subsequently, a number of other experiments in model membrane systems were performed and the results from some of these measurements are given in Table 4.6. In the pure lipid model systems, the summation of the data suggests that flip-flop does not occur on any reasonable time scale, in pure phospholipid bilayers.

The next issue is whether the same conclusion pertains to biological membranes, since so many other properties of phospholipid bilayers mimic properties of biological membranes. Several lines of evidence suggest that transmembrane phospholipid movement in biological membranes is facilitated. Phospholipid biosynthesis is asymmetric. Therefore, during synthesis some phospholipids must move from one side of the membrane to the other. Direct measures of transmembrane movement of phospholipids in biological membranes also suggest that such transmembrane movement must occur in time scales much shorter than suggested by the model membrane work. Some results are summarized in Table 4.6.

From these data, one might hypothesize that the other major component of membranes, protein, plays a role. In fact, one experiment, using the glycoprotein glycophorin from the human erythrocyte, demonstrated that transmembrane movement of a phospholipid did occur much more rapidly

TABLE 4.6
Rate of Phospholipid Flip-Flop in Membranes

Model systems	Rate ($T_{1/2}$)
Phospholipid vesicles	Too slow to measure
Influenza virus	Too slow to measure
Erythrocytes	Hours
Bacterial membranes	Minutes

in the presence of the membrane protein recombined in a model membrane than in the absence of the membrane protein.[73] (However, this result may be dependent on lipid composition.) How this might occur is then an interesting question. Perhaps the membrane protein does not pack well in the lipid bilayer and therefore creates discontinuities in the bilayer structure. Such discontinuities would then provide a mechanism for transmembrane phospholipid movement by lowering the activation energy barrier to the process. Membrane discontinuities formed by domains of phospholipids unstable to the lamellar phase structure could serve a similar function.

The issue of a phospholipid gradient across a biological membrane is not addressed by the above discussion. In cell membranes, two kinds of mechanisms for transmembrane of phospholipids have been described. One is facilitated diffusion, such as in endoplasmic reticulum, where particular ER proteins play a role in reducing the energy barrier for transmembrane movement of phospholipids. The other mechanism is an active, energy-requiring transport mechanism, which has been best studied in the erythrocyte membrane. This mechanism utilizes ATP and membrane proteins to "pump" specific phospholipids from one side of a membrane to another. In the erythrocyte membrane, in particular, there appear to be "pumps" for the amino phospholipids (PE and PS) that utilize ATP to maintain the PE and PS in the inner leaflet of the phospholipid bilayer of the erythrocyte membrane (see Section III,K).

J. Movement of Phospholipids between Membranes

For reasons similar to those expressed above, against transmembrane lipid movement, movement of phospholipids from one membrane to another is considered unlikely. Experimental evidence confirms that movement is slow; many hours are required to demonstrate significant exchange of phospholipids between membranes.

However, there is a class of protein that can lower significantly the energy barrier to such intermembrane phospholipid movement. These phospholipid exchange proteins promote the movement of phospholipids between membranes, frequently with some specificity with respect to phospholipid headgroup. Most of them seem to perform their function by carrying the lipids from one membrane to another, where the carried lipid is exchanged for another of the same kind from the acceptor membrane. The phospholipid exchange proteins are not, therefore, capable of a promoting a net flux of phospholipid from one membrane to another. A preparation of a nonspecific exchange protein that catalyzes exchange without specificity for headgroup has been described. This preparation can be useful in modifying membrane phospholipid content.[74]

In the absence of such proteins, therefore, one would expect little phospholipid exchange between membranes of the cell. However, little data are available and the *in vivo* role for the phospholipid exchange proteins is not known.

K. Transmembrane Lipid Asymmetry

The two halves of a phospholipid bilayer potentially constitute two independent domains. They could, for example, have different physical properties or even different lipid compositions. In fact, an interesting problem develops when considering the transmembrane distribution of phospholipids in biological membranes. In some cases the two sides of the membrane do not have the same lipid composition, yet the same membranes exhibit flip-flop of phospholipids across the membrane. The mammalian erythrocyte is an example. These results suggest on the one hand that the distribution of phospholipids across the membrane is asymmetric and at the same time that there is communication between both sides of the membrane. Therefore in this section both the question of transmembrane distribution of phospholipids and maintenance of that distribution in the presence of finite flip-flop rates must be considered.

Transmembrane distribution of phospholipids has been examined in both model systems and in biological membranes. One of the best characterized examples in model systems is that of sonicated vesicles containing PC and PE. The method used was to chemically label the amino group of PE selectively from one side of the membrane. It was possible to choose to label either the outside of the vesicle or the inside of the vesicle. The labeled PE could then be separated from the unlabeled PE by thin-layer chromatography and the ratio of labeled to unlabeled PE obtained. By this method the transmembrane distribution of PE could be determined. Figure 4.31 shows the results as a function of PE content of the vesicle. Clearly, over most of the range of PE concentrations, the PE favors the inside of the vesicle. However, at low concentrations of PE, the PE favors the outside of the vesicle. (The dashed line represents an even distribution. About 70% of the total vesicle phospholipid is on the outside surface of such a small sonicated PC vesicle.) This asymmetric distribution is likely due to the relatively hydrophobic nature of the PE headgroup. Exposure to the aqueous phase of the PE headgroup is less on the inside of the vesicle membrane than on the outside of the membrane. Thus the distribution of PE may be driven by entropy. In large unilamellar vesicles there is no asymmetry, corresponding to the much longer radius of curvature.

In another study, cholesterol above 30 mol% was found to distribute preferentially toward the inside of a small sonicated vesicle. One could argue that the cross-sectional area of the headgroup of cholesterol is much

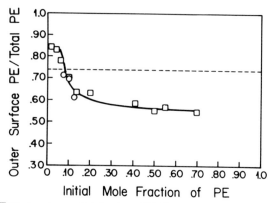

Fig. 4.31. Transmembrane distribution of PE in small sonicated vesicles as a function of mole fraction PE in PC, determined by TNBS labeling. Reprinted with permission from B. S. Litman, *Biochemistry* **13** (1974): 2844, American Chemical Society.

smaller than that for PC and that this will allow it to partition favorably toward the inside of the vesicle. Alternatively, cholesterol distribution may be driven by entropy as noted for PE.[75]

Even the effects of unsaturation of the hydrocarbon chains can cause a preferential partitioning of phospholipids across a sonicated vesicle membrane. These experiments were performed with PC, in which only the hydrocarbon chain composition was changed. The distribution was determined by using NMR. ^{13}C-labeled PC was made, and the distribution of this labeled PC was determined by shift reagent analysis. In this approach, a reagent that can change the chemical shift of a nucleus to which it can closely approach is added to the outside of the vesicle. This shifts (changes the chemical shift of) only the resonance from PC molecules on the outside of the vesicle and allows the assessment of the transmembrane distribution of the lipid. In this case an increase in unsaturation causes a partitioning of the phospholipid toward the outside membrane surface.[76]

These represent only a few examples of the measurements in model membrane systems. The driving force for asymmetry appears to arise predominantly from packing constraints and related hydration in the sonicated vesicles, although the details are not understood. These factors do not generally apply to biological membranes or even to large unilamellar vesicles.

A well-studied example of asymmetry in a biological membrane is the mammalian erythrocyte membrane. In addition to chemical labeling techniques, phospholipase treatment also has been used in this system to assess transmembrane phospholipid distribution. This relies on selective hydrolysis of the phospholipid for which the lipase is specific; hydrolysis should occur only on the side of the membrane to which the lipase has access.

A summary of results for the erythrocyte membrane is presented in Fig. 4.32. Clearly there is a significant asymmetry in the distribution of phospholipids across this membrane. The outside leaflet of the erythrocyte membrane consists primarily of PC and sphingomyelin. The inside leaflet of the erythrocyte membrane is enriched in PE and PS.

This is a striking asymmetry in phospholipid distribution. Since the erythrocyte membrane is a plasma membrane, do other cell (including nucleated cells) plasma membranes exhibit a similar asymmetry? Unfortunately there is no clear answer to this question yet.

A second question is how such asymmetry is maintained in the erythrocyte. In pure phospholipid bilayers that are not constrained by a small radius of curvature, such asymmetry is not observed. Furthermore, transmembrane movement of phospholipids can occur in biological membranes.

The major role in maintaining the asymmetry of the erythrocyte membrane appears to be "flipases," which specifically move PE and PS (and not PC or sphingomyelin) to the inner leaflet of the membrane bilayer, utilizing ATP as an energy source. These flipases are likely enzymes that exist as integral membrane proteins and directly translocate specific phospholipids through interactions between the phospholipid headgroups and the enzyme.[77, 78]

The asymmetry of biological membranes, where it exists, could also be established by the biosynthetic process. If so, then this asymmetry could be maintained in the membrane if the only mechanism for transmembrane

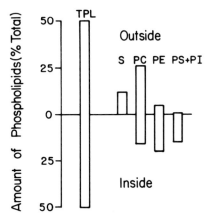

Fig. 4.32. Transmembrane distribution of lipids in the erythrocyte membrane, as determined by preferential hydrolysis by phospholipases. Reprinted with permission from A. J. Verkleij, R. F. A. Zwaal, R. Roelofson, P. Comfurius, D. Kastelijn, and L. L. M. van Deenen, *Biochim. Biophys. Acta* **323** (1973): 178.

movement involved exchange of two phospholipids of the same headgroup class. If the presence of membrane proteins facilitates such a process, then some control of this process could be achieved, as in the case of phospholipid exchange proteins, which frequently exhibit a specificity for a particular phospholipid headgroup class.

IV. SUMMARY

The complex behavior of lipids of biological membranes is in part due to the largely two-dimensional world described by cell membranes. These membrane lipids can form a variety of phase structures, including lamellar, hexagonal II, and micellar phases. Furthermore transitions between phases can take place, as in the gel to liquid crystalline phase transition or the lamellar to H_{II} phase transition. Lateral diffusion of lipids in the plane of the membrane occurs rapidly, but does not inhibit, under certain conditions, lateral phase separations of lipids. Lateral phase separations can lead to complex phase behavior of the membrane. The phase behavior of the lipids can profoundly affect the properties of the membrane in which these lipids reside. The details of these membrane properties are described by the concepts of motional rate and motional order of various segments of the phospholipid molecules. The composition of the membrane can, in turn, modulate the phase behavior of the membrane. Thus the immediate membrane environment can affect membrane properties, and obviously membrane properties can affect the cellular environment. It is also clear that the membrane interior is not well described by a single parameter, such as fluidity, but is a complex, highly anisotropic medium. The membrane cannot be modeled by an isotropic hydrophobic system, like a simple organic phase. Although the membrane is largely two-dimensional, some movement in a third direction is possible, such as transmembrane lipid movement (flip-flop) and lipid exchange between membranes. Furthermore, the two sides of a biological membrane are distinct, and may not have the same lipid composition. These properties of lipids and lipid bilayers can be used to understand the role lipids play in modulating membrane protein activity.

REFERENCES

1. Luzzati, V., In *Biological Membranes* (D. Chapman, ed.) (New York: Academic Press, 1968).
2. Levine, Y. K., "X-ray diffraction studies of the membranes," *Prog. Surf. Sci.* **3** (1973): 279–352.

3. Huang, C., "Mixed chain phospholipids and interdigited bilayer systems," *Klin. Wochemschrift* **68** (1990): 149–165.
4. Slater, J. L., and C. Huang, "Lipid bilayer interdigitation," *The Structure of Biological Membranes* (P. L. Yeagle, ed.) (Boca Raton: CRC Press, 1992), 175–210.
5. Ellena, J. F., W. C. Hutton, and D. S. Cafiso, "Elucidation of cross-relaxation pathways in phospholipid vesicles utilizing two-dimensional H-1 NMR spectroscopy," *J. Am. Chem. Soc.* **107** (1985): 1530–1537.
6. Becker, R., A. Helenius, and K. Simons, "Solubilization of the semliki forest virus membrane with sodium dodecyl sulfate," *Biochemistry* **14** (1975): 1835.
7. Strasberg, S. M., and P. R. C. Harvey, "Biliary cholesterol transport and precipitation," *Hepatology* **12** (1990): 1S–5S.
8. Reiss-Husson, F., "Structure des phases liquide-cristallines de differents phospholipides, monoglycerides, spingolipides, anhydres ou en presence d'eau," *J. Mol. Biol.* **25** (1967): 363–382.
9. Cullis, P. R., and B. de Kruijff, "Lipid polymorphism and the functional roles of lipids in biological membranes," *Biochim. Biophys. Acta* **507** (1978): 207–218.
10. Gruner, S. M., "Intrinsic curvature hypothesis for biomembrane lipid composition: A role for nonbilayer lipids," *Proc. Natl. Acad. Sci. U.S.A.* **82** (1985): 3665–3669.
11. Siegel, D. P., J. Banschbach, and P. L. Yeagle, "Stabilization of H-II phases by low levels of diglycerides and alkanes: An NMR, DSC and X-ray diffraction study," *Biochemistry* **28** (1989): 5010–5018.
12. Brown, P. M., J. Steers, S. W. Hui, P. L. Yeagle, and J. R. Silvius, "Role of head group structure in the phase behavior of amino phospholipids. 2. Lamellar and nonlamellar phases of unsaturated phosphatidylethanolamine analogues," *Biochemistry* **25** (1986): 4259–4267.
13. Hitchcock, P. B., R. Mason, K. M. Thomas, and G. G. Shipley, "Structural chemistry of 1,2 dilauroyl-DL-phosphatidylethanolamine: Molecular conformation and intermolecular packing of phospholipids," *Proc. Natl. Acad. Sci. U.S.A.* **71** (1974): 3036–3040.
14. Rand, R. P., N. Fuller, V. A. Parsegian, and D. C. Rau, "Variation in hydration forces between neutral phospholipid bilayers: evidence for hydration attraction," *Biochemistry* **27** (1988): 7711–7722.
15. Yeagle, P. L., and A. Sen, "Hydration and the lamellar to hexagonal (II) phase transition of phosphatidylethanolamine," *Biochemistry* **25** (1986): 7518–7522.
16. Cullis, P. R., A. J. Verkleij, and Ph. J. T. Ververgaert, "Polymorphic phase behaviour of cardiolipin as detected by P-31 NMR and freeze fracture techniques: Effects of calcium, dibucaine and chloropromazine," *Biochim. Biophys. Acta* (1978): 11–20.
17. Wieslander, A., J. Ulmius, G. Lindblom, and K. Fontell, "Water binding and phase structures for different *Acholeplasma laidlawii* membrane lipids studied by deuteron NMR and X-ray diffraction," *Biochim. Biophys. Acta* **512** (1978): 241–253.
18. Albert, A. D., A. Sen, and P. L. Yeagle, "The effect of calcium on the bilayer stability of lipids from bovine rod outer segment disk membranes," *Biochim. Biophys. Acta* **771** (1984): 28–34.
19. Hui, S. W., T. P. Stewart, A. J. Verkleij, and B. deKruijff, "'Lipidic particles are intermembrane attachment sites," *Nature (London)* **290** (1981): 427–428.
20. Verkleij, A. J., C. Mombers, J. Leunissen-Bijvelt, and P. H. J. Th. Ververgaert, "Lipidic intramembraneous particles," *Nature (London)* **279** (1979): 162–163.
21. Hui, S. W., T. P. Stewart, P. L. Yeagle, and A. D. Albert, "Bilayer to non-bilayer transition in mixtures of phosphatidylethanolamine and phosphatidylcholine: Implications for membrane properties," *Arch. Biochem. Biophys.* **207** (1981): 227–240.
22. Cullis, P. R., and B. de Kruijff, "Polymorphic phase behaviour of lipid mixtures as

detected by P-31 NMR: Evidence that cholesterol may destabilize bilayer structure in membrane systems containing phosphatidylethanolamine," *Biochim. Biophys. Acta* 507 (1978): 207–218.

23. Allegrini, P. R., G. van Scharrenburg, G. H. de Haas and J. Seelig, "H-2 and P-31 NMR studies of bilayers composed of 1-acyllysophosphatidylcholine and fatty acids," *Biochim. Biophys. Acta* **731** (1983): 448–455.
24. Jain, M. K., M. K. Jain, C. J. A. van Echteld, F. Ramirez, J. deGier, G. H. de Haas, and L. L. M. van Deenen, "Association of lysophosphatidylcholine with fatty acids in aqueous phase to form bilayers," *Nature (London)* **284** (1980): 486–487.
25. Hui, S. W., and L. T. Boni, "Lipidic particles and cubic phase lipids," *Nature (London)* **296** (1982): 175–176.
26. Lindblom, G., and L. Rilfors, "Cubic phases and isotropic structures formed by membrane lipids—Possible biological relevance," *Biochim. Biophys. Acta* **988** (1989): 221–256.
27. Siegel, D. P., "Inverted micellar intermediates and the transitions between lamellar, cubic, and inverted hexagonal lipid phases," *Biophys. J* **49** (1986): 1155–1170.
28. Lewis, R. N. A. H., and R. N. McElhaney, "The mesomorphic phase behavior of lipid bilayers," In *The Structure of Biological Membranes* (P. L. Yeagle, ed.) (Boca Raton: CRC Press, 1992), 73–156.
29. Huang, C., "Empirical estimation of the gel to liquid crystalline phase transition temperatures for fully hydrated saturated phosphatidylcholines," *Biochemistry* **30** (1991): 26–30.
30. Barton, P. G., and F. D. Gunstone, "Hydrocarbon chain packing and molecular motion in phospholipid bilayers formed from unsaturated lecithins." *J. Biol. Chem.* **250** (1975): 4470–4476.
31. McElhaney, R. "The effect of alterations in the physical state of the membrane lipids on the ability of acholeplasma laidlawii B. to grow at various temperatures," *J. Mol. Biol.* **84** (1974): 145.
32. Barenholz, Y., and T. E. Thompson, "Sphingomyelins in bilayers and biological membranes," *Biochim. Biophys. Acta* **604** (1980): 129–158.
33. Seelig, J., and A. Seelig, "Lipid conformation in model membranes and biological membranes," *Q. Rev. Biophys.* **13** (1980): 19–61.
34. Hauser, H., and G. Poupart, "Lipid structure," In *The Structure of Biological Membranes* (P. L. Yeagle, ed.) (Boca Raton: CRC Press, 1992).
35. Pearson, R. H., and I. Pascher, "The molecular structure of lecithin dihydrate," *Nature (London)* **281** (1979): 499–501.
36. Seelig, J., *et al.*, "Deuterium and phosphorus NMR and fluorescence depolarization studies of functional reconstituted sarcoplamic reticulum membrane vesicles," *Biochemistry* **20** (1981): 3922–3932.
37. Ernst, J., W. S. Sheldrick, and J.-H. Fuhrhop, "Structures of the essential unsaturated fatty acids," *Z. Naturforsch. B* **34** (1979): 706–711.
38. Baenziger, J. E., H. C. Jarrell, and I. C. P. Smith, "Molecular motions and dynamics of a diunsaturated chain in a lipid bilayer. Implications for the role of polyunsaturation in biological membranes," *Biochemistry* **31** (1992): 3377–3385.
39. Lewis, B. A., S. K. Das Gupta, and R. G. Griffin, "Solid state NMR studies of the molecular dynamics and phase behavior of mixed chain phosphatidylcholines," *Biochemistry* **23** (1984): 1988–1993.
40. Yeagle, P. L., and R. B. Martin, "Hydrogen-bonding of the ester carbonyls in phosphatidylcholine bilayers," *Biochem. Biophys. Res. Commun.* **69** (1976): 775–780.
41. Yeagle, P. L., "Phospholipid headgroup behavior in biological assemblies," *Accounts Chem. Res.* **11** (1978): 321–326.

42. Buldt, G., H. U. Gally, A. Seelig, J. Seelig, G. Zaccai, "Neutron diffraction studies on selectively deuterated phospholipid bilayers," *Nature (London)* **271** (1978): 182–184.
43. Yeagle, P. L., W. C. Hutton, C-h. Huang, and R. B. Martin, "Phospholipid headgroup conformations; intermolecular interactions and cholesterol effects," *Biochemistry* **16** (1977): 4344–4349.
44. Browning, J. L., and J. Seelig, "Bilayers of phosphatidylserine: A deuterium and phosphorus NMR study," *Biochemistry* **19** (1980): 1262–1270.
45. Renou, J.-P., *et al.*, "Glycolipid membrane surface structure: Orientation, conformation, and motion of a disaccharide headgroup," *Biochemistry* **28** (1989): 1804–1814.
46. Jarrell, H. C., A. J. Wand, J. B. Giziewicz, and I. C. P. Smith, "The dependence of glyceroglycolipid orientation and dynamics on head-group structure," *Biochim. Biophys. Acta* **897** (1987): 69–82.
47. Zaccai, G., J. K. Blasie, and B. P. Schoenborn, "Neutron diffraction studies on the lcation of water in lecithin bilayer model membranes," *Proc. Natl. Acad. Sci. U.S.A.* **72** (1975): 376–380.
48. Berliner, L., *Spin Labeling* (New York: Academic Press, 1975).
49. Smith, R. L., and E. Oldfield, "Dynamic structure of membranes by deuterium NMR," *Science* **225** (1984): 280–288.
50. Gaber, B. P., P. Yager, and W. L. Petcolas, "Conformational nonequivalence of chains 1 and 2 of dipalmitoyl phosphatidylcholine as observed by Raman spectroscopy," *Biophys. J.* **24** (1978): 677–688.
51. Huang, C., J. T. Mason, F. A. Stephenson, and I. W. Levin, "Raman and P-31 NMR spectroscopic identification of a highly ordered lamellar phase in aqueous dispersions of 1-stearoyl-2-acetyl-*sn*-glycerol-3-phosphorylcholine," *J. Phys. Chem.* **88** (1984): 6454–6458.
52. Mendelson, R., and J. Maisano, "Use of deuterated phospholipids in Raman spectroscopic studies of membrane structure. I. Multilayers of dimyristoyl phosphatidylcholine (and its-d54 derivative) with distearoyl phosphatidylcholine," *Biochim. Biophys. Acta* **506** (1978): 192–201.
53. Mantsch, H. H., C. Madec, R. N. A. H. Lewis, and R. N. McElhaney, "An infrared spectroscopic study of the thermotropic phase behavior of phosphatidylcholines containing o-cyclohexyl fatty acyl chains," *Biochim. Biophys. Acta* **980** (1989): 42–49.
54. Seelig, J., "Deuterium magnetic resonance: theory and application to lipid membranes," *Q. Rev. Biophys.* **10** (1977): 353–418.
55. Seelig, A., and J. Seelig, "The dynamic structure of fatty acyl chains in a phospholipid bilayer measured by deuterium magnetic resonance," *Biochemistry* **13** (1974): 4839–4845.
56. Dufourc, E. J., I. C. P. Smith, and H. C. Jarrell, "A H-2 NMR analysis of dihydrosterculoyl-containing lipids in model membranes: Structural effects of a cyclopropane ring," *Chem. Phys. Lipids* **33** (1983): 153–177.
57. Kelusky, E. C., E. J. Dufourc, and I. C. P. Smith, "Direct observation of molecular ordering of cholesterol in human erythrocyte membranes," *Biochim. Biophys. Acta* **735** (1983): 302–304.
58. Yeagle, P. L., A. D. Albert, K. Boesze-Battaglia, J. Young, and J. Frye, "Cholesterol dynamics in phosphatidylcholine bilayers," *Biophys. J.* **57** (1990): 413–424.
59. Stockton, B. W., and I. C. P. Smith, "A deuterium NMR study of the condensing effect of cholesterol on egg phosphatidylcholine bilayer membranes. I. Perdeuterated fatty acid probes," *Chem. Phys. Lipids* **17** (1976): 251.
60. Lange, A., D. Marsh, K. H. Wassmer, P. Meier, and G. Kothe, "Electron spin resonance study of phospholipid membranes employing a comprehensive line-shape model," *Biochemistry* **24** (1985): 4383–4392.

61. Meier, P., E. Ohmes, and G. Kothe, "Multipulse dynamic NMR of phospholipid membranes," *J. Chem. Phys.* **85** (1986): 3598–3614.
62. Yeagle, P. L., W. C. Hutton, C.-H. Huang, and R. B. Martin, "Headgroup conformation and lipid–cholesterol association in phosphatidylcholine vesicles: A P-31 [H-1] nuclear Overhauser effect study," *Proc. Natl. Acad. Sci. U.S.A.* **72** (1975): 3477–3481.
63. Milburn, M. P., and K. R. Jeffrey, "Dynamics of the phosphate group in phospholipid bilayers: A P-31 nuclear relaxation time study," *Biophys. J.* **52** (1987). 791–799.
64. Frye, J., A. D. Albert, B. S. Selinsky and P. L. Yeagle, "Cross polarization P-31 nuclear magnetic resonance of phospholipids," *Biophys. J.* **48** (1985): 547–552.
65. Yeagle, P. L., A. D. Albert, K. Boesze-Battaglia, J. Young, and J. Frye, "Cholesterol dynamics in membranes," *Biophys. J.* **57** (1990): 413–424.
66. Schindler, H., and J. Seelig, "EPR spectra of spin labels in lipid bilayers. II. Rotation of steroid spin probes," *J. Chem. Phys.* **61** (1974): 2946–2951.
67. Shinitzky, M., and Y. Barenholz, "Fluidity parameters of lipid regions determined by fluorescence depolarization," *Biochim. Biophys. Acta* **515** (1978): 367–394.
68. Shimshick, E. J., and H. M. McConnell, "Lateral phase separations in binary mixtures of cholesterol and phospholipids," *Biochem. Biophys. Res. Commun* **53** (1973): 446–451.
69. Sankaram, M. B., and T. E. Thompson, "Cholesterol-induced fluid-phase immiscibility in membranes," *Proc. Natl. Acad. Sci. U.S.A.* **88** (1991): 8686–8690.
70. Thompson, T. E., "Organization of glycosphingolipids in bilayers and plasma membranes of mammalian cells," *Annu. Rev. Biophys. Biophys. Chem.* **14** (1985): 361–386.
71. Almeida, P. F. F., W. L. C. Vaz, and T. E. Thompson, "Lateral diffusion in the liquid phases of DMPC/cholesterol lipid bilayers," *Biochemistry* **31** (1992): 6739–6747.
72. Kornberg, R. D., and H. M. McConnell, "Inside–outside transitions of phospholipids in vesicle membranes," *Biochemistry* **10** (1971): 1111–1120.
73. deKrujiff, B., E. J. J. van Zoelen, and L. L. M. van Deenen, "Glycophorin facilitates the transbilayer movement of phosphatidylcholine in vesicles," *Biochim. Biophys. Acta* **509** (1978): 537–542.
74. Crain, R. C., and D. B. Zilversmit, "Two non-specific phospholipid exchange proteins from beef liver: purification and characterization," *Biochemistry* **19** (1980): 1433–1447.
75. Huang, C-H., J. P. Sipe, S. T. Chow, and R. B. Martin, "Differential interaction of cholesterol with phosphatidylcholine on the inner and outer surfaces of lipid bilayer vesicles," *Proc. Natl. Acad. Sci. U.S.A.* **71** (1974): 359–362.
76. Yeagle, P. L., W. C. Hutton, R. B. Martin, B. Sears, and C-H. Huang, "Transmembrane asymmetry of vesicle lipids," *J. Biol. Chem.* **251** (1976): 2110–2112.
77. Morrot, G., A. Zachowski, and P. F. Devaux, "Partial purification and characterization of the human erythrocyte $Mg^{2(+)}$-ATPase. A candidate aminophospholipid translocase," *FEBS Lett.* **266** (1990): 29–32.
78. Herrmann, A., and P. F. Devaux, "Alteration of the aminophospholipid translocase activity during in vivo and artificial aging of human erythrocytes," *Biochim. Biophys. Acta* **1027** (1990): 41–46.

5

Cholesterol and Cell Membranes

Cholesterol poses an interesting paradox. Normal mammalian cell growth and function is dependent on cholesterol. Cholesterol is therefore an essential lipid in the membranes of mammalian cells. Many cells synthesize cholesterol to satisfy that requirement. In fact, the role of cholesterol is so important that some cells not capable of synthesizing cholesterol make other sterols, or structural equivalents, which apparently play similar roles. For example, in plant cells, cholesterol is not found. However, other sterols, such as stigmasterol or sitosterol, are found. Yeasts make and use ergosterol. *Tetrahymena* do not synthesize a sterol, but do synthesize tetrahymenol. Tetrahymenol is not structurally a sterol but apparently functions in a manner similar to that of sterols. The analog between sterols and tetrahymenol is most dramatically evidenced by the ability of *Tetrahymena* to take up cholesterol from the medium and substitute cholesterol for tetrahymenol in its membranes.

There is a dark side to the cholesterol story, however. Cholesterol often plays a lethal role, particularly in human physiology. High levels of serum cholesterol can lead to the development of plaques on arterial walls. When these plaques occlude coronary arteries feeding heart muscle, some heart muscle can die as a result. The myocardial infarct, or heart attack, that results can kill.

In this chapter, the emphasis will be on understanding the essential role of cholesterol in cholesterol-requiring cells, with limited treatment of the sterols of nonmammalian cells for comparison. The avilable knowledge on cholesterol in membranes will be summarized and then a discussion will follow on a current hypothesis for the essential role of cholesterol in cholesterol-requiring cells (and other sterols in other cells).[1, 2]

I. GENERAL FEATURES OF THE CHOLESTEROL MOLECULE

The structure of cholesterol appears in Fig. 2.15. Note that the structure consists of four fused rings, referred to as the A,B,C and D rings, reading left to right. The most common conformation of this steroid ring system is planar. One face of the plane, the alpha face, is flat. The opposite side of this sterol is not flat, due to two protruding methyl groups.

This is not the only conformation available to the sterol, however. From crystal structures, it is evident that the A ring can adopt an alternate conformation in the crystal, indicating some conformational flexibility in the A ring. Much more conformational flexibility is enjoyed by the tail of cholesterol, as is also revealed in the crystal structures shown in Fig. 5.1.[3]

In addition to an approximately planar steroid ring system, and the hydrophobic tail just referred to, cholesterol possesses a 3β-hydroxyl function. All three features are important for characteristic cholesterol-like behavior.

Although most of the molecule is hydrophobic, the 3β-hydroxyl is polar and gives the molecule an amphipathic character. Cholesterol is surface active, orienting in a phospholipid bilayer with its polar hydroxyl facing the aqueous phase and the hydrophobic steroid ring parallel to, and buried

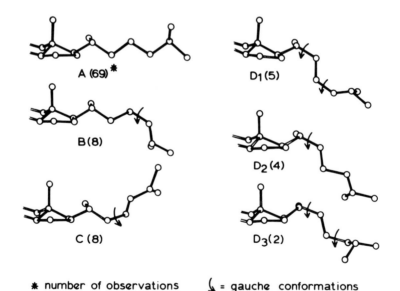

A (69) *

D₁ (5)

B(8)

D₂ (4)

C (8)

D₃(2)

* number of observations   = gauche conformations

Fig. 5.1. Conformation of the tail of cholesterol derived from X-ray crystal structures. Figure courtesy of Dr. J. Griffin and Dr. W. Daux.

in, the hydrocarbon chains of the phospholipids. The hydrophobic character of cholesterol makes it highly insoluble in water, but cholesterol is accommodated in membranes readily at molar ratios with respect to the phospholipids of 1:1 and even higher.

Evidence has been offered that cholesterol, when in a hydrophobic medium, forms dimers with itself. Dimerization may alter the free energy difference characterizing the partitioning of cholesterol between an aqueous phase and a hydrophobic phase.

Cholesterol is synthesized by a pathway involving enzymes whose membrane bound members are located in the endoplasmic reticulum. Therefore cholesterol is synthesized in a membrane whose relative cholesterol content is low compared to that of plasma membranes. Why are different membranes in the same cell so different in their cholesterol content?

II. CHOLESTEROL LOCATION IN CELLS

A. Distribution of Cholesterol among Cell Membranes

Cholesterol is disproportionately distributed among the membranes of cells. For example, plasma membranes have a much higher cholesterol content (35–45 mol%) than many intracellular membranes (for example, endoplasmic reticulum membranes contain about 10–12 mol% cholesterol). The inner membrane of mitochondria contains even less cholesterol. Studies on the distribution of cholesterol in intact cells (fibroblasts, Chinese hamster ovary cells, and rat hepatocytes) using oxidation of cholesterol by cholesterol oxidase indicated that 80 to 90% of total cell cholesterol is located in the cell plasma membrane. Thus the disproportionation of cellular cholesterol toward the plasma membrane is extreme.

Within plasma membranes, cholesterol may not be uniformly distributed. For example, in the rat intestinal mucosa, cholesterol is found in higher concentrations in the basolateral membranes than in brush border membranes.[4] Filipin studies have suggested that cholesterol is disproportionally distributed within the Golgi stacks, with greater cholesterol contents found in the portion of the Golgi located near the plasma membrane than in the Golgi elements located near the endoplasmic reticulum.[5] This distribution corresponds to the division of the Golgi apparatus into cis medial, and trans Golgi.

Cholesterol distribution can change as the state of the membrane changes. Disk membranes containing the photopigment rhodopsin are stacked in the outer segment of retinal rod cells. New disks form from evaginations of the plasma membrane at the base of the outer segment.

Old disks are phagocytosed from the apical tip of the outer segment by the pigmented epithelium. The disks thus go through an aging process from biogenesis to phagocytosis and move along the axis of the outer segment in the process. When these disks form, they are similar in content to the plasma membrane from which they were made. Old disks are low in cholesterol content, suggesting a time-dependent expulsion of cholesterol from the disk membrane during disk aging.[6]

B. Cholesterol Movement between Membranes

The process of cholesterol exchange between membrane vesicles provides a means to assess the kinetics and mechanism of cholesterol movement between membranes. The movement of cholesterol can be manifest as exchange, in which case no net movement of cholesterol occurs, or as a flux of cholesterol from one membrane to another. This latter phenomenon has been exploited to alter the cholesterol content of cell membranes. One example involves the human erythrocyte membrane. Vesicles enriched in cholesterol (to cholesterol/phospholipid ratios greater than the erythrocyte membrane) were incubated with erythrocytes. This led to an increase in the erythrocyte plasma membrane cholesterol content by net cholesterol influx during the incubation. Likewise when vesicles without cholesterol were incubated with erythrocytes, the erythrocytes were depleted of cholesterol through a net cholesterol flux to the acceptor vesicles.[7]

Studies on the kinetics of the movement of cholesterol from one membrane to another have revealed the mechanism of that movement. Several studies with small phosphatidylcholine vesicles indicated cholesterol can move between vesicles by transfer through the aqueous phase.[8] Perhaps most dramatic was the observation that cholesterol could transfer between two vesicle populations separated by a membrane impermeable to the vesicles.[9] However, transfer can also be enhanced significantly by collision of donor and acceptor membranes.[10]

It is interesting that cholesterol transfers through the aqueous phase even though the hydrophobic effect determines that the solubility of cholesterol in water is vanishingly small. Not surprisingly then, agents that increase the critical micelle concentration apparently enhance the exchange rates.[11]

In vitro experiments allow for a role for collision between membranes to contribute to the pathway of cholesterol movement between membranes. Thus in laboratory experiments, cholesterol movement can occur by two competing mechanisms. Intracellularly, the dominant mechanism, however, is likely vesicular transport of cholesterol from one site to another, in which cholesterol is carried as a membrane component.

C. Factors Controlling the Distribution of Cholesterol among Membranes

How is the distribution of cholesterol among membranes established and maintained? At least three possibilities should be considered. First, cholesterol distribution could be maintained by a thermodynamic equilibrium. In this model the composition of each of the cell membranes determines the partition coefficient for cholesterol. Factors that might influence the partition coefficient for cholesterol include the lipid composition of the membranes and the nature of the proteins contained in the membranes.

Second, cholesterol distribution could be maintained by a flux of cholesterol from the site of synthesis to the plasma membrane. In this case, an energy-driven movement of cholesterol would obscure the underlying thermodynamic factors. The distribution of cholesterol might then be described by a "steady-state" equation, rather than an equilibrium equation.

Third, cholesterol distribution could be established by an energy-driven sorting process coincident with transport vesicle formation. Transport vesicles from the ER carrying newly synthesized cholesterol must undergo a sorting process so that these vesicles do not carry away from the ER intrinsic ER proteins. That sorting process could also include cholesterol (although cholesterol is not included in the sorting process (for phospholipids and proteins) in disk membrane biogenesis in the rod outer segment[12]).

Although any one of these factors could dominate, one must also consider whether all are important. Further, one must consider whether the biosynthetic pathway of cholesterol is organized among intracellular organelles with an end point physically located at or near the plasma membrane of the cell, thus determining the plasma membrane location of the majority of the cellular cholesterol.

Some answers to the questions posed above can be deduced. Coated vesicles pinch off from the plasma membrane. When they fuse with Golgi or endoplasmic reticulum, the coated vesicles join a membrane of significantly lower content than the plasma membrane, and yet do not, on a time average, cause the target membrane to become equal in cholesterol content with the cell plasma membrane. This would suggest that cholesterol is sorted and transferred back to the plasma membrane from the intracellular membrane to maintain the cholesterol distribution normally observed.

Likewise, in the case of newly formed rod outer segment disks that were considered earlier in this chapter, the cholesterol content of the disks, although initially resembling the plasma membrane (from which the disks originated), eventually drops to a value similar to those of other intracellular membranes. This would suggest that much of the cholesterol in the new disks is transferred back to the plasma membrane and from there to extracellular cholesterol carriers. This observation would be consistent with a thermodynamic partitioning of cholesterol among membranes that was dependent on the lipid and protein content of these membranes. However, alternative explanations cannot be ruled out.

An important experiment that is helpful in this regard was reported by Silbert and colleagues.[13] Preparations of various biological membranes were incubated with each other, and the membrane cholesterol content was monitored as a function of time. The results suggested that membranes originally high in cholesterol content appeared to remain relatively high in cholesterol content during the incubation, whereas membranes originally low in cholesterol tended to remain low in cholesterol. Thus thermobynamics may have a role to play in determining cholesterol distribution among membranes (although the same study suggested that other factors may also be important).

If thermodynamics does play a role, then one must inquire what membrane components contribute to the thermodynamic partitioning of cholesterol among membranes. Some data are available on the role of lipid composition. Several experiments have suggested that cholesterol has a greater affinity for phosphatidylcholine than for phosphatidylethanolamine. Membranes containing substantial phosphatidylethanolamine provide a thermodynamically unfavorable environment for cholesterol. In a partitioning experiment, this causes cholesterol to leave membranes high in PE content for membranes low in PE content. In a companion partitioning experiment, cholesterol in a PC membrane tends to stay in the PC membrane when incubated with PE membranes containing no cholesterol (see Fig. 5.2).[14]

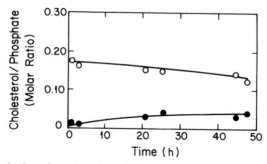

Fig. 5.2. Incubation of small sonicated PC vesicles containing cholesterol with large unilamellar PE vesicles containing no cholesterol. As a function of time, the vesicle populations were separated and the cholesterol–phospholipid mole ratio was determined. It is evident that cholesterol prefers to stay in the PC vesicle.

   A basis for understanding these results can be found in the discussion in Chapter 4 on the thermodynamics of the interaction between water and phosphatidylethanolamine. Phosphatidylethanolamine binds fewer water molecules than phosphatidylcholine, and cholesterol binds even fewer water molecules than does phosphatidylethanolamine. For that reason there is an unfavorable entropy term describing the interaction of water with the surface of a phosphatidylethanolamine bilayer. Consider the addition of cholesterol to the PE membrane. Cholesterol will occupy surface area without binding a large compensating complement of water molecules. The PE–cholesterol membrane surface will cause water molecules to become ordered because of the hydrophobic effect. Therefore, per unit surface area, the unfavorable entropy associated with the surface interfacial structure will become a more severe problem without compensation from favorable entropy terms derived from water binding to the lipid. Therefore, cholesterol partitioning into PE bilayers can be predicted to be unfavorable in agreement with the experimental result.

   Interestingly, one would therefore predict that significantly changing the nature of the aqueous phase would change the partitioning of cholesterol between membranes of phosphatidylcholine and phosphatidylethanolamine. In fact, such expectations are fulfilled by experiment. In the presence of the chaotropic agent guanidine hydrochloride, cholesterol partitions equally well into phosphatidylcholine and phosphatidylethanolamine membranes (see Fig. 5.3).

   Returning one last time in this discussion to the interesting example of the membranes of the retinal rod outer segment, it is interesting to note that rod outer segment disk membranes are high in PE. Recent studies have revealed that the PE/PC mole ratio for the plasma membrane is much

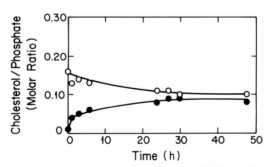

Fig. 5.3. Incubation as in Fig. 5.2, but in the presence of 5 *M* guanidinum hydrochloride (GuHCl). In the presence of the chaotropic agent, cholesterol distributes readily between the PC and the PE membranes with no evidence for preference.

lower than for the disk membranes.[12] Furthermore, the disk membranes exhibit their characteristically high PE content in all disks, including the newest disks. Apparently, phospholipid sorting occurs concurrent with, or as a part of, disk biogenesis. One might therefore expect that, based on the studies just described, cholesterol would move back to the plasma membrane (with its much lower PE content) with time following rod outer segment disk biogenesis. Again, recent studies have revealed that as disks age, their cholesterol content decreases from a value in the newest disks similar to the plasma membrane (cholesterol/phospholipid mole ratio of 0.4) to a much lower value in the oldest disks (cholesterol/phospholipid mole ratio of 0.05), as shown in Fig. 5.4. This is as the partitioning model for cholesterol predicted. Phospholipid composition therefore likely affects cholesterol partitioning among some cell membranes.

However, little evidence exists for a general role for membrane lipids in control of the partitioning of cholesterol among cell membranes; the rod outer segment disk membrane is unusually high in PE content. Note that there are also potential roles for hydrocarbon chain length and extent of unsaturation in the partitioning of cholesterol among membranes, although this has not yet been systematically studied.[15] Furthermore, the role of protein has not yet been adequately examined. All that is known

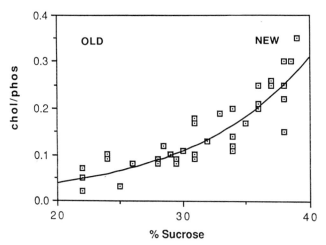

Fig. 5.4. Relationship between the cholesterol content of retinal rod outer segment disks and disk age. The disks were separated according to cholesterol content on a sucrose density gradient after treatment with digitonin. The digitonin forms a 1:1 complex with cholesterol and proportionally alters membrane density. Disk age was determined in a separate experiment. New disks are highest in cholesterol content. Reproduced with permission from K. Boesze-Battaglia, T. Hennessey, and A. D. Albert, *J. Biol. Chem.* **264** (1989): 8151–8155.

so far is that both band 3 and glycophorin of the human erythrocyte membrane, reconstituted into phosphatidylcholine membranes, incorporate cholesterol in exchange incubations at levels greater than pure phosphatidylcholine membranes with which the reconstituted protein-containing membranes have been incubated. In contrast, the Ca^{2+}-ATPase, reconstituted and subjected to the same experimental conditions, has no discernable effect.

D. Intracellular Cholesterol Movement

None of these data rule out a role in the observed distribution of cholesterol among membranes for a flux of cholesterol through the cell. Consider what is known about the movement of cholesterol from the site of synthesis to the plasma membrane. Measurements suggest that the half-time for this process is of the order of minutes.[16, 17] The rate of cholesterol movement is different from the rate of movement of newly synthesized glycoproteins to the plasma membrane and, oddly enough, different from the movement of newly synthesized lipid to the plasma membrane. Taken together, these data suggest that cholesterol transport to the plasma membrane may be by a membranous route, perhaps involving vesicular transport. Recent efforts have been aimed at defining those vesicle-mediated pathways.

The question of a role for protein in intracellular cholesterol movement has not been satisfactorily answered. A protein, called sterol carrier protein, has been identified in a number of tissues. It binds cholesterol, and may also bind fatty acids. Although its presence is not a matter of controversy, it is uncertain what role this protein plays biologically.

Details of the pathway of cholesterol movement remain among the important and exciting questions of membrane biology.

E. Transbilayer Distribution of Cholesterol

So far this discussion has concentrated on the distribution of cholesterol among cell membranes. A related topic is the transmembrane distribution of cholesterol. This question has not been extensively examined. In model membranes consisting of egg phosphatidylcholine and cholesterol, it appears that at higher cholesterol contents, cholesterol prefers the inside of SUV.[18] This is likely due to packing constraints within the bilayer (due to the small radius of curvature of the SUV) that do not apply widely to cell membranes.

Some determinations of the transbilayer distribution of cholesterol have been made in biological membranes. In the human erythrocyte membrane, cholesterol is nearly symmetrically distributed.[19] In mycoplasma, cholesterol distribution is approximately symmetric, with perhaps a small excess on the outside of the membrane.[20] On the other hand, fluorescence quenching studies using the fluorescent sterol probe dehydroergosterol suggest that the membrane cholesterol is disproportionately distributed toward the outside of LM plasma membranes.[21] Therefore the transmembrane distribution is dependent on the membrane.

F. Rate of Transbilayer Movement of Cholesterol

In Chapter 4, the conclusion reached was that phospholipid flip-flop was thermodynamically unfavorable, and did not occur in phospholipid vesicle systems on the time scale of hours or days. It was further noted that in biological membranes, flip-flop does occur, both passively and actively. With this in mind and since the polar moiety of cholesterol (the hydroxyl) is much smaller than the polar headgroup found on phospholipids, one might expect that transmembrane movement of cholesterol would occur more readily than for phospholipids. In fact, this does prove to be the case. In small sonicated vesicles, cholesterol oxidase has been used to examine the rate of transmembrane movement. A half-time for transmembrane movement of cholesterol of less than 1 min was determined.[22] The same approach indicated that the half-time of transmembrane movement of cholesterol in the human erythrocyte membrane was less than 3 sec.[23]

With such rapid transmembrane movement, cholesterol could assist in shape changes that must be endured by the erythrocyte during circulation. A shape change for the erythrocyte requires the mass of membrane material on each side of the plasma membrane to change slightly to accommodate changes in curvature. Rapid transmembrane movement of cholesterol could satisfy the mass changes required, thereby maintaining the integrity of the plasma membrane.

III. LOCATION OF CHOLESTEROL IN THE MEMBRANE

X-ray diffraction and neutron diffraction data have provided a clear picture of the location of cholesterol in the membrane.[24, 25] The data show that the cholesterol molecule is located so that the hydroxyl is in the immediate vicinity of the phospholipid ester carbonyl, as shown in Fig. 5.5.

Fig. 5.5. Space-filling models showing the most likely position of cholesterol in a membrane relative to the phospholipids.

IV. CHOLESTEROL EFFECTS ON THE PHYSICAL PROPERTIES
OF MEMBRANES

Cholesterol has a wide variety of effects on the physical properties of membranes. A number of sophisticated techniques have been employed to examine these effects.

One of the earliest studies of the effect of cholesterol on lipid properties demonstrated that cholesterol caused the phase transition of sphingomyelin to "disappear." ^{1}H NMR spectra in the same study indicated that cholesterol was creating an intermediate fluid state of the membrane. Thus cholesterol can induce a liquid crystal state in lipids that would otherwise occupy a gel state.[26]

A. Effects on Membrane Ordering

^{2}H NMR studies of deuterium-labeled phospholipids and ESR studies of spin-labeled lipids have revealed that the degree of motional ordering of the lipid hydrocarbon chains in membranes is highly dependent on the position of the probe (whether deuterium label or spin label) in the chain. There is an apparent plateau in the order parameter describing the degree of motional order in the portion of the hydrocarbon chain nearest the glycerol. In contrast, near the center of the bilayer, the hydrocarbon chain segments experience considerable motional freedom. For example, the terminal methyl groups experience nearly isotropic motion in the bilayer center or, in other words, can move in all directions with nearly equal ability (see Chapter 4 for details).

The effect of cholesterol on this order profile is dramatic (Fig. 4.27). Cholesterol increases the motional order of the plateau region (or, in other words, decreases the number of degrees of freedom of motion), with little effect on the properties of the bilayer center. This effect is different from that which occurs when the membrane enters a gel state, when all positions in the hydrocarbon chains experience an increase in motional order.

Increasing the motional order of the plateau region corresponds to a decrease in trans–gauche isomerizations in the plateau region. However, the average number of gauche bonds for the whole hydrocarbon chain of dipalmitoylphosphatidylcholine changes slightly, if at all.[27] There may be an increase in the number of gauche bonds in the part of the chain near the bilayer center to compensate. The packing of the cholesterol molecule in the membrane may require such an effect. The steroid nucleus is planar and relatively rigid. Therefore the portion of the phospholipid hydrocarbon chain encountering the steroid nucleus would be hindered in its motion. The cholesterol side chain is not quite as free to move and fill up space as is the hydrocarbon chain of the phospholipid. ^{2}H NMR data indicate that the tail of cholesterol is more ordered than the corresponding region of the phospholipid hydrocarbon chains.[28] Therefore the end of the phospholipid hydrocarbon chain may need to compensate for the cholesterol, particularly when the phospholipid is longer than the cholesterol, by an increase in the number of gauche bonds near the chain terminus. Cholesterol thus increases the anisotropy of the interior of the lipid bilayer.

These results may provide clues to one of the important roles of cholesterol in biological membranes. Propagation of the special ordering effect, which is unique to cholesterol among common membrane lipids, may affect the conformational motility of membrane proteins and thus affect their function (see Chapter 7 for a description of this mechanism).

The effect of cholesterol on the ordering of the phospholipid hydrocarbon chains would predict that inclusion of cholesterol will induce the upper portion of the lipid hydrocarbon chains to adopt more trans configurations about the carbon–carbon bonds and become effectively longer. This increase in effective length then must lead to an increase in thickness of the bilayer. Cholesterol does increase the bilayer thickness of membranes consisting of phospholipids with hydrocarbon chains up to 16 carbon atoms in length. However, for a phospholipid with an 18-carbon chain, an interesting deviation from this behavior is noted. Cholesterol actually decreases the width of the bilayer. Cholesterol is not as long as an 18 : 0 chain. One way to alleviate packing problems at the center of the bilayer is for the phospholipid hydrocarbon chain to fill up the volume in the bilayer center not occupied by the sterol tail by becoming more disordered. But to do so, the effective length of the 18 : 0 chain must decrease. Therefore the bilayer thickness must decrease also.[29]

The ability of a sterol to have such effects is dependent on its structure. Significant changes in structure will inhibit the ability of the sterol to produce this ''special ordering'' effect. Several studies have demonstrated what structural changes are permitted and what changes are not.[30, 31] In general, the structural requirements for sterol-like behavior include a flat fused ring system, a β-hydroxyl at position 3, and a cholesterol-like tail. These agree fairly well with the requirements for effective reduction in membrane permeability (see below).

B. Effects on Membrane Permeability

The effects of sterols on membrane permeability, determined by measuring glucose release from liposomes, led to the identification of important structural features of the cholesterol molecule.[32] Permeability was measured by trapping glucose within liposomes consisting of phosphatidylcholine and a prescribed amount of sterol. The liposomes were then washed and the amount of glucose leaking out was assayed as a function of time. By using sterols of different structure, it was discovered that a planar ring system, a 3β-hydroxyl, and a side chain are all required for a sterol-mediated reduction in membrane permeability. The effects of cholesterol can be seen in Fig. 5.6. Significant deviations from these structural features defeat the effectiveness of the sterol in inhibiting the permeability of a phospholipid bilayer. In some cases, the permeability can even be increased by some sterols and related compounds.

Permeability of small molecules across membranes can be thought to occur by jumping between defects in the bilayer structure as the small molecules transit the membrane (see Chapter 4, Section III,6). Kinks in the hydrocarbon chains (caused by trans–gauche isomerizations) could provide such defects. Cholesterol apparently reduces the incidence of these kinks, at least in the plateau region (with respect to membrane order) of the membrane. This leads to a reduction in the number of defects, and a consequent decrease in the permeability. The model may also explain the decrease in solubility of hydrophobic molecules in the membrane with an increase in sterol content.

Litman and co-workers have approached this issue from a different perspective, using the properties of a fluorescent probe to calculate the effective "free volume" in a membrane.[33] The concept of free volume is related to the defects in packing just discussed. The addition of cholesterol reduces the "free volume" in the membrane. Figure 5.7 shows an example of the influence of cholesterol on this "free volume" parameter.

A similar effect is noted on the incorporation of an integral membrane

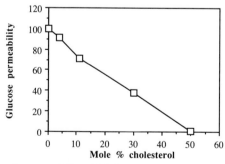

Fig. 5.6. Decrease in permeability of PC liposomes to glucose by increased cholesterol. Data regraphed from R. A., Demel, K. R. Bruckdorfer, and L. L. M. van Deenen, *Biochim. Biophys. Acta* **255** (1972): 321–330.

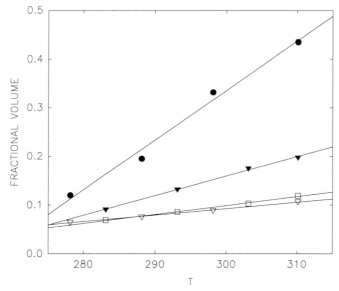

Fig. 5.7. Fractional volume as a function of temperature. (●) palmitoylarachidonoyl PC (PAPC); (▽) PAPC + 30 mole % cholesterol; (▼) PAPC + rhodopsin (PC/rhodopsin = 100/1) (□) PAPC + rhodopsin (PC/rhodopsin = 100/1) + 30 mol% cholesterol. Data taken from M. Straume and B. J. Litman, *Biochemistry* **26** (1987): 5121; M. Straume and B. J. Litman, *Biochemistry* **27** (1988) 7723. Figures provided by M. Straumme and B. J. Litman.

protein in the membrane bilayer.[34] A membrane protein, on incorporation into a lipid bilayer, reduces the available "free volume," analogous to the incorporation of cholesterol into the membrane. This observation can be understood by a requirement of the membrane protein on incorporation into a membrane for some response by the bilayer to accommodate the volume occupied by the protein. Scavenging free volume from packing defects in the lipids may provide part of the volume required by a membrane protein undergoing normal "breathing modes" as the protein samples a limited set of conformations within the bilayer, the extent of which is determined, in part, by the ambient temperature. More details are discussed in Chapter 7.

One would already anticipate that the influence of cholesterol on the properties of the lipid bilayer may be antagonistic to the need for "free volume" by the integral membrane protein. In such a case, cholesterol might be expected to inhibit the function of such a membrane protein Section V).

C. The Condensing Effect of Cholesterol

The concept of the condensing effect is derived from studying mixed monolayers of cholesterol and phospholipids (see Chapter 3, Section II,5). The condensing effect is related to the ordering of membranes by cholesterol. Cholesterol decreases the surface area per molecule occupied by saturated and monounsaturated phospholipids in monolayers at the air–water interface.[35] The role of the ester carbonyls of the phospholipids on the condensing effect has been examined with X-ray diffraction.[36] Although the condensing effect is not as complete with the non-ester-containing phospholipids, some condensing does occur, as measured by bilayer thickness. Interestingly, cholesterol has no condensing effect on 18 : 2,18 : 2 PC or on 18 : 3,18 : 3 PC.[37]

D. Effects of Cholesterol on Phospholipid Headgroups

Phospholipid headgroups in membranes consisting of phosphatidylcholine, phosphatidylethanolamine, and/or sphingomyelin interact with neighboring phospholipids via intermolecular headgroup interactions. These intermolecular interactions are disrupted by the insertion of cholesterol into the membrane through a spacing effect.[38]

E. Effects of Cholesterol on Lateral Diffusion

Measurements indicate that cholesterol has only modest effects on lateral diffusion of phospholipids in most circumstances. The only profound effects were on lateral diffusion below the gel to liquid crystalline phase transition temperature. Cholesterol, because it inhibits the formation of gel state lipid, increases the diffusion rate of the lipids under the same conditions.[39]

The effect of cholesterol on the lateral diffusion of proteins has also been examined. The lateral diffusion of M-13 coat protein was measured in phospholipid bilayers as a function of cholesterol concentration.[40] Lateral diffusion of M-13 coat protein was little affected by the addition of cholesterol in the liquid crystal state, as in the case of lateral diffusion of phospholipid. The only dramatic effects of cholesterol were observed below the phase transition temperature of the phospholipid due to cholesterol inhibition of gel phase formation.

F. Effects of Cholesterol on Phase Transitions

One of the best known of the effects of cholesterol on the properties of phospholipid bilayers is the dramatic change in the enthalpy and cooperativity of the gel to liquid crystalline phase transition in phospholipid bilayers. Cholesterol causes the elimination of the sharp, highly cooperative phase transition in dipalmitoylphosphatidylcholine bilayers at about 20–25 mol% cholesterol. Furthermore the size of the cooperative unit for the sharp transition also decreases as cholesterol content of the membrane increases. Similar phenomena are observed for cholesterol in other phospholipids.[1]

When the effects of cholesterol on the phase transition behavior of sphingomyelin are measured, the observed behavior is somewhat analogous to the effects of cholesterol on the phospholipid gel to liquid crystalline phase transitions. Calorimetry measurements indicate that the system is complex in the presence of cholesterol, and much hysteresis is observed during the measurements.

This effect of cholesterol has been suggested to be important biologically. Since biological membranes should, for the most part, be in the liquid crystal state for proper function, a role for cholesterol in maintaining this state is easy to imagine. However, most plasma membranes would be in the liquid crystal state at 37°C without the cholesterol. Furthermore, the introduction of a double bond in the hydrocarbon chain of the membrane lipid is sufficient to keep the membrane bilayer in the liquid crystal state. Simple desaturation is metabolically much cheaper than synthesizing cholesterol, so inhibiting gel phase in membranes seems unlikely as a major role for cholesterol in cell biology.

Below the phase transition temperature of the host phospholipid, cholesterol may create two domains. A lateral phase separation occurs in the presence of cholesterol forming cholesterol-rich regions and cholesterol-poor regions.[41] Data also suggested that cholesterol induces domain formation in membranes in the liquid crystal state. This is a question worthy of further examination.

V. ROLE OF STEROLS IN STEROL-REQUIRING CELLS

Cholesterol has long been known to be an essential component of mammalian cells.[42] Yet despite much study, the role cholesterol plays in mammalian cell biology has remained a mystery. Without cholesterol, mammalian cells cannot experience normal growth. Mammalian cells, in fact, are capable of making their own cholesterol.

Many steps are involved in cholesterol biosynthesis. Even after the synthesis of lanosterol, 18 enzymatically catalyzed steps remain to produce cholesterol.[43] Much valuable cellular energy is therefore utilized in the complex biosynthetic pathway to produce the particular chemical structure of cholesterol. Why this obtains is not fully understood. Bloch has suggested that evolutionary pressure for a more biologically competent sterol led to the development of the pathway from lanosterol to cholesterol. However the molecular details describing why the cholesterol structure is required for biological competence (for example, in mammalian cells) have not yet been fully described.

A. Structural Requirements for Sterols in Cell Biology

Some studies on specificity for sterol structure by sterol-requiring cells are available. *In vitro* experiments have shown that lanosterol cannot fully substitute for cholesterol as the essential sterol for mycoplasma cell function.[44] In particular, *Mycoplasma mycoides* can be adapted to grow on low cholesterol media. However, they cannot grow in the total absence of cholesterol in the medium, since they do not make their own cholesterol and cholesterol is required for cell growth and function. Supplementation in the medium of lanosterol will not support cell growth in the absence of cholesterol. However, cell growth will occur at nearly the same rate if cells fed low cholesterol levels supplemented with high (relatively) lanosterol, as in cells fed high (relatively) cholesterol levels. Thus cholesterol appears both adequate and necessary for minimal cellular function in these mycoplasma. For optimum cellular function, higher membrane sterol content is required but without the structural specificity associated with the requirement for minimal cellular function. This sterol synergism has pointed to a special role for cholesterol in supporting cell growth in which the particular chemical structure of cholesterol is required for mycoplasma.

The requirement of cholesterol for normal function of mycoplasma is mirrored in yeast by an analogous requirement for ergosterol. Yet one sterol cannot substitute for the other; cholesterol cannot fully substitute for ergosterol in yeast, and ergosterol cannot support normal mammalian cellular function.[45] Anaerobic growth of *Saccharomyces cerevisiae* requires ergosterol supplement to the culture medium. Supplementation only with cholesterol will not support normal growth. Yet cholesterol is more effective at modifying the properties of lipid bilayers than is ergosterol. These data point to a specific recognition of sterol structure, crucial to normal cellular function, that is different in yeast than it is in mycoplasma.

The requirement for cholesterol in mammalian cells is not well understood due to the paucity of available mammalian sterol auxotrophes. Therefore other methods have been used to explore the role of cholesterol. In mammalian cells, inhibition of cellular cholesterol biosynthesis inhibits cell growth. In addition to the role of metabolic products from mevalonate other than cholesterol in cell growth, cholesterol itself played a role, since the addition of cholesterol in the culture medium would restore cell growth.[46]

Data such as these suggest that a unique sterol structure leads to biological competence in each sterol-requiring organism. What defines that biological competence is central to the discussion in this chapter. From this analysis emerges a hypothesis for the essential role of cholesterol in mammalian (and other cholesterol-requiring) cells.

B. Modulation of Biological Function of Membranes by Cholesterol

The modulation by cholesterol of the function of membrane proteins has been examined, both by modifying the cholesterol content of the native membranes in which the protein is found, and by reconstituting the membrane proteins into membranes of defined lipid content. Studies of this sort have led to three different classes of observations.

1. An increase in the level of cholesterol in the membrane leads to a proportionate decrease in membrane protein function. An example can be seen in Fig. 5.8. In this example, the equilibrium constant for the Meta I–Meta II transition of rhodopsin was measured as a function of the cholesterol content of reconstituted membranes.[34] The data show an inverse relationship between the function of the membrane protein and the cholesterol content of the membrane. For rhodopsin, cholesterol apparently acts as a negative modulator, or an inhibitor, at high cholesterol levels.

2. An increase in the level of cholesterol in the membrane leads to a proportionate increase in membrane protein function. This can be seen at low membrane cholesterol levels for the Na^+,K^+-ATPase in Fig. 5.9. From a cholesterol/phospholipid mole ration of 0 to about 0.35, an increase in cholesterol in the membrane leads to an increase in the ATP hydrolyzing activity of this enzyme in the modified native membrane.[47]

3. Some membrane functions appear to be insensitive to cholesterol levels in the membrane. As membrane cholesterol content was varied over a wide range, no alteration in sarcoplasmic reticulum Ca^{2+} ATPase activity was observed.

The challenge is to find an explanation for these varied observations. There are at least two general classes of interactions in which cholesterol can engage while in a membrane. One is best represented by the discussion above on "free volume" (see Section IV,B).

The other interaction between cholesterol and membrane components

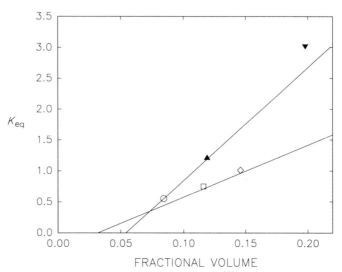

Fig. 5.8. K_{eq} (Meta I : Meta II equilibrium for rhodopsin) as a function of fractional volume. (◇) Egg PC + rhodopsin (PC/rhodopsin = 100/1), 0 mol% cholesterol; (□) same + 15 mol% cholesterol; (○) + 30 mol% cholesterol; (▼) palmitoylarachidonoyl PC (PAPC) + rhodopsin (PAPC/rhodopsin = 100/1), 0 mol % cholesterol; (▲), same + 30 mol % cholesterol (all at 37°C). Data obtained from D. C. Mitchell, M. Straume, J. L. Miller, and B. J. Litman, *Biochemistry* **29,** (1990): 9143; D. C. Mitchell, M. Straume, and B. J. Litman, *Biochemistry* **31** (1992): 662. Figures provided by D. Mitchell, M. Straume, and B. J. Litman.

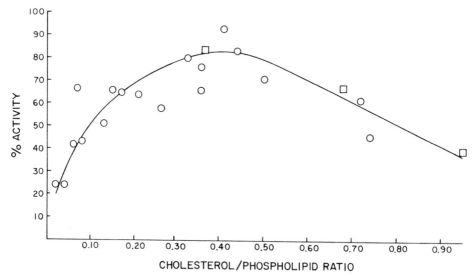

Fig. 5.9. Na^+,K^+-ATPase activity from bovine kidney cortex, as a function of membrane cholesterol. Apparently, membrane cholesterol stimulates activity at low contents and inhibits activity at high contents. The inhibition is likely due to the cholesterol-induced increase in membrane order. The stimulation may be due to a direct sterol–protein interaction. (□) Erythrocyte; (○) kidney.

that could be important to membrane function is a direct interaction between the sterol and the membrane protein. Studies have shown an apparent binding of cholesterol to some integral membrane proteins, including glycophorin[48] and band 3[49] from the human erythrocyte membrane (see Chapter 7).

The possibility that cholesterol can bind directly to membrane proteins suggests a mechanism of sterol modulation of membrane proteins in a manner analogous to that of effectors of water-soluble proteins. That is, binding of sterol to a sterol-specific site on a membrane protein would lead to a modulation of the function of that membrane protein. One would predict that such a mechanism would also give rise to a structural specificity in the sterol modulation of membrane proteins. The degree of such specificity would be governed by nature of the sterol–protein interactions.

C. Hypothesis for a Specific Sterol Requirement for Cellular Function

This discussion has led to the conclusion that cholesterol likely modulates the function of biological membranes by more than one mechanism. These are summarized in the following.

1. Cholesterol alters the bulk biophysical properties of membranes. Cholesterol increases the orientational order of the lipid hydrocarbon chains of membranes and reduces the "free volume" available to membrane proteins for conformation changes that may be required for membrane protein function. In this role, cholesterol likely inhibits membrane function.

2. Cholesterol may bind directly to membrane proteins and regulate their function. In this role cholesterol may stimulate or may inhibit membrane function.

The discussion can now return to the specific cases provided above for cholesterol modulation of membrane proteins. The first case was that of bovine rhodopsin reconstituted into membranes containing varying levels of cholesterol. Increasing cholesterol levels inhibited the Meta I to Meta II transition.

Comparison of Fig. 5.7 and Fig. 5.8 shows that the relationships described for cholesterol effects on "free volume" and cholesterol inhibition of the Meta I to Meta II transition are remarkably similar. These data suggest that (a) free volume may be required by the protein for function; (b) cholesterol reduces the free volume available to the protein in the bilayer; and (c) cholesterol thereby inhibits the function of rhodopsin in the membrane. It has been shown, for example, that activation of the GTP cascade by retinal rod outer segment plasma membranes is inhibited by the high cholesterol content of that membrane.[50]

The second case provided above for cholesterol modulation of membrane function was that of the Na^+, K^+-ATPase. Cholesterol stimulates the Na^+, K^+-ATPase at low to moderate membrane cholesterol levels (see Fig. 5.9). Lanosterol and ergosterol were shown to be unable to substitute fully for cholesterol in this stimulation of the enzyme.[47]

The influence of several other sterols on the activity of the Na^+, K^+-ATPase provided greater insight into the specificity of the stimulation for sterol structure.[51] For each of these sterols, previous measurements provided a measure of the ability of these sterols to reduce the packing defects and favor an all-*trans* conformation of the carbon–carbon single bonds in the lipid hydrocarbon chains. One such parameter is the reduction in the area per headgroup for the phospholipids in monolayers. This is a measure of the condensing effect of sterols as described earlier.

Figure 5.10 shows a plot of the extent of the condensing effect of the sterols (reduction in area per lipid headgroup at a fixed sterol level in the monolayer of phospholipid) as a function of the ability of each of these sterols to support ATP hydrolyzing activity of the Na^+, K^+-ATPase. Each point represents a different sterol. What is evident from this graph is that no simple direct proportionality exists between the ability of the sterol to "condense" the phospholipids in a monolayer and the ability of the sterols to support activity of the enzyme under otherwise constant conditions in a reconstituted system.

These data indicate that the activation of the Na^+, K^+-ATPase by sterol

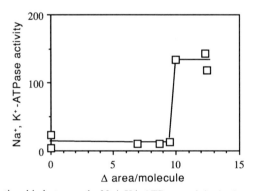

Fig. 5.10. Relationship between the Na^+, K^+-ATPase activity in the presence of various sterols and the condensing effect of cholesterol on phosphatidylcholine bilayers, represented as the additional reduction in area per headgroup due to the presence of the sterol in the membrane at 50 mol%. The sterols represented in this graph are epicholesterol, ergosterol, stigmasterol, 7-dehydrocholesterol, cholesterol, and cholestanol. Reprinted with permission from P. L. Yeagle, *Biochemie* **73** (1991): 1303–1310.

is not a process that is modulated by the condensing or ordering of the membrane lipids by cholesterol. Not only is there no correlation in Fig. 5.10 with enzyme activity, but the small amounts of cholesterol that are required to activate some of the enzymes (see Fig. 5.9) do not have an equivalently dramatic effect on the ordering of the hydrocarbon chains of the membrane lipids at those same low cholesterol levels. At the membrane cholesterol levels that do significantly affect the bulk properties of the lipid bilayer, one observes inhibition of enzyme activity (see below).

What is suggested by the graph in Fig. 5.10 is that activation of the Na^+,K^+-ATPase by sterol at low to moderate sterol levels in the membrane is highly structurally specific. Such structural specificity can be best explained by a direct interaction of sterol with membrane protein. The shape of the activation curve of the Na^+,K^+-ATPase is similar to that for a binding isotherm.

With this model, the data on cholesterol modulation of the Na^+,K^+-ATPase can be explained by the competition of two effects. One is stimulation of the enzyme by cholesterol at low to moderate cholesterol levels. Stimulation in this model would result from a direct interaction of the sterol with the protein. The other is inhibition of the enzyme by cholesterol at high membrane cholesterol levels. Inhibition in this model would result from a restriction of conformational changes by the enzyme required for function due to a reduction in free volume available within the lipid bilayer by the presence of the cholesterol.

These two effects of cholesterol on the Na^+,K^+-ATPase compete to produce a region of membrane cholesterol levels that support maximal activity. Interestingly for kidney Na^+,K^+-ATPase, maximal activity is observed at the cholesterol level of the native membrane. This is appropriate for an organ in which the Na^+,K^+-ATPase is very important to the function. Thus cholesterol homeostasis may be important to ion transport in the kidney. In contrast in the human erythrocyte, where high fluxes of sodium and potassium are not as crucial to cellular function, the membrane cholesterol level determines operation of the erythrocyte Na^+,K^+-ATPase at submaximal activity.

For those enzymes in which only inhibition by cholesterol is observed, one could postulate that there are no sites for effective interaction between sterol and protein. In this case, the only effect would be the influence of cholesterol on the bulk properties of the bilayer, thereby indirectly affecting the ability of membrane proteins to undergo conformational changes.

Based on the above analysis, it is possible to formulate a hypothesis for the essential role of cholesterol in cholesterol-requiring cells: the essential role of cholesterol in cell function is to activate membrane enzymes that are necessary for cellular function and growth. Cholesterol carries out this function through specific binding to particular integral membrane enzymes. In mammalian cells, the sterol–protein interaction accommodates cholesterol perferentially. Therefore cholesterol is required for mammalian cell function. Interestingly, some plasma membrane enzymes would be expected to exhibit reduced activity at their site of synthesis in the ER.

VI. SUMMARY

Consideration of recent research on the behavior and consequences of cholesterol in cell membranes helps to focus attention on several important issues concerning the roles of cholesterol in mammalian cells. Most cellular cholesterol is found in the plasma membrane. Membrane cholesterol levels influence membrane permeability and stability. Cholesterol does this by its special ordering effect on the membrane lipids. Membrane cholesterol modulates plasma membrane protein activity. Positive modulation likely occurs through structurally specific, direct sterol–protein binding, favoring an active state of the enzyme. Negative modulation likely occurs through a reduction in free volume within the bilayer, necessary for required enzyme conformational changes, or possibly through direct sterol–protein binding. Cholesterol, either directly or indirectly, plays a role in gene expression. For example, LDL cholesterol can apparently inhibit expression of the structural gene for the LDL receptor. There is currently no information on mechanisms for cholesterol-mediated gene regulation.

REFERENCES

1. Yeagle, P. L., "Cholesterol and the cell membrane," *Biochim. Biophys. Acta Biomembrane Rev.* **822** (1985): 267–287.
2. Yeagle, P. L., *The Biology of Cholesterol* (Boca Raton: CRC Press, 1988), 242.
3. Duax, W. L., Z. Wawrzak, J. F. Griffin, and C. Cheer, "Sterol conformation and molecular properties," In *Biology of Cholesterol* (P. L. Yeagle, ed.) (Boca Raton: CRC Press, 1988).
4. Chapelle, S., and M. G. Baillien, "Phospholipids and cholesterol in brush border and basolateral membranes from rat intestinal mucosa," *Biochim. Biophys. Acta* **753** (1983): 269–271.
5. Orci, L., R. Montesano, P. Meda, F. Malaisse-Lagae, D. Brown, A. Perrelet, and P. Vassalli, "Heterogeneous distribution of filipin–cholesterol complexes across the cisternae of the Golgi apparatus," *Proc. Natl. Acad. Sci. U.S.A.* **78** (1981): 293–297.
6. Boesze-Battaglia, K., S. J. Fliesler, and A. D. Albert, "Relationship of cholesterol content to spatial distribution and age of disk membranes in retinal rod outer segments," *J. Biol. Chem.* **265** (1990): 18867–18870.

7. Lange, Y., H. B. Cutler, and T. L. Steck, "The effect of cholesterol and other intercalated amphipaths on the contour and stability of the isolated red cell membrane," *J. Biol. Chem.* **255** (1980): 9331–9336.

8. McLean, L. R., and M. C. Phillips, "Mechanism of cholesterol and phosphatidylcholine exchange or transfer between unilamellar vesicles," *Biochemistry* **20** (1981): 2893–2900.

9. Backer, J. M., and E. A. Dawidowicz, "Mechanism of cholesterol exchange between phospholipid vesicles," *Biochemistry* **20** (1981): 3805–3810.

10. Steck, T., F. Kezdy, and Y. Lange, "An activation–collision mechanism for cholesterol transfer between membranes," *J. Biol. Chem.* **263** (1988): 13023–13031.

11. Bruckdorfer, K. R., and M. K. Sherry, "The solubility of cholesterol and its exchange between membranes," *Biochim. Biophys. Acta* **769** (1984): 187–196.

12. Boesze-Battaglia, K., and A. D. Albert, "Phospholipid distribution in bovine rod outer segment membranes," *Exp. Eye Res.* **54**, (1992): 821–823.

13. Wattenberg, B. W., and D. F. Silbert, "Sterol partitioning among intracellular membranes," *J. Biol. Chem.* **258** (1983): 2284–2289.

14. Yeagle, P. L., and J. Young, "Factors contributing to the distribution of cholesterol among phospholipid vesicles.," *J. Biol. Chem.* **261** (1986): 8175–8181.

15. Lange, Y., J. S. D'Alessandro, and D. M. Small, "The affinity of cholesterol for phosphatidylcholine and sphingomyelin," *Biochim. Biophys. Acta* **556** (1979): 388–398.

16. Lange, Y., and H. J. G. Matthies, "Transfer of cholesterol from its site of synthesis to the plasma membrane," *J. Biol. Chem.* **259** (1984): 14624–14630.

17. Kaplan, M. R., and R. D. Simoni, "Transport of cholesterol from the endoplasmic reticulum to the plasma membrane," *J. Cell Biol.* **101** (1985): 446–453.

18. Huang, C. H., J. P. Sipe, S. T. Chow, and R. B. Martin, "Differential interaction of cholesterol with phosphatidylcholine on the inner and outer surfaces of lipid bilayer vesicles," *Proc. Natl. Acad. Sci. U.S.A.* **71** (1974): 359–362.

19. Blau, L., and R. Bittman, "Cholesterol distribution between the two halves of the lipid bilayer of human erythrocyte membranes," *J. Biol. Chol.* **253** (1978): 8366.

20. Bittman, R., and S. Rottem, "Distribution of cholesterol between the outer and inner halves of the lipid bilayer of *Mycoplasma* cell membranes," *Biochem. Biophys. Res. Commun.* **71** (1976): 318–324.

21. Schroeder, F., "Use of a fluorescent sterol to probe the transbilayer distribution of sterols in biological membranes," *FEBS Lett.* **135** (1981): 127–130.

22. Backer, J. M., and E. A. Dawidowicz, "Transmembrane movement of cholesterol in small unilamellar vesicles detected by cholesterol oxidase," *J. Biol. Chem.* **256** (1981): 586–588.

23. Lange, Y., J. Dolde, and T. L. Steck, "The rate of transmembrane movement of cholesterol in the human erythrocyte," *J. Biol. Chem.* **256** (1981): 5321–5323.

24. Franks, N. P., "Structural analysis of hydrated egg lecithin and cholesterol bilayers. I. X-ray diffraction," *J. Mol. Biol.* **100** (1976): 345–358.

25. Worcester, D. L., and N. P. Franks, "Structural analysis of hydrated egg lecithin and cholesterol bilayers. II. Neutron diffraction," *J. Mol. Biol.* **100** (1976): 359–378.

26. Oldfield, E., and D. Chapman, "Molecular dynamics of cerebroside–cholesterol and sphingomyelin–cholesterol interactions: Implications for myelin membrane structure," *FEBS Lett.* **21** (1972): 303–306.

27. Bush, S. F., R. G. Adams, and I. W. Levin, "Structural reorganizations in lipid bilayer systems: Effect of hydration and sterol addition on Raman spectra of dipalmitoylphosphatidylcholine multilayers," *Biochemistry* **19** (1980): 4429–4435.

28. DuFourc, E. J., E. J. Parish, S. Chitrakorn, and I. C. P. Smith, "Structural and dynamical details of the cholesterol–lipid interaction as revealed by deuterium NMR," *Biochemistry* **23** (1984): 6062–6071.

29. McIntosh, T. J., "The effect of cholesterol on the structure of phosphatidylcholine bilayers," *Biochim. Biophys. Acta* (1978): 43–58.

30. Ahmad, P., and A. Mellors, "NMR studies on liposomes: Effects of steroids on lecithin fatty acyl chain mobility," *J. Membrane Biol.* **41** (1978): 235–247.

31. Yeagle, P. L., R. B. Martin, A. K. Lala, H. Lin, and K. Bloch, "Differential effects of cholesterol and lanosterol on artificial membranes," *Proc. Natl. Acad. Sci. U.S.A.* **74** (1977): 4924–4926.

32. Demel, R. A., K. R. Bruckdorfer, and L. L. M. van Deenen, "The effect of sterol structure on the permeability of liposomes to glucose, glycerol and Rb^+," *Biochim. Biophys. Acta* **255** (1972): 321–330.

33. Straume, M, and B. J. Litman, "Influence of cholesterol on equilibrium and dynamic bilayer structure of unsaturated acyl chain phosphatidylcholine vesicles as determined from higher order analysis of fluorescence anisotropy decay," *Biochemistry* **26** (1987): 5121–5126.

34. Mitchell, D. C., M. Straume, J. L. Miller, and B. J. Litman, "Modulation of metarhodopsin formation by cholesterol-induced ordering of bilayer lipids," *Biochemistry* **29** (1990): 9143.

35. deKruijff, B., R. A. Demel, A. J. Slotboom, L. L. M. van Deenen, and A. F. Rosenthall, "The effect of the polar headgroup on the lipid–cholesterol interaction: A monolayer and differential scanning calorimetry study," *Biochim. Biophys. Acta* **307** (1972): 1–19.

36. Schwarz, F. T., F. Paltauf, and P. Laggner, "Studies on the interaction of cholesterol with diester- and diether-lecithin," *Chem. Phys. Lipids* **17** (1976): 423.

37. Demel, R. A., W. S. M. G. van Kessel, and L. L. M. van Deenen, "The properties of polyunsaturated lecithins in monolayers and liposomes and the interactions of these lecithins with cholesterol," *Biochim. Biophys. Acta* **266** (1972): 26–40.

38. Yeagle, P. L., W. C. Hutton, C-H. Huang, and R. B. Martin, "Phospholipid headgroup conformations: Intermolecular interactions and cholesterol effects," *Biochemistry* **16** (1977): 4344–4349.

39. Rubenstein, J. L. R., B. A. Smith, and H. M. McConnell, "Lateral diffusion in binary mixtures of cholesterol and phosphatidylcholines," *Proc. Natl. Acad. Sci. U.S.A.* **76** (1979): 15–18.

40. Smith, L. M., J. L. R. Rubenstein, J. W. Parce, and H. M. McConnell, "Lateral diffusion of M-13 coat protein in mixtures of phosphatidylcholine and cholesterol," *Biochemistry* **19** (1980): 5907.

41. Shimshick, E. J., and H. M. McConnell, "Lateral phase separations in binary mixtures of cholesterol and phospholipids," *Biochem. Biophys. Res. Commun.* **53** (1973): 446–451.

42. Yeagle, P. L., "Modulation of membrane function by cholesterol," *Biochemie* **73** (1991): 1303–1310.

43. Faust, J. R., J. M. Trzaskos, and J. L. Gaylor, "Cholesterol biosynthesis," In *Biology of Cholesterol* (P. L. Yeagle, ed.) (Boca Raton: CRC Press, 1988), 19–38.

44. Dahl, J. S., C. E. Dahl, and K. Bloch, "Sterol in membranes: Growth characteristics and membrane properties of *Mycoplasma capricolum* cultured on cholesterol and lanosterol," *Biochemistry* **19** (1980): 1467–1472.

45. Dahl, C., and J. Dahl, "Cholesterol and cell function," In *Biology of Cholesterol* (P. L. Yeagle, ed.) (Boca Raton: CRC Press, 1988), 147–172.

46. Esfahani, M., L. Scerbo, and T. M. Devlin, "A requirement for cholesterol and its structural features for a human macrophage-like cell line," *J. Cell. Biochem.* **25** (1984): 87–97.

47. Yeagle, P. L., D. Rice, and J. Young, "Effects of cholesterol on (Na,K)-ATPase ATP hydrolyzing activity in bovine kidney," *Biochemistry* **27** (1988): 6449–6452.

48. Yeagle, P. L., "Incorporation of the human erythrocyte sialglycoprotein into recombined membranes containing cholesterol," *Membrane Biol.* **78** (1984): 201–210.

49. Klappauf, E., and D. Schubert, "Band 3 from human erythrocyte membranes strongly interacts with cholesterol," *FEBS Lett.* **80** (1977): 423–425.

50. K. Boesze-Battaglia and A. Albert, "Cholesterol modulation of photoreceptor function in bovine rod outer segments," J. Biol. Chem. **265** (1990): 20727–20730.

51. Vemuri, R., and K. D. Philipson, "Influence of sterols and phospholipids on sarcolemmal and sarcoplasmic reticular cation transporters," *J. Biol. Chem.* **264** (1989): 8680–8685.

6

Membrane Proteins

Two of the three major components of biological membranes, water and polar lipids, alone are sufficient for the construction of a semipermeable barrier that exhibits many of the attributes of a cell membrane. Bilayers made from lipids and water can create separate compartments, each with its own identity, as in the case of cells and organelles. Vesicles made of just these two components can be sealed against leakage of protons and other ions, as are biological membranes. Lipid bilayer membranes can fuse; biological membranes can fuse. Lipid bilayers can hold an electrical potential as can cellular membranes. Lipid vesicles of the right composition can even provide specific binding sites for lectins or antibodies. These and other properties of biological membranes can be mimicked by lipid bilayer vesicles or liposomes.

However, there are many properties of cell membranes that cannot be mimicked by lipid bilayers. Energy-driven transport of ions across membranes, receptor-mediated events, synthesis of membrane components, secretion, and ATP synthesis are only a few of a myriad of membrane-associated functions that lipid bilayers are incapable of performing on their own.

Membrane proteins are required to provide these additional functions to the membranes of cells. Protein structures can be extraordinarily versatile in their structure–function relationships, thus imparting enormous additional functionality to cell membranes. However, in most cases, membrane proteins cannot function properly on their own. They require lipid bilayers to complement their own intricate structures to exhibit correctly the functions associated with the membranes that contain the proteins. This will become more clear later in this and later chapters in this book as the functions of cell membranes, as expressed by the association of its constituents, are explored.

Just as each cell membrane is distinguishable by its lipid content (see Chapter 2), each cell membrane is also distinguishable by its protein composition. Thousands of different proteins are found as constituents of biological membranes. This variety is reflected in the abundance of different functionalities found in the myriad of membranes in cells. Not only the relative abundance, but also the absolute presence or absence of particular protein components, distinguishes individual membranes. Sodium docecyl sulfate (SDS) polyacrylamide gels of purified membranes frequently provide a "fingerprint" of that membrane; different membranes have different "fingerprints" on the SDS gels.

The covalent structure of membrane proteins is no different than soluble proteins. Membrane proteins are constructed of linear sequences of amino acids covalently bonded together by amide bonds. These sequences are prescribed most immediately by linear sequences of nucleic acids in the mRNA using standard translation codes. Thus in these aspects there is nothing to distinguish membrane proteins from nonmembrane proteins.

The proteins of cell membranes are distinguished from nonmembranous proteins by their association with lipid bilayers to form the functioning cell membrane. The association may be loose or tight. The proteins may be incorporated into the structure of the lipid bilayer or may be simply associated with a lipid or protein component on the surface of the membrane. Based on the nature of the association with the membrane, membrane proteins have been classified to understand more easily the kinds of protein–membrane association possible.[1, 2]

I. CLASSIFICATION OF MEMBRANE PROTEINS

Membrane proteins have been divided into two major classes, based on the nature of their association with the membrane: peripheral membrane proteins and integral membrane proteins.[3] Peripheral membrane proteins can be described functionally as proteins that are associated with membranes, but are readily removed by washing the membranes, changing the ionic strength, or changing the pH. These proteins are generally not integrated into the hydrophobic region of the membrane lipid bilayer.

Peripheral membrane proteins can be further subdivided into two subclasses. One subclass will be referred to as associated membrane proteins. An example is cytochrome c, which is associated with the inner mitochondrial membrane. It binds to some of the integral membrane proteins of that membrane through portions of those integral membrane proteins that are exposed on the surface of the membrane. The second subclass is referred to as skeletal. The skeletal membrane proteins form membrane skeletons, which are cytoplasmic protein networks intimately associated with the membrane. A good example of skeletal membrane proteins is spectrin of the human erythrocyte membrane.

A portion of the mass of integral membrane proteins is integrated into the structure of the lipid bilayer. Integral membrane proteins can, generally, be removed from the membrane only with detergents or chaotropic agents that disrupt the membrane structure or the water structure. Therefore, a portion of the integral membrane protein must be thermodynamically compatible with the hydrophobic interior of the membrane. The amino acids used to build these proteins can be subdivided into hydrophobic and hydrophilic amino acids, depending upon the nature of the amino acid side chain. The side chain of isoleucine, for example, is hydrophobic, whereas the side chain of glutamic acid is clearly hydrophilic. Thus the free energy of transfer of the former from hydrocarbon to aqueous phase is much less favorable than the free energy of transfer of the latter. One might expect a preponderance of the hydrophobic amino acids in the intramembranous portions of the integral membrane proteins. This does not necessarily lead to a preponderance of hydrophobic amino acids in the overall integral membrane structure, however. Rather, some interesting topological consequences for the structure of integral membrane proteins follow (to be examined shortly).

The class of integral membrane proteins can be further subdivided into two subclasses. Transmembrane proteins constitute one of these sub-classes. As their name implies, transmembrane proteins span the lipid bilayer of the membrane. At least some mass of the protein therefore appears on both sides of the membrane. An example of a transmembrane protein is human erythrocyte glycophorin. Glycophorin is a glycoprotein of approximately 30,000 Da. Of that mass, about 55% is carbohydrate. This protein is known not only to span the bilayer, but to have the carbohydrate bearing portion on the outside of the erythrocyte, and another portion of the peptide on the inside of the erythrocyte.

The other subclass of integral membrane proteins can be referred to as anchored proteins. These proteins also bury a portion of their mass in the hydrophobic interior of the lipid bilayer. However, these proteins are not transmembrane. The portion buried in the membrane therefore serves as an anchor to hold the protein in the membrane. An example of this class is a membrane bound form of acetylcholine esterase.

Figure 6.1 helps to define schematically the structures associated with each of the classes.

A. Cytochrome c, an Associated Membrane Protein

Cytochrome c is a small, water soluble protein of molecular weight about 12,000. It is a peripheral membrane protein since it can be readily

INTEGRAL MEMBRANE PROTEINS

Transmembrane

Anchored

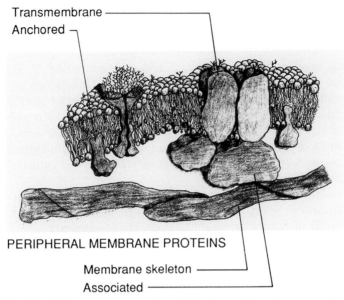

PERIPHERAL MEMBRANE PROTEINS

Membrane skeleton

Associated

Fig. 6.1.   Schematic representation of the classes of membrane proteins.

stripped from the inner mitochondrial membrane where it is found. Cytochrome c is associated with protein binding sites on the membrane surface. It is approximately spherical in shape. The topology for the amino acids of this protein is typical of that normally found for water soluble proteins. Polar amino acid residues largely reside on the outer surface of the protein, whereas hydrophobic amino acids are largely located in the interior of the protein. The hydrophobic effect is satisfied by this topology.

Cytochrome c is functionally involved in the electron transport chain of the mitochondria, which is intimately involved in the production of ATP. The role of cytochrome c is to carry electrons from one complex of the electron transport system to another. The electron donor is cytochrome c-1, an integral membrane protein of the inner mitochondrial membrane. The electron acceptor is the protein complex referred to as cytochrome-c oxidase, which is also located in the inner mitochondrial membrane. Thus electron transport can occur by cytochrome c binding to cytochrome c-1, picking up an electron, then disengaging itself and subsequently binding to cytochrome oxidase where it gives up that electron. Cytochrome c contains a heme group and it is this heme group that actually carries the electron. The rest of the protein creates the appropriate pocket in which the heme sits and offers the recognition sites for binding to the cytochrome c-1 and cytochrome-c oxidase. These sites involve some electrostatic interactions made possible by discrete basic patches on the surface of cytochrome c.

B. Spectrin, a Membrane Skeleton Protein

There appears to be an intricate network, consisting of protein and attached to the plasma membrane of the erythrocyte, which gives shape to the red cell. This network can sometimes be preserved after dissolving the remainder of the membrane. The term membrane skeleton is used to describe this network. Considerable data suggest that other cells, including nucleated cells, have membrane skeletons too.[4] These membrane skeletons may also constitute a connection between the cell cytoskeleton and the plasma membrane.

A protein typical of the membrane skeleton is human erythrocyte spectrin. Spectrin comes in at least two forms, one of molecular weight 240,000 and another of molecular weight 220,000. The fundamental unit is a dimer. Spectrin generally assumes an elongated shape on the inside of the human erythrocyte plasma membrane. The length of the dimer is about 1000 Å and the diameter is about 50 Å. These units associate to form a tetramer with binding sites for actin on the ends of the tetramers. In the presence of band 4.1 of the erythrocyte (a protein of about 80,000 molecular weight that also binds to spectrin) a tight complex is formed between spectrin, actin, and band 4.1 to link the spectrin tetramers in a lattice.

This spectrin network is connected to the plasma membrane by protein links. Figure 6.2 shows spectrin membrane skeleton represented as analogous to a rope net underneath the membrane. That network does not float freely. Rather the network is attached to the membrane by discrete protein bridges. One bridge is provided by a combination of band 3 (the anion transport protein) and ankyrin.[5] Band 3 is an integral membrane protein of the erythrocyte membrane of molecular weight about 90,000. Band 3 spans the membrane and exposes a considerable portion of its mass on the cytoplasmic surface of the membrane.

Ankyrin has a molecular weight of about 210,000. Ankyrin has two interesting binding sites. One site exhibits affinity for a site on the spectrin network (the β subunit of spectrin). The other site exhibits affinity for band 3. Thus band 2.1 links the membrane skeleton to the membrane.

Another bridge from the membrane skeleton of the erythrocyte to the membrane involves a second abundant integral membrane protein of the erythrocyte glycophorin (see below). A portion of the glycophorin polypeptide protrudes from the cytoplasmic surface of the erythrocyte membrane. Band 4.1 can bind directly to glycophorin and simultaneously to spectrin, thus forming a second link between the membrane skeleton and the membrane. The binding of band 4.1 to glycophorin is modulated by

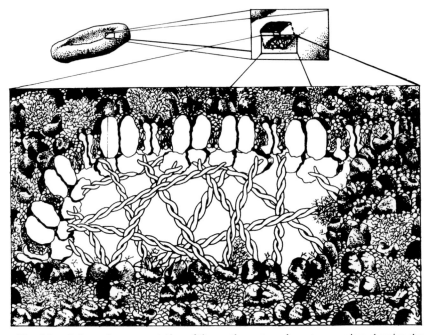

Fig. 6.2. Schematic representation of the erythrocyte and a cut-away view showing the spectrin network underlying the plasma membrane. Figure courtesy of W. Scherer.

the state of phosphorylation of phosphatidylinositol, bound specifically to glycophorin in the erythrocyte membrane.[6, 7]

As mentioned above, the membrane skeleton of the mammalian erythrocyte is the best studied of the membrane skeletons and provides something of a model for the membrane skeletons for other cells. It appears that other cells also have membrane skeletons with similar components. Interestingly, the spectrin-like proteins that are found in other cells do not react with antibodies for human erythrocyte spectrin. But antibodies from avian spectrins do cross react with spectrin-like proteins of non-erythroid cells. Likewise, analogies to band 3 and 4.1 also appear in non-erythroid cells. The details of the structure and function of the membrane skeletons of non-erythroid cells have yet to be determined. However, they may exhibit additional functions, over the erythroid models, including connection and communication between cell cytoskeleton and extracellular matrices.

C. Acetylcholinesterase, an Anchored Protein

A good example of anchored proteins can be found in those proteins that are covalently bound to a phospholipid to anchor them to the membrane.[8]

Membrane bound acetylcholinesterase is an example of an anchored protein whose sequence is known.[9] This enzyme is bound to membranes at the cholinergic synapses where its primary role is termination of nerve transmission after a stimulated impulse, which results in acetylcholine release from the presynaptic membrane.

The most common membrane bound form of this enzyme is covalently bound at its carboxyl terminal to phosphatidylinositol. This acetylcholine esterase can be released from the membrane by treatment with PI-specific phospholipase C.[10] The phospholipid serves as a hydrophobic tail to anchor the protein to the membrane.

Most of the structure of membrane bound acetylcholinesterase consists of a polypeptide with structural characteristics typical of water-soluble proteins (about 62 kDa from *Torpedo californica*). In fact, after release by phospholipase C, the protein behaves as a water soluble enzyme as dimer. The soluble form contains a functional active site for hydrolysis and a binding site for the positively charged choline.[11] The addition of a hydrophobic lipid anchor can be thought of as a means to bind a soluble enzyme to a membrane.

A number of proteins are found in this category of anchored membrane proteins: some forms of alkaline phosphatase, Thy-1 antigen, and 5'-nucleotidase are among the proteins that are anchored to the membrane by covalent attachment to a membrane lipid.

D. Glycophorin, an Integral Membrane Protein

Glycophorin was one of the first membrane proteins for which a complete amino acid sequence was obtained, and one of the first to be shown to span the membrane.[12] As a transmembrane protein, a portion of the peptide is found on each side of the lipid bilayer. Glycophorin has a molecular weight near 30,000, of which about 14,000 is protein and the remainder carbohydrate. Glycophorin is actually the name for a family of proteins, glycophorin A, B, and C.

The sequence of glycophorin is given in Fig. 6.3. On the extracellular portion of the peptide are found the attachment sites for the large carbohydrate chains that are covalently bound to the protein. The sugar–amino acid links occur on serine and threonine residues. The branched-chain carbohydrate is built largely of subunits of galactose, mannose, glucose, fucose, and N-acetylneuraminic acid, an acidic sugar. The N-acetylneuraminic acid (sialic acid), gives the surface of the protein a negative charge. Glycophorin carries the M and N antigens of the corresponding blood groups. It provides a binding site for influenza virus on the red cell, one of whose envelope glycoproteins recognizes the neuraminic acid.

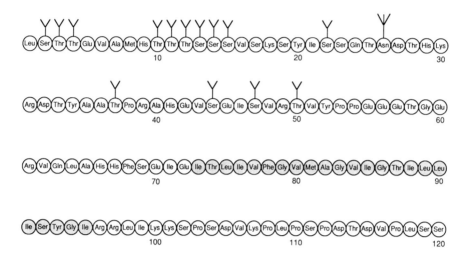

Fig. 6.3. Amino acid sequence of human erythrocyte glycophorin. Data derived from M. Tomita, H. Furthmayr, and V. T. Marchesi, *Biochemistry* **17** (1978): 4756, American Chemical Society.

Glycophorin binds wheat germ agglutinin, but not concanalin A (Con A), through its carbohydrate. Glycophorin is one of the two major integral membrane proteins of the human erythrocyte, but its function is as yet unknown.

Return to Fig. 6.3 and examine the primary structure in detail. Other than the high concentrations of threonine and serine for carbohydrate attachment, there is nothing unusual about the amino acid composition of the extracellular portion of the peptide.

However, the interior of the sequence of glycophorin introduces a fundamental fact about integral membrane protein topology. Within the primary structure is found a linear sequence of 23 amino acids, all of which are hydrophobic. The sequence begins with the hydrophobic amino acid isoleucine at position 73. Following that isoleucine are 22 more hydrophobic amino acids ending in an isoleucine. On the amino terminus of this hydrophobic sequence there is a polar, charged amino acid, glutaminic acid, and on the carboxyl terminus of the hydrophobic sequence are two arginines, all highly polar, charged amino acids. Therefore, in the middle of this transmembrane protein is a linear sequence of hydrophobic amino

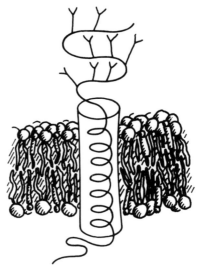

Fig. 6.4. Schematic representation of the disposition of human erythrocyte glycophorin in the membrane.

acids, 23 residues in length, which is flanked by regions with a high concentration of charge.

This transmembrane protein has a primary sequence perfectly adapted to the requirements of an integral membrane protein. Some regions of the polypeptide chain form the same kind of protein structues common to nonmembranous water-soluble proteins. One would expect to find these regions on either side of the membrane lipid bilayer in contact with the aqueous phase. Connecting these two regions in glycophorin is a hydrophobic segment suited to span the hydrophobic interior of the bilayer. The linear hydrophobic sequence is just long enough to span the membrane in the form of an α helix. From this one can envision a model for the disposition of glycophorin in the membrane as shown in Fig. 6.4.

There is some additional experimental evidence to help complete this model for glycophorin. Trypsin treatment of glycophorin produces seven

---

Fig. 6.5. (A) Freeze-fracture electron micrograph of reconstituted membranes containing human erythrocyte glycophorin, egg phosphatidycholine, and cholesterol showing particles corresponding to glycophorin in the membrane. Bar: 100 nm. Micrograph courtesy of S. W. Hui. (B) Freeze-fracture electron micrograph of human erythrocyte membranes showing particles corresponding to integral membrane proteins in the membrane. The E and P faces are evident, the former with the higher particle density. Bar: 100 nm. Micrograph courtesy of S. W. Hui.

peptides. Six of these peptides are water-soluble. The seventh is extremely hydrophobic and will readily incorporate itself into a membrane. This last peptide contains the linear sequence of hydrophobic amino acids that span the membrane in the intact protein.[12] Freeze-fracture electron microscopy shows particles in the fracture face from membrane proteins, indicative of their transmembrane orientation (see Fig. 6.5).

The topology of the amino acid distribution in integral membrane proteins reflects the structure of the lipid bilayer the protein inhabits. If the overall amino acid composition of membrane proteins is not very different from water soluble proteins, and if these proteins have, as in the case of glycophorin, linear sequences of hydrophobic amino acids, one must expect to find concentrated clusters of polar and charged amino acids at the interface between the linear hydrophobic sequences and the remainder of the protein primary sequence. The primary sequence of glycophorin is a good example of this topology. Note that this region of glycophorin is also at the interface of the polar and hydrophobic regions of the lipid bilayer in which it resides. Therefore, these clusters of charges are in the same region of the membrane as the polar headgroups of the lipids (see Chapter 7).

This topology firmly anchors the transmembrane proteins in the membrane. The transmembrane protein will not be readily removed from the membrane, since such removal would require passage through the hydrophobic interior of the membrane, of a large region of the protein with a polar surface, as well as expose the hydrophobic transmembrane segment to the water. Furthermore, this topology will precisely position glycophorin in the membrane. This is because the hydrophobic sequence ends abruptly with highly charged amino acids on both ends. Thus there is no room for "slippage" (or displacement along the normal to the bilayer surface) of the protein in the membrane, without violating the hydrophobic effect. Instead, glycophorin has a single well-defined transmembrane location.

II. MEMBRANE PROTEIN CONFORMATION

A. Primary Structure

The structural properties of glycophorin appear to mirror the structures of other membrane proteins. However, it was difficult to obtain detailed structural information about membrane proteins because membrane proteins are not amenable to normal sequencing techniques (the hydrophobic transmembrane sequences are insoluble in aqueous media used in normal sequence analysis by degradation). This situation changed significantly with the advent of gene cloning. Now amino acid sequences of membrane proteins are more likely to be interpreted from the sequence of the message (mRNA) or the sequence of the structural gene (DNA) than be obtained from classical sequencing techniques.

Several groups of investigators have introduced useful techniques to analyze the primary sequence of membrane proteins.[13] If a linear sequence of purely hydrophobic amino acids is found in a primary sequence, it is of course readily identifiable. However, suppose a mixture of hydrophobic and hydrophilic amino acids is present in a portion of a sequence. How does one determine whether such a sequence is sufficiently hydrophobic to be incorporated into the interior of a membrane?

One means of answering this question is the following. Suppose one examines a sequence, several amino acids at a time. Within that sequence one can add up the free energy of transfer from aqueous to hydrocarbon phase for each of the individual amino acids. An overall free energy of transfer can then be calculated that gives a measure of the ability of this sequence to enter a hydrophobic phase (i.e., the interior of the membrane). What is often done is then to assign that calculated value to the point of the primary sequence that corresponds to the middle of the segment in question. The calculation is then continued point by point throughout the sequence.

1. RHODOPSIN

One method of representation of the primary sequence is given in Fig. 6.6. for bovine rhodopsin. Rhodopsin is the photosensitive pigment of retinal rod cells and is an intergal transmembrane protein. Rhodopsin is about 38,000 kDa and is a glycoprotein, containing a Con A binding site. Rhodopsin is found in the disk membrane of the retinal rod cell and is a photon counter responsible for black and white, and night vision. Little of its mass is outside the membrane. It contains the chromophore retinal, which isomerizes from the 11-*cis* to the all-*trans* form upon absorption of a photon of light. Note that there are approximately seven troughs in the graph in Fig. 6.6. Each of the troughs corresponds to a hydrophobic sequence. Therefore, there are about seven separate hydrophobic sequences in this primary structure.[14] According to the discussion above, this would suggest that there might be seven transmembrane segments in rhodopsin. The first transmembrane segment starts in the vicinity of amino acid 40, the second near 80, the third near 115, the fourth near 155, the fifth at 202, the sixth at 251, and the seventh near 280. Circular dichroism measurements provide supporting evidence for this model, indicating that the transmembrane segments are in the form of an α helix.

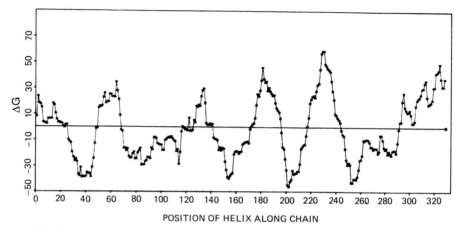

Fig. 6.6. Relative hydrophobicity plot of rhodopsin. Helix length = 20. The troughs represent hydrophobic helices. Plot courtesy of D. Engelman.

2. BAND 3

Band 3, a major transmembrane protein of the erythrocyte membrane, is the anion transport protein of the human erythrocyte.[15] It facilitates the rapid exchange of bicarbonate ion and chloride ion across the red cell membrane. Because of the realtively high concentration of carbonate hydratase inside the erythrocyte, a rapid conversion of carbon dioxide to bicarbonate can occur, and the resulting high intracellular bicarbonate

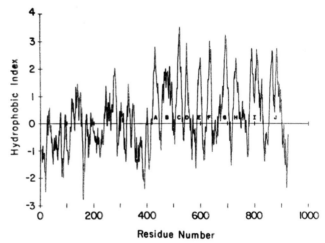

Fig. 6.7. Relative hydrophobicity plot of band 3. The peaks represent hydrophobic helices. Reprinted with permission from R. R. Kopito and H. F. Lodish, *Nature (London)* **316** (1985): 234. Plot courtesy of R. R. Kopito.

concentration can be alleviated by band 3-facilitated exchange of intracellular bicarbonate for extracellular chloride. Thus tissue-generated carbon dioxide can be transported in the blood in the more soluble bicarbonate form for expulsion in the lungs by a reversal of the same mechanism.

The protein has a molecular weight of about 90,000.[16] It has a small amount of carbohydrate on the extracellular surface of the protein. The oligosacharrides are heterogeneous in structure. They contain galactose, mannose, *N*-acetylglucosamine, *N*-acetylgalactosamine, and fucose. They do not contain any sialic acid. The carboxyl terminus of the protein is the portion found outside the cell.

Some aspects of band 3 structure have been worked out from proteolytic studies. Proteolysis to a core of about 17,000 Da leaves a membrane bound peptide that contains the binding site for specific inhibitors of anion transport. This small transmembrane peptide also appears capable of facilitating anion transport.[17]

The amino acid sequence of band 3 has been determined.[18] A plot of the relative hydrophobicity of the segments is given in Fig. 6.7. As in the case of rhodopsin, several hydrophobic sequences are observed in the transmembrane domain of the protein. These hydrophobic sequences likely represent hydrophobic transmembrane helices.

However, a few hydrophilic residues are found in these domains that might be involved in the anion transport process. Perhaps they are involved in the structure of a polar channel for the passage of anions.

3. Na$^+$,K$^+$-ATPase

The Na$^+$,K$^+$-ATPase is an enzyme capable of hydrolyzing ATP and simultaneously pumping sodium out of the cell and potassium into the cell.[19] The protein consists of two subunits. The α subunit contains the active site for ATP hydrolysis and the ouabain (specific inhibitor) binding site. The molecular weight for the α subunit is near 110,000. The β subunit is glycosylated and its molecular weight is about 55,000. No function is yet known for the β subunit.

The ATP binding site and phosphorylation site are on the cytoplasmic surface of the α subunit and the binding site for cardiac glycosides is on the extracellular surface of the protein. The amino acid sequence for this protein has been determined. A "hydrophobicity plot" for this protein appears in Fig. 6.8.[20] Several regions emerge in this plot that are good candidates for hydrophobic transmembrane helices.

This protein appears to be organized into three major domains. In the central region of the protein sequence lies a domain that apparently contains no transmembrane helices. This domain is likely found outside the membrane, on the cytoplasmic surface of the plasma membrane. In this

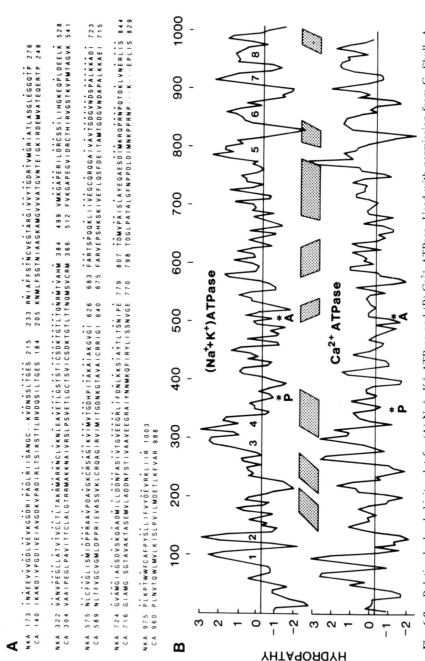

Fig. 6.8. Relative hydrophobicity plot for (A) Na$^+$, K$^+$-ATPase and (B) Ca^{2+}-ATPase. Used with permission from G. Shull. A. Schwartz, and J. B. Lingrell, *Nature (London)* **316**, (1985): 691. Figure courtesy of G. Shull.

domain is found the active site for ATP hydrolysis. This domain is flanked on either side by two other domains containing hydrophobic transmembrane sequences that firmly anchor the protein to the membrane.

The Na^+,K^+-ATPase is specifically inhibited by ouabain, which binds tightly to the enzyme. The ouabain binding site appears to be in the region of residues 395–415, which is between the third and fourth putative transmembrane helices in the amino terminal hydrophobic domain of Na^+,K^+-ATPase and near the junction of that hydrophobic domain and the extramembranous domain. The phosphorylation site is near the ouabain binding site.

4. Ca^{2+}-ATPase

The Ca^{2+}-ATPase of sarcoplasmic reticulum has a molecular weight of about 110,000. It transports calcium ions into the lumen of the sarcoplasmic reticulum at the expense of ATP.[21] Much of the mass of the protein is in the membrane.[22]

Figure 6.8 shows that there is considerable similarity in the primary structure of the Ca^{2+}-ATPase and Na^+,K^+-ATPase.[23] The center of the primary sequence in both cases contains the cytoplasmic extramembraneous portion of these pump proteins. A set of hydrophobic transmembrane domains flank the cytoplasmic portion of the primary sequence on both sides in these proteins.

These similarities suggest a homology between the primary sequence of the two proteins. In fact about 30% homology is observed. The homology is even greater if one allows conservative replacements. It has been suggested that these two transport proteins may have a common ancestral gene. The homology is particularly apparent in the sequence containing the catalytic site for ATP hydrolysis. Homology is notably lacking in the region of the Na^+,K^+-ATPase sequence that contains the ouabain binding site. Correspondingly, the Ca^{2+}-ATPase is not sensitive to ouabain and does not bind ouabain. The structural organization of the Ca^{2+}-ATPase has been described as depicted in Fig. 6.9.[23]

B. Secondary and Tertiary Structure

Employing a linear sequence of hydrophobic amino acids is only part of the solution to accommodating a protein to a membrane. The hydrophobic amino acid side chains are compatible with the hydrophobic interior of the lipid bilayer. The absence of charged amino acids in the transmembrane segment makes it thermodynamically favorable to insert this sequence into the membrane. However, the amide bond connecting these amino acids is still polar. The carbonyl is polar and is a good hydrogen bond acceptor; the amino hydrogen is polar and is a good hydrogen bond donor.

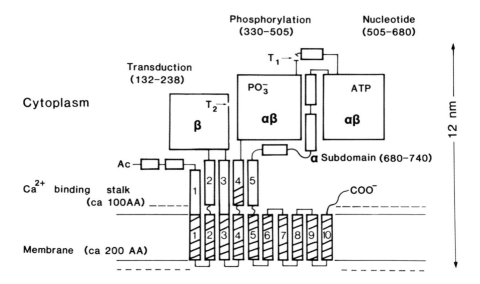

Fig. 6.9. Possible structure for the Ca^{2+}-ATPase, derived from the primary structure using structure prediction techniques. Reproduced with permission from D. H. MacLennan, C. J. Brandl, B. Korczak, and N. M. Green, *Nature (London)* **316** (1985): 696. Courtesy of D. MacLennan.

These polar structures must be reconciled to the hydrophobic membrane interior.

Referring again to soluble proteins provides a useful model. When a stretch of amino acids must cross the hydrophobic interior of a globular protein, one finds that the carbonyls and amino hydrogens are involved in hydrogen bonds. This is because the amide bonds are polar, and are not otherwise compatible with the interior of the globular protein. Hydrogen bonding the amino hydrogens to carbonyls of the peptide bonds "neutralizes" their polarity. One of the most common means of achieving this is to form an α helix. The α helix thus becomes an excellent candidate for the secondary structure of the hydrophobic sequences of transmembrane proteins.

1. GLYCOPHORIN

Circular dichroism is a spectroscopic technique that is exquisitely sensitive to the secondary structure of proteins. To a first approximation, it is possible to obtain estimates of protein secondary structure from the circular dichroism spectra exhibited by the protein. This measurement for the transmembrane segment of glycophorin yields results consistent with an α-helical conformation.[24] The water-soluble trypsin-generated fragments of glycophorin do not have helix as their dominant structural feature by the same analysis. However, the circular dichroism of the individual tryptic peptides adds up to the circular dichroism of the whole protein. This suggests a domain structure for the membrane protein.

When a peptide containing 19–23 amino acids adopts an α-helical conformation, the distance from end to end of that helix is just that required to span the hydrophobic region of a typical lipid bilayer in a biological membrane. Therefore the hydrophobic α helix of glycophorin is admirably suited to the structural requirements of this transmembrane protein. Furthermore, the transmembrane segment of glycophorin is conformationally rigid. The motionally restricted region includes the hydrophobic segment and some of the polar regions on either side. This structural feature suggests a rigid anchoring of this transmembrane protein in the membrane.

2. BACTERIORHODOPSIN

Just how good are these structure predictions? Do membrane proteins really have hydrophobic transmembrane α helices? There is a paucity of membrane protein structures, due to the difficulty in crystalizing membrane proteins into single crystals. The few available examples will be examined here.

One example of a high-resolution structure determination of a transmembrane membrane protein can be found in bacteriorhodopsin. Bacteriorhodopsin functions as a light-driven proton pump for the *Halobacterium halobium*. The pigment is organized into a specialized patch of membrane referred to as the purple membrane because of its characteristic color. Structure prediction techniques, using the primary sequence, predict seven α helices in this integral membrane protein. Circular dichroism studies indicate that the protein is mostly α helix in conformation.

The structure, shown in Fig. 6.10 shows that these predictions reflect accurately what is observed in the three-dimensional image.[25] This structure determination represents a truly remarkable achievement by Henderson and colleagues, exploiting the two-dimensional arrays formed by bacteriorhodopsin with modern image enhancement techniques and electron diffraction, since single crystals were not available for normal diffraction techniques. Seven α helices span the membrane, although not all of them align perpendicular to the membrane surface. The retinal is bound between two of the helices in the interior of the membrane. Considerable similarity has been noted between this structure and that of bovine rhodopsin.[41]

An α helix will exhibit a dipole moment, with a partial positive charge at the amino terminus and a partial negative charge at the carboxyl termi-

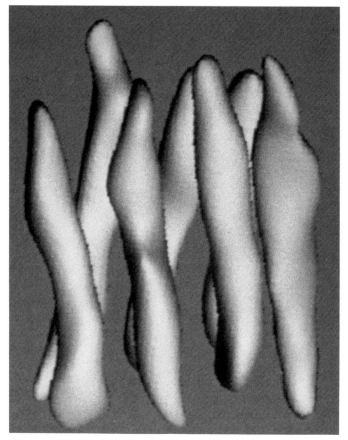

Fig. 6.10. A computer representation of the structure of bacteriorhodopsin. The image is the result of averaging the results from two separate structure determinations. The work is the result of R. Henderson and J. M. Baldwin. Courtesy of R. Henderson.

nus. The bundle of helices may accentuate that dipole moment, depending on the directionality of the helices relative to each other. This may prove important to protein function in membranes with a transmembrane electrical potential.

A closer examination of this protein is in order. The seven helices of the protein are arranged as a bundle. End on, these transmembrane helices can be described as inhabiting a group of three and a group of four, with the groups immediately adjacent to each other. This organization raises the question whether interhelical interactions occur, since each helix has contacts with other helices roughly along the helix axis. Many of the helix–helix contacts involve hydrophobic amino acids and are governed by van der Waals interactions proportional to their surface area. However, interactions between polar amino acids may also occur along helix–helix contacts.[26] These polar interactions involve only protein, presumably shielded from lipid hydrocarbon–protein interactions by an amphipathic organization of the helix; that is, a helix with a polar face and with a nonpolar face. The possibility of polar interactions of this sort could provide structural integrity as well as a pathway for transport of polar species. The concept of bundles of helices may well be general for membrane proteins with more than one transmembrane helix.

The structure obtained at 7 Å resolution (Fig. 6.10) provides a view of the tertiary structure of bacteriorhodopsin in the relative disposition of the transmembrane helices. The regions of polypeptide connecting the helices do not show up well in the structure determination because they are disordered (that is, these regions experience more than one conformation and they do not appear in the structure determination). However, one should not lose sight of the fact that these helices are connected to each other in a particular order by the covalent structure of the primary sequence.

3. BACTERIAL PORIN

At this point it is appropriate to question whether the α helix is the only secondary structure to be expected in the hydrophobic regions of a membrane protein. In water-soluble proteins one other common element of secondary structure fulfills the requirements that result from inserting a peptide sequence into the hydrophobic core of a protein. That is β-sheet structure. All the peptide bond carbonyls and amide nitrogens can be involved in hydrogen bonds in the β-sheet structure, as well as in the α helix. Thus the polarity problems arising from insertion of the peptide bond into the hydrophobic interior of the membrane can be solved.

The minimum domain size for an intramembranous β sheet would be much larger than the minimum domain size for an intramembranous protein structure made of an α helix. The runs of the β sheet could form a β barrel, perhaps for transport purposes. One example of such a structure is provided by the porins from the outer membrane of bacteria.[27] The porins are responsible for the formation of pores in the outer membrane through which solutes (<500 kDa, in most cases) may pass. One example of a porin from *Escherichia coli* is about 37 kDa. This protein forms trimers in the membrane, each monomer potentially providing a channel for passive transport. The predominant secondary structure is β sheet, likely in the form of a β barrel.

4. Photosynthetic Reaction Center

The crystal structures of the photosynthetic reaction center from *Rhodo-bacter sphaeroides* and *Rhodopseudomonas viridis* are the only available high-resolution crystal structures of integral membrane proteins. Their structure determination led to a Nobel prize. A figure representing this crystal structure is given in Fig. 6.11.[28]

This crystal structure confirms some of the structural patterns for membrane proteins discussed above, which were deduced from less direct evidence. The transmembrane portions of this protein are indeed α helices. These α helices stretch from one side of the membrane to another and may be tilted with respect to the bilayer normal, as in the case of bacterio-

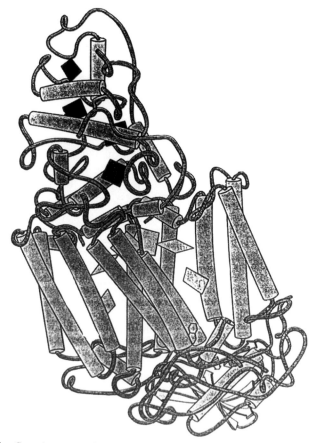

Fig. 6.11. Crystal structure of photosynthetic reaction center. From Stryer, L."Biochemistry, Third Edition (1988)." Copyright © 1988 by Lubert Stryer. Reprinted with permission of W. H. Freeman and Co., New York.

rhodopsin. The extensive extramembraneous domains of this protein are organized following structural principles common to water-soluble proteins.

One of the unanswered questions from this crystal structure is the relationship between the thickness of the lipid bilayer and the thickness of the α-helical bundle. Does the projection of the α helix onto the bilayer normal match the thickness of the hydrophobic portion of the lipid bilayer? What portion of the protein sequence contacts the lipid headgroups of the lipid bilayer? These are important questions for the interaction between lipids and proteins and the control of protein function by the lipid bilayer (see Chapter 7).

Unfortunately, no other crystal structures are currently available for membrane proteins. Therefore, most information about membrane protein structure will continue to be obtained from less direct methods.

C. Domain Structure of Membrane Proteins

Since different structural features are associated with the extramembraneous portions of integral membrane proteins compared to the intramembraneous portions of those proteins, one might expect different physical behavior for those regions. Furthermore, structurally distinct domains are found in integral membrane proteins, and these domains tend to maintain their structure even when separated from each other by proteolysis. A schematic representation of the organization of domains in membrane proteins is presented in Fig. 6.12.

1. CYTOCHROME b_5

Cytochrome b_5 is a small protein of molecular weight about 17,000. Like cytochrome c it is involved in electron transfer. Also like cytochrome c the electron transfer is carried out with the help of a heme group bound in a hydrophobic pocket on the protein. This protein is found in the endoplasmic reticulum. The electron transfer is used for the desaturation of fatty acids. One such system is the stearoyl-CoA desaturase, which converts stearic acid to oleic acid.

If the ER membranes containing this protein are treated with trypsin, a water-soluble peptide is released, about 10,000 in molceular weight. This peptide accounts for a large part of the mass of cytochrome b_5. Contained in this peptide is the heme. This peptide is capable of carrying out electron transfer reactions. The full cytochrome b_5 is not required for the fundamental chemistry with which this protein is involved. The crystal structure of this peptide reveals a compact globular protein with a topology of hydrophobic and polar amino acids similar to that found in cytochrome c.

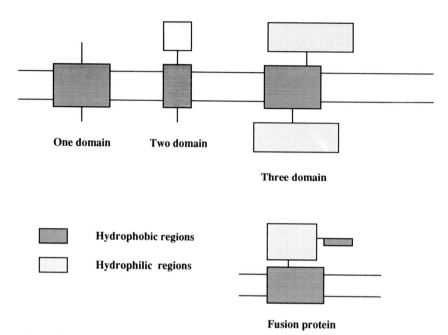

Fig. 6.12. Schematic representation of the topology of domains in membrane proteins.

The remainder of the protein is left in the membrane after proteolysis. It is not removed from the membrane, suggesting that it is a highly hydrophobic peptide. The sequence for cytochrome b_5 has been determined. The portion of the sequence corresponding to the peptide that the trypsin cleavage leaves in the membrane is mostly hydrophobic as expected.

An interesting denaturation experiment has been performed with cytochrome b_5. In experiments of this type, a chaotropic agent is added to the medium that is capable of altering water structure sufficiently to perturb the thermodynamic balance between a fully folded protein and one that retains no organized structure, or that is denatured. Guanidine hydrochloride (GuHCl) is a classic chemical used for such measurements in studies of water-soluble proteins.[29]

The chaotropic agent alters the ΔG of transfer of an amino acid side chain from the interior of a protein to direct contact with the aqueous phase. The unfavorable nature of this ΔG for hydrophobic amino acids stabilizes the native structures of proteins. If the concentration of the chaotropic agent is high enough to alter sufficiently the water structure, the hydrophobic effect is partially defeated. This can lead to an unfolding of the protein since hydrophobic amino acids need not, under these conditions, be as protected from the aqueous phase.

In denaturation experiments with guanidine hydrochloride, one usually measures some parameter that is indicative of the native state of the protein. This might be a spectral property, for example. What one usually finds for water-soluble proteins is a transition from the native to the denatured state with increasing GuHCl concentration. This normally is observed as one smooth transition over a fairly narrow range of GuHCl concentration, indicating a cooperativity in the transition.

For cytochrome b_5, a transition is observed in the region of 2–3 M GuHCl, which is common for water-soluble proteins (Fig. 6.13). Additionally, there is a second transition in the region of 5–6 M GuHCl. A reasonable assignment for these two transitions can be suggested based on the observation that the trypsin-released water-soluble fragment exhibits a transition in the region 2–3 M GuHCl. It would appear that the extramembraneous portion of cytochrome b_5 denatures separately from the intramembraneous portion. From these data comes the idea that the protein consists of two rather separate structural domains, and that the transmembrane region is structurally more stable with respect to the aqueous phase.

2. RHODOPSIN

Another example of the domain structure of integral membrane proteins is provided by rhodopsin. Treatment of rhodopsin with papain or thermolysin cleaves the protein into two major pieces that remain associated with the membrane. These two domains retain their structural integrity even

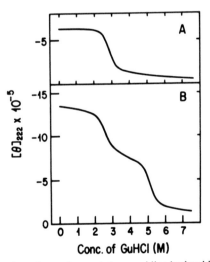

Fig. 6.13. Denaturation of cytochrome b_5 by guanidine hydrochloride (A) Water-soluble portion only; (B) whole protein. These data suggest that the water-soluble portion behaves in a manner typical of water-soluble proteins, whereas the intramembranous segment is more stable. Reprinted with permission from S. Tajima, K. Enomoto, and R. Sato, *Arch. Biochem. Biophys.* **172** (1976): 90.

when cleaved from each other, as defined by circular dichroism studies. When in the membrane, the two domains tend to stay together.[30] These results suggest not only that structural domains exist in this membrane protein, but also that noncovalent interdomain interactions exist that help to stabilize the tertiary structure of the membrane protein. In fact, some molecular modeling of putative interhelical interactions has been done for the protein bacteriorhodopsin. Phospholipid–protein interactions may also play a role.

Domain structures can be observed in the "hydrophobicity plots" in Figs. 6.7 and 6.8. The transmembrane region consists of a series of hydrophobic helices, all within localized regions of the polypeptide sequence. The extramembranous portion of the protein occurs in another section of the polypeptide. The proteins appear to be divided by their linear sequence into domains, each with their own integrity.

3. Hydroxymethylglutaryl-CoA (HMG-CoA) Reductase

Consider one more example of a transmembrane protein, the HMG-CoA reductase. The molecular weight of this protein is about 97,000. It is glycosylated even though it is located in the endoplasmic reticulum, where much of cholesterol biosynthesis takes place. Specific proteolysis releases a fragment of molecular weight 50,000 that is water soluble. The remainder of the protein is hydrophobic and stays integrated in the membrane.

Its properties indicate that it is an integral membrane protein. It is topologically analogous to glycophorin, discussed earlier. The size of the hydrophobic domain of HMG-CoA reductase is much larger than is required to span the membrane only once. The sequence shows that this protein is likely organized with a transmembrane region consisting of more than one hydrophobic α helix. These putative hydrophobic transmembrane segments appear to be located all in one region of the primary sequence in contrast to the ion pumps.

This protein is worth a second look, from the point of view of the construction of the structural gene. The map of the gene reflects, in part, the ultimate topography of the protein. The transmembrane domain maps to one portion of the gene, whereas the extramembraneous region of the protein maps to another portion of the gene.[31]

However, an even finer division of the structure is evident in the gene map. Many of the hydrophobic transmembrane sequences map to unique sections of the gene and are separated from the coding region for other hydrophobic transmembrane sequences by introns. This structure has given rise to speculation that the transmembrane region of membrane proteins may turn out to be conserved, and possibly repeated from one membrane protein to another.

D. Post-translational Modification of Membrane Proteins

Five common kinds of post-translational modification merit brief mention here. One is glycosylation, as is found for glycophorin. Carbohydrate may be bonded through the alcohol of serine or threonine (O-linked glycosylation). Alternatively, glycosylation may occur through the side chain of asparagine for N-linked glycosylation. Many of the plasma membrane proteins are glycosylated in this manner.

The second is acylation. In this case, fatty acids are covalently attached to membrane proteins. Acylation was discovered through the observation that some fatty acids copurifying with membrane proteins could not be removed by organic solvent extraction, by detergent treatment, or by phospholipase treatment. They could, however, be removed by hot methanol–HCl. Some of these fatty acids are acylated by an ester bond to the protein and other fatty acids are attached to the protein by an amide bond. A number of examples of acylated proteins now exist. Specific examples include the G protein of VS virus envelope and rhodopsin. Proteins such as the β-adrenergic receptor and the influenza HA have been suggested to require acylation for proper function. Isoprenyl chains have also been found bound to membrane proteins.

The third type of post-translational modification is limited proteolysis of the protein. An example of this is provided by M13 coat protein. This post-translation modification may be important in insertion of the M13 protein into a membrane after its synthesis. Limited proteolysis is required for the attachment of lipid to anchored proteins. As will be seen in the last chapter in this book, proteolysis is an integral part of the maturation of a class of membrane proteins during biosynthesis.

A fourth type of post-translational modification is phosphorylation. As in soluble proteins, membrane proteins will serve as substrate for protein kinases. One example is the β-adrenergic receptor, which when phosphorylated on serines and/or threonines by β-adrenergic receptor kinase, is rendered less able to participate in the signal transduction process.

The last post-transitional modification that will be maintained is cross-linking of polypeptide chains by disulfide bonds. For example, the fusion protein of Sendai virus consists of two polypeptides (in mature form) linked by a disulfide bond. Only one polypeptide integrates into the membrane. Therefore reduction of the disulfide bond leads to dissociation of the two subunits, with subsequent loss of one subunit. Activity is also lost.

E. Motional Characteristics of Proteins in Membranes

Some membrane proteins undergo lateral diffusion in the plane of the membrane.[32] The diffusion coefficients for such motion tend to represent slower lateral diffusion than observed for lipids. Table 4.5 presents representative examples. Lateral diffusion can be limited by interaction with other proteins, including those of the membrane skeleton. For example, band 3 proteins bound to the membrane skeleton (only a portion of the total band 3 in the membrane) are rendered immobile, relative to band 3 proteins not bound to the membrane skeleton.

Integral membrane proteins also can rotationally diffuse about an axis perpendicular to the surface of the membrane. The rate of rotational diffusion is slower than that for phospholipids, and generally in the range 10–100 kHz. However, not all membrane proteins diffuse in this manner. For example, bacteriorhodopsin in the purple membrane of *H. halobium* is immobile in the native membrane. Only in reconstituted systems does bacteriorhodopsin exhibit axial diffusion.

Finally there is internal motion within the structure of the protein. However, there may be a difference between the extramembraneous regions and the intramembranous regions. The latter may exhibit much less internal motion than the former.

F. Membrane Asymmetry

There is a contrast between the transmembrane distribution of lipids in membranes and the transmembrane distribution of proteins. Lipids are not symmetrically distributed across biological membranes. However, in almost no case is one class of lipid found only on one side of the membrane. In other words, lipid asymmetry is seldom absolute as it has been measured to date.

In contrast, protein asymmetry is absolute in native membranes. All the proteins of a particular type are oriented in the membrane in the same way. For example, consider the F_1,F_0-ATPase of mitochondria. This protein complex consists of two parts. F_0 is the integral membrane protein complex that spans the inner mitochondrial membrane. F_1 is noncovalently attached to F_0 and protrudes considerably from the membrane surface. In electron micrographs of mitochondria one can see knobs, corresponding to F_1, on the matrix side of the inner mitochondrial membrane; in other words, the distribution of this protein represents absolute asymmetry. This asymmetry is functionally important for membrane proteins.

G. Quaternary Structure or the State of Oligomerization of Membrane Proteins in Membranes

As in the case of water-soluble proteins, the organization of subunits, including identical subunits, of membrane proteins may be important to the function of those proteins in biological membranes. Addressing this question requires a means of determining the molecular weight of functional membrane proteins. However, most of the methods used classically require the proteins to be solubilized. With membrane proteins this usually necessitates the use of detergents. Furthermore, in one of the important techniques for determination of oligomeric state, sedimentation equilibrium, the amount of detergent binding to the membrane protein affects the measurement. Therefore one must separately determine the binding of the detergent to the protein and the partial specific volumes of the detergent and the protein in order to determine the molecular weight. Although these obstacles have been overcome, only the molecular weight in a detergent micelle, and not the molecular weight in a membrane, can be measured. This last problem also applies to SDS gel electrophoresis, which is often used to esimtate molecular weight because of its ease of use. These problems become even more severe when considering a multisubunit protein and attempting to determine the stoichiometry of the subunits.

To assess the effective molecular weight in a membrane (or the extent of aggregation of protein monomers), chemical cross-linking studies have been performed. Crosslinking provides an effective nearest neighbor analysis. A second approach is radiation inactivation, which is effectively a target size analysis based on destruction of membrane protein activity due to collision with accelerated paraticles. The larger the protein, the less flux of such a beam of accelerated particles is required to inactivate the protein.

From analyses such as these, some information on the state of aggregation of integral membrane proteins in biological membranes has been determined. Evidence obtained so far suggests that membrane proteins can exist as monomers and dimers as well as higher oligomeric states. Oligomerization may be the dominant form found in biological membranes. Rhodopsin of the retinal rod outer segment disk membrane may be an example of a protein that can exist as a monomer in the membrane. Ca^{2+}-ATPase of the sarcoplasmic reticulum is an example of a protein that can exist as a dimer in the membrane.[33] The Na^+,K^+-ATPase is a dimer of dimers ($\alpha_2\beta_2$). The state of aggregation of some membrane proteins may be dependent on the medium in which the membrane is placed. The state of aggregation may modulate the activity of the protein.

In the case of glycophorin, the state of oligomerization is maintained even in SDS–polyacrylamide gels where dimers are observed. Such oligomerization may well be driven, in part, by helix–helix interactions. In the case of glycophorin, the transmembrane helix, when isolated from the rest of the protein, exists as an oligomer in the membrane.

A second important type of interaction is between different proteins that are involved in the same functional process. Two possibilities exist: one is that the different proteins involved in a particular function are always associated in a large protein complex; the other is that proteins must laterally diffuse and collide to react.

1. CYTOCHROME OXIDASE

In the case of a protein complex, such as the cytochrome c oxidase complex of the inner mitochondrial membrane, several polypeptides are involved. Some of these subunits contain heme groups and are involved in electron transport. Cytochrome oxidase is involved in an oxidation–reduction reaction and, on movement of electrons, transports protons across the membrane.[34]

The important observation for the present discussion is that these proteins remain associated in the membrane. Thus the chemical reactions involved do not require lateral diffusion of the membrane proteins to proceed. The complex creates an efficient mechanism to carry out the pertinent chemical reaction. Consequently the subunits of this complex cannot be readily shared with other proteins outside the complex for participation in other reactions.

2. CYTOCHROME-b_5 REDUCTASE

The second case of membrane protein interaction is that of proteins that must collide through the process of lateral diffusion to react. A good example of this case is the reaction of cytochrome b_5 and cytochrome-b_5 reductase. The structure of cytochrome b_5 consists of two domains. One is the hydrophilic extramembraneous region that contains the heme group involved in electron transport. The other is the hydrophobic intramembranous domain that anchors the protein to the membrane.

Cytochrome-b_5 reductase exhibits a similar domain structure. The total molecular weight is about 43,000, which is substantially larger than that of cytochrome b_5. Trypsin releases a soluble domain, just as in the case of cytochrome b_5, with a molecular weight of about 34,000. Left in the membrane is a hydrophobic domain similar in size to the hydrophobic domain for cytochrome b_5. Thus the structural model for cytochrome-b_5 reductase can be similar to the model for cytochrome b_5 with allowance for the larger protein size.

How do these proteins interact in the membrane? Some interesting experiments have addressed this question.[35] Cytochrome b_5 and cytochrome-b_5 reductase were purified separately and then reconstituted into the same membrane. A reconstitution procedure was employed that regenerated proteins capable of transferring electrons, yet properly bound to the membrane. This experimental protocol permits the investigator to vary the ratio between the lipids and the proteins in the membrane. By so doing, one can change the effective concentration of proteins in the membrane. Changing the concentration of proteins in the two-dimensional regime of the membrane is roughly equivalent to changing the concentration of the reacting species of a chemical reaction in three dimensions.

If a collision between cytochrome b_5 and its reductase is all that is needed to effect electron transfer, then one might expect this reaction to be first order in each of the reacting species. Normally the kinetics of such a reaction cannot be analyzed for integral membrane proteins, because such proteins are not in solution. However, the present example does allow one to vary "concentration," in a manner analogous to varying the concentration of species in solution. Thus increasing the "concentration" of one of the proteins in the membrane while keeping the "concentration" of the other constant should lead to an increase in reaction rate. Such dependence of reaction rate on "concentration" was observed. Therefore, in contrast to the case of cytochrome oxidase complex, this reaction requires the lateral diffusion and subsequent collision of the cytochrome b_5 and cytochrome-b_5 reductase. One consequence is that inhibition of lateral diffusion should inhibit the reaction. Thus a phase transition of the membrane to the gel state will decrease the reaction rate. Another consequence of this model is that these proteins could be shared with other enzymes in other reaction pathways. Thus the cytochrome b_5 and its reductase could donate electrons to more than one class of desaturase.

H. Organization of Proteins in Membranes

1. SARCOPLASMIC RETICULUM MEMBRANE

The sarcoplasmic reticulum membrane (SR membrane) is a specialized muscle membrane system. It is functionally and morphologically divided into the longitudinal tubules and the terminal cisternae. Calcium is apparently released from the terminal cisternae after stimulation by a nerve impulse. The calcium release is subsequently involved in contraction of muscle fibers. Following contraction, the calcium is pumped back into the lumen of the SR, in concert with muscle relaxation. The sequestration of the calcium occurs (though not exclusively) in the longitudinal tubules (Fig. 6.14).[36]

A number of membrane proteins are required for the proper function of the SR. However, many details remain to be elucidated. The one

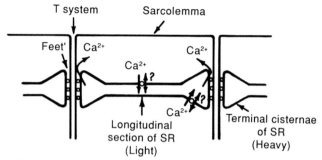

Fig. 6.14. Schematic representation of the function of muscle sarcoplasmic reticulum. Reprinted with permission from B. Meissner, *Mol. Cell. Biochem.* **55** (1983): 65, Martinus Nijhoff.

function that is reasonably understood is the calcium pumping function. Calcium is pumped into the lumen of the SR at the expense of ATP hydrolysis by the calcium ATPase of the SR.

This polypeptide contains both the ATP hydrolysis function and the calcium transport function. A single protein is capable of ATP hydrolysis, without the assistance of another unit of the same protein. It is not clear whether a monomer of the protein is capable of calcium transport. However, studies in the native membrane suggest that the calcium pump protein exists in an aggregated state. Evidence has been accumulated for a dimeric form of the protein and in some cases the protein may exist as a functional tetramer in the membrane.

Biological membranes are complex and generally contain a number of different proteins. Such is the case with the SR membrane. Two major fractions of SR are obtained.[37] One is the heavy SR, derived largely from the terminal cisternae; the other is the light SR, derived largely from the longitudinal tubules.[38] The major band in each fraction arises from the calcium pump protein. The calcium pump protein appears to function somewhat differently in the two membranes.

A number of other proteins are in the membrane. Included in the heavy SR are proteins that bind calcium, such as calcequestrin. Calcequestrin is more concentrated in the heavy SR. Two glycoproteins, one of molecular weight about 160,000 and the other near 50,000, are found in the membrane. Although their function is not known, one or both may associate with the pump protein. How all these proteins function in the sarcoplasmic reticulum is as yet ill-defined. But this brief discussion serves to emphasize that membrane proteins do not operate in the absence of other membrane proteins when in the biological membrane.

2. ROD OUTER SEGMENT DISK MEMBRANE

The photosensitive disk membranes of the vertebrate rod cells are found in the outer segment of the retinal rod cell. The major membrane protein of these membranes is the pigment protein rhodopsin.

Rhodopsin contains the chromophore 11-*cis*-retinal. This molecule undergoes a photoisomerization when a photon of light is absorbed. A conformation change of the protein results, and the rhodopsin enters an activated state. During this process, the visual pigment goes through a defined series of spectral intermediates.

The light-activated rhodopsin binds a G_t protein (transducin) tightly in a manner characteristic of a peripheral, associated membrane protein.[39] G_t protein exchanges GDP for GTP. G_t is similar to the G_s protein of the receptor-activated adenylate cyclase system. This G_t protein can then activate a phosphodiesterase. Activation of the phosphodiesterase results in a reduction in cGMP in the rod outer segment where cGMP is otherwise relative high (as high as mM). Eventually sodium channels in the plasma membrane of the rod cell are closed by this light-activated process, leading to a hyperpolarization of the plasma membrane.

Light-activated protein kinase activity is also found in the rod outer segment. These kinases will phosphorylate rhodopsin.[40] Again, the complexity of the environment in which a membrane protein must operate is apparent. Further details of this system will be discussed in Chapter 10.

3. INNER MITOCHONDRIAL MEMBRANE

The inner mitochondrial membrane is required for the cellular synthesis of ATP by a pathway commonly known as oxidative phosphorylation. The details of oxidative phosphorylation are well described in introductory biochemistry texts and therefore lie outside the scope of this book. However, the rough outline of the process will be presented, as it is an elegant example of the complex functioning of a membrane. It also is an example of an important cellular function that could not occur without a membrane.

The most favored view of the synthesis of ATP is referred to as the Mitchell hypothesis, after the individual who was awarded the Nobel prize for its formulation. In its simplest form the hypothesis states that the energy stored in a transmembrane proton gradient is utilized for the synthesis of ATP. This energy derived both from the difference in concentration of the species, the proton, across the membrane and from the transmembrane distribution of electrical charge. Thus both a difference in chemical potential and an electrical potential across the membrane are important to function.

A number of membrane proteins are involved in the production of the proton gradient. The gradient of protons occurs at the expense of a series of oxidation–reduction reactions in which electrons are transmitted via proteins from species such as NADH to O_2. The free energy change resulting from the electron transfers is captured in a proton gradient.

The F_1, F_0-ATPase of the inner mitochondrial membrane then provides the key step in converting the proton gradient into the synthesis of ATP. The F_1, F_0-ATPase is a large structure, consisting of a number of subunits. The F_1 extramembranous domain of the enzyme contains the active site where ATP synthesis (or under certain conditions, ATP hydrolysis) occurs. The F_0 domain is largely within the inner mitochondrial membrane and is responsible for proton permeability of the membrane. The details of the mechanism for the synthesis of ATP utilizing the proton gradient are, for the most part, unknown. The molecular organization of the mitochondrial membrane is represented in Fig. 6.15.

III. SUMMARY

Membrane proteins can be divided into two classes; Peripheral membrane proteins and integral membrane proteins. Peripheral membrane proteins include both associated proteins and proteins of the membrane skeleton. Integral membrane proteins include both transmembrane proteins and proteins that are simply anchored to the membrane.

Peripheral membrane proteins have sites with affinity for binding sites

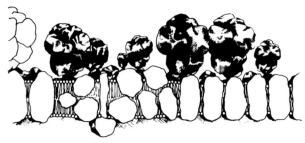

Fig. 6.15. Representation of the molecular organization of the inner mitochondrial membrane. The largest extramembranous protein masses are the F_1 subunit of the mitochondrial ATPase responsible for the synthesis of ATP using the energy stored in the proton gradient. The F_1 subunit is bound to the intramembraneous F_0 subunit. The F_1 subunit can be seen as "knobs" in electron micrographs of the mitochondria. The upper portion represents the matrix side of the inner mitochondrial membrane. Note that there is very little lipid bilayer in this membrane because of the high protein content. Drawing courtesy of W. Scherer.

on the membrane. Frequently the binding sites are on integral membrane proteins, although binding of proteins to the surface of the lipid bilayer is also possible. Peripheral membrane proteins can be removed from the membrane with mild treatments. Integral membrane proteins incorporate some of their mass into the hydrophobic interior of the membrane. They may span the membrane (transmembrane) or they may penetrate only from one side of the membrane (anchored).

Integral membrane proteins organize their primary sequence to accommodate the hydrophobic interior of the membrane. Usually linear sequences of hydrophobic amino acids in an α-helical conformation are thought to provide the means of integrating the protein into the membrane. In two cases, bacteriorhodopsin and the photosynthetic reaction center, three-dimensional structural determinations have demonstrated this model for membrane protein structure to be correct.

Because of these structural constraints on integral membrane proteins, such proteins appear to be organized into domains. Some domains are extramembranous, resembling water soluble proteins in their structure. These domains are often easily removed by limited proteolysis from the membrane. Other domains are clearly intramembraneous, containing the membrane-spanning hydrophobic α helices. These domains sometimes reflect domains in the structural gene coding for the protein. Interactions between elements of the intramembraneous domains may be important to membrane protein structure and function.

Membrane proteins often are found in dimeric, or higher oligomeric, states in biological membranes. Membrane proteins in biological membranes are absolutely asymmetric with respect to their orientation in the membrane. That means, for example, that all the carboxyl terminal domains of a particular membrane protein protrude from the same side of the membrane.

Although membrane proteins are often considered purified proteins, frequently in simple reconstituted systems they derive from complex biological membranes. Many different proteins reside in the same membrane and may play interdependent roles in the function of that membrane.

REFERENCES

1. Reithmeier, R. A. F., and C. M. Deber, "Intrinsic membrane protein structure: principles and prediction," In *The Structure of Biological Membranes* (P. L. Yeagle, ed.) (Boca Raton: CRC Press, 1992).
2. McCarthy, M. P., "The three-dimensional structure of membrane proteins," In *The Structure of Biological Membranes* (P. L. Yeagle, ed.) (Boca Raton: CRC Press, 1992).

3. Singer, S. J., and G. L. Nicholson, "The fluid mosaic model of the structure of cell membranes," *Science* 175 (1972): 720–731.
4. Repasky, E. A., and C. C. Gregorio, "Plasma membrane skeletons," In *The Structure of Biological Membranes* (P. L. Yeagle, ed.) (Boca Raton: CRC Press, 1992), 449–506.
5. Bennett, V., "The membrane skeleton of human erythrocytes and its implication for more complex cells," *Annu. Rev. Biochem.* 54 (1985): 273–304.
6. Anderson, R. A., and R. E. Lovrien, "Glycophorin is linked by band 4.1 protein to the human erythrocyte membrane skeleton," *Nature (London)* 307 (1984): 655–658.
7. Anderson, R. A., and V. T. Marchesi, "Associations between glycophorin and protein 4.1 are modulated by polyphosphoinositides: A mechanism for membrane skeleton regulation," *Nature (London)* 318 (1985): 295–298.
8. Low, M. G., "The glycosyl-phosphatidylinositol anchor of membrane proteins," *Biochim. Biophys. Acta* 988 (1989): 427–454.
9. MacPhee-Quigley, K., P. Taylor, and S. Taylor, "Primary structures of the catalytic subunits from two molecular forms of acetylcholinesterase," *J. Biol. Chem.* 260 (1985): 12185–12189.
10. Perelman, A., and E. Brandan, "Different membrane-bound forms of acetylcholinesterase are present at the cell surface of hepatocytes," *Eur. J. Biochem.* 182 (1989): 203–207.
11. Sussman, J., *et al.,* "Atomic structure of acetylcholinesterase from *Torpedo californica:* A prototypic acetylcholine-binding protein," *Science* 253 (1991): 872–879.
12. Tomita, M., H. Furthmayr, and V. T. Marchesi, "Primary structure of human erythrocyte glycophorin A. Isolation and characterization of peptides and complete amino acid sequence," *Biochemistry* 17 (1978): 4756–4770.
13. Kyte, J., and R. F. Doolittle, "A simple method for displaying the hydropathic character of a protein," *J. Mol. Biol.* 157 (1982): 105–132.
14. Hargrave, P. A., J. H. McDowell, D. R. Curtis, J. K. Wang, E. Juszczak, S. L. Fong, J. K. M. Rao, and P. Argos, "The structure of bovine rhodopsin," *Biophys. Struct. Mech.* 9 (1983): 235–244.
15. Jennings, M. L., "Structure and function of the red blood cell anion transport protein," *Annu. Rev. Biophys. Chem.* 18 (1989): 397.
16. Tanner, M. J. A., P. G. Martin, and S. High, "The complete amino acid sequence of the human erythrocyte membrane anion transport protein deduced from the cDNA sequence," *Biochem. J.* 256 (1988): 703.
17. Grinstein, S., S. Ship, and A. Rothstein, "Anion transport in relation to proteolytic dissection of band 3 protein," *Biochim. Biophys. Acta* 507 (1978): 294–304.
18. Kopito, R. R., and H. F. Lodish, "Primary structure and transmembrane orientation of the murine anion exchange protein," *Nature (London)* 316 (1985): 234–238.
19. Glynn, I., and C. Ellory (eds.) *The Sodium Pump* (Cambridge: Company of Biologists Limited, 1985).
20. Kawakami, K., S. Noguchi, M. Noda, H. Takahashi, T. Ohta, M. Kawamura, H. Nojima, K. Nagano, T. Hirose, S. Inayama, H. Hayashida, T. Miyata, and S. Numa, "Primary structure of the alpha-subunit of *Torpedo californica* (Na + + K +)ATPase deduced from cDNA sequence," *Nature (London)* 316 (1985): 733–736.
21. Meissner, G., "Calcium transport and monovalent cation and proton fluxes in sarcoplasmic reticulum vesicles," *J. Biol. Chem.* 256 (1981): 636–643.
22. Herbette, L., P. Defoor, S. Fleischer, D. Pascolini, A. Scarpa, and J. K. Blasie, "The separate profile structures of the functional calcium pump protein and the phospholipid bilayer within isolated sarcoplasmic reticulum membranes determined by X-ray and neutron diffraction," *Biochim. Biophys. Acta* 817 (1985): 103–122.
23. MacLennan, D. H., C. J. Brandl, B. Korczak, and N. M. Green, "Amino-acid sequence

of a Ca2+ +Mg2+-dependent ATPase from rabbit muscle sarcoplasmic reticulum, deduced from its complementary DNA sequence," *Nature (London)* **316** (1985): 696–700.

24. Schulte, T. H., and V. T. Marchesi, "Conformation of the human erythrocyte glycophorin A and its constitutent peptides," *Biochemistry* **18** (1979): 275–280.

25. Leifer, D., and R. Henderson, "Three dimensional structure of orthorhombic purple membrane at 6.5 resolution," *J. Mol. Biol.* **163** (1983): 451–466.

26. Engelman, D., "An implication of the structure of bacteriohopsin globular membrane proteins are stabilized by polar interactions," *Biophys. J.* **37** (1982): 187–188.

27. Benz, R., "Structure and selectivity of porin channels.," *Curr. Topics Membr. Transp.* **21** (1984): 199–219.

28. Deisenhofer, J., O. Epp, K. Miki, R. Huber, and H. Michel, "X-ray structure analysis of a membrane protein complex: electron density may at 3 Å resolution and a model of the chromophores of the photosynthetics reaction center from Rhodopseudomonas viridis," *J. Mol. Biol.* **180** (1984): 385–398.

29. Tajima, S., K. Enomoto, and R. Sato, "Denaturation of cytochrome b5 by guanidine hydrochloride: Evidence for independent folding of the hydrophobic and hydrophilic moieties of the cytochrome molecule," *Arch. Biochem. Biophys.* **172** (1976); 90–97.

30. Albert, A. D., and B. J. Litman, "Independent structural domains in the membrane protein bovine rhodopsin," *Biochemistry* **17** (1978): 3893–3900.

31. Chin, D. J., G. Gill, D. W. Russell, L. Liscum, K. L. Luskey, S. K. Basn, H. Okayama, P. Berg, J. L. Goldstein, and M. S. Brown, "Nucleotide sequence of 3-hydroxy-3-methyl-glutaryl coenzyme A reductase, a glycoprotein of endoplasmic reticulum," *Nature (London)* **308** (1984): 613–617.

32. Edidin, M., "Translational diffusion of membrane proteins," In *The Structure of Biological Membranes* (P. L. Yeagle, ed.) (Boca Raton: CRC Press, 1992).

33. Hymel, L., A. Maurer, C. Berenski, C. Y. Jung, and S. Fleischer, "Target size of calcium pump protein from skeletal muscle sarcoplasmic reticulum," *J. Biol. Chem.* **259** (1984): 4890–4895.

34. Malmstrom, B. G., "Cytochrome c oxidase as a proton pump. A transition state mechanism," *Biochim. Biophys. Acta* **811** (1985): 1–12.

35. Strittmatter, P., and M. J. Rogers, "Apparent dependence of interactions between cytochrome b5 and cytochrome b5 reductase upon translational diffusion in dimyristoyl lecithin liposomes," *Proc. Natl. Acad. Sci. U.S.A.* **72** (1975): 2658–2661.

36. Meissner, B., "Monovalent ion and calcium ion fluxes in sarcoplasmic reticulum," *Mol. Cell. Biochem.* **55** (1983): 65–82.

37. Louis, C. F., P. A. Nash-Adler, G. Fudyma, M. Shigekawa, A. Akowitz, and A. M. Katz, "A comparison of vesicles derived from terminal cisternae and longitudinal tubules of sarcoplasmic reticulum isolated from rabbit skeletal muscle," *Eur. J. Biochem.* **111** (1980): 1–9.

38. Winkle, W. B. Van, R. J. Bick, D. E. Tucker, C. A. Tate, and M. L. Entman, "Evidence for membrane microheterogeneity in the sarcoplasmic reticulum of fast twitch skeletal muscle," *J. Biol. Chem.* **257** (1982): 11689–11695.

39. Bennett, N., M. Michel-Villary, and H. Kuhn, "Light-induced interaction between rhodopsin and the GTP-binding protein: Metarhodopsin II is the major photoproduct involved," *Eur. J. Biochem.* **127** (1982): 97–103.

40. Aton, B., and B. J. Litman, "Activation of rod outer segment phosphodiesterase by enzymatically altered rhodopsin: a regulatory role for the carboxyl terminus of rhodopsin," *Exp. Eye Res.* **38** (1984): 547–559.

41. Schertler, G. F. X., C. Villa, and R. Henderson, "Projection structure of rhodopsin," *Nature* **362** (1993): 770.

7

Lipid–Protein Interactions and the Roles of Lipids in Biological Membranes

Having studied each of the major components of the membranes of cells separately, it is now possible to examine the interactions among these components. If one considers just the lipids and the proteins of cell membranes, three pairwise interactions are important to membrane structure: protein–protein interactions, lipid–protein interactions, and lipid–lipid interactions. These interactions are linked by the structure of the membrane. Therefore changes in one kind of interaction will affect the other interactions. (The complexity of the membrane structure becomes apparent when one considers the effect of water interactions with the other two components from this point of view.)

Protein–protein interactions refer predominantly to the state of oligo-merization of the membrane proteins. As noted in Chapter 6, examples are known of membrane proteins being in a monomeric state, dimeric state, and higher oligomeric states. These protein–protein interactions are influenced by the interactions between the protein surfaces in contact with each other. Hydrophobic regions (transmembrane regions) of the proteins may be in contact or the extramembranous regions of the proteins may be in contact, or portions of both may be in contact. The energies of such interactions have not been well described. However, contact between extramembranous regions of the membrane proteins are likely influenced by the same forces governing oligomerization of water-soluble proteins. These forces can include electrostatic forces between concentra-tions of charges on the protein surfaces, ion pair formation as between subunits of hemoglobin, dispersive forces, and the hydrophobic effect due to hydrophobic patches on the surfaces of the extramembranous portions of the membrane proteins. Contact between transmembrane regions of the membrane proteins usually involves contact between hydrophobic surfaces. This may require a tilt of the α-helical transmembrane segments, with respect to the bilayer normal, to allow close contact between these portions of the protein (due to the projections of the mass of amino acid side chains perpendicular to the helix, which spiral around the helix due to the direction of the polypeptide chain). This line of thought could also lead to the idea that a conformational change in the membrane protein could change the tilt of the transmembrane helices and thus alter the ability of the protein to oligomerize. In addition, at least in the case of bacteriorhodopsin, some polar interactions have been suggested to be involved in helix–helix interactions.

The interplay between these interactions determines the structure of the membrane. If protein–protein interactions are favored over lipid–protein interactions, for example, then protein oligomerization will occur in the membrane, as is frequently observed. Likewise if lipid–lipid interactions are favored over lipid–protein interactions, then protein oligomerization will also occur. Therefore, one needs to consider all these interactions to understand membrane structure. In particular, one needs to be aware that changes in one set of interactions will lead to changes in other sets of interactions. For example, unless there are binding sites on proteins for lipids, lipid–protein interactions can be relatively unfavorable (see below), compared to lipid–lipid interactions, thus leading to the widespread observation of dimerization by membrane proteins in membranes. If lipid–lipid interactions are favored by the state of the system, as in the gel state of the lipids, then membrane proteins can phase separate into protein-rich regions (as is observed experimentally).

The discussion of the interactions among membrane components will begin here with a consideration on theoretical grounds of interactions that have not yet been examined: the interactions between the lipid and the protein components at the lipid–protein interface.

I. THEORETICAL CONSIDERATIONS OF LIPID–PROTEIN INTERACTIONS

A. Contacts between Lipid Hydrocarbon Chains and Integral Membrane Proteins

Transmembrane regions of integral membrane proteins are characterized by linear sequences of hydrophobic amino acids. Those hydrophobic sequences are most likely α helices. Therefore, the hydrophobic hydrocarbon chains of the lipids must coexist with the surface of hydrophobic α helices in the membrane interior. This structure satisfies the hydrophobic effect and maintains the integrity of the hydrophobic interior of the membrane.

Consider first the protein surface involved in these interactions. Likely all the polar carbonyls and amide hydrogens of the peptide bonds are involved in hydrogen bonds within the α helix. By engaging in hydrogen bonds, the polarity of the peptide bond is reduced. As noted in Chapter 6, the hydrogen bonding of these polar groups allows the peptide bonds to be immersed in the hydrophobic region of the bilayer. This hydrogen bonding scheme has the interesting side effect of making the carbonyls and amines unavailable (or unnecessary) for interaction with lipid molecules.

Protruding from the α helix are the side chains of the hydrophobic amino acids. The bulk of the side chain mass is oriented roughly perpendicular to the helix axis. For an amino acid like glycine, the side chain does not perturb the cylindrical shape of the α helix. Even for alanine, little mass protrudes from the helix axis. However, for an amino acid like tryptophan, the side chain is large. Its mass protrudes a substantial distance from the helix axis.

There is considerable variation in amino acid side chain size in the amino acids of a typical hydrophobic transmembrane sequence. Now imagine this sequence distributed in three dimensions along an α helix. On the scale of the lipid molecules, the surface is rough. The length of a tryptophan side chain is equivalent to perhaps four methylene segments of a hydrocarbon chain. This length corresponds to about a quarter of the total length of a typical lipid hydrocarbon chain. Therefore within the context of lipid–protein interactions, the transmembrane region of a membrane protein is not well represented by a smooth cylinder.

These structural features lead to some interesting problems for the coexistence of lipids and proteins in membranes. The transmembrane segment of membrane proteins can be expected to be conformationally quite rigid. Some flipping and spinning of amino acid side chains can take place. However, because of the rigidity of the peptide backbone, the rough protein surface cannot be smoothed out by protein motion. Therefore, the hydrocarbon chains of the lipids, which are conformationally flexible, must be expected to accommodate themselves to the rough protein surface and fill in the spaces between the amino acid side chains.

Because of such mismatches in packing of the membrane components, one might expect to find increased permeability at the lipid–protein interface. Any inability of the lipid hydrocarbon chains to fill fully and constantly the awkward spaces in the protein surface will lead to transient defects in the membrane structure. According to the model developed previously for membrane permeability to small solutes, an increase in structural defects in the membrane should lead to an increase in membrane permeability. Thus it would not be surprising to find an increase in membrane permeability when a protein is incorporated into a membrane. In reconstituted systems, this has been experimentally observed.

Presumably the transmembrane movement of phospholipids, or flip-flop, is also enhanced by the formation of defects in the bilayer structure. Thus one might also expect to see an increase in the rate of transmembrane movement of phospholipids in membranes in the presence of membrane proteins. This has been observed as a result of the incorporation of a membrane protein into a phospholipid bilayer in reconstituted systems.

The difficulty of packing lipid molecules and protein molecules in the membrane creates a free energy problem for the interaction between lipid hydrocarbon chains and the protein surface. The lipid chains must conform to a rough, but structured, surface. This means that the chains must adopt a conformation complementary to the protein surface. To do so is to restrict the conformations available to the chains during the time they interact with the membrane protein. A restriction in the number of conformations available to the lipid chain thermodynamically implies an ordering of the chain. This represents an unfavorable entropy contribution to lipid–protein interactions.

Recent calculations have suggested that, for the hydrocarbon chains, lipid–lipid interactions would be favored over lipid–protein interactions.[1] In these calculations, aggregation of hydrophobic helices in a lipid matrix is energetically favored over mixing of the lipids and the protein helices. Experimental systems consisting of hydrophobic helices of transmembrane proteins and lipids exhibit aggregation into bundles of helices large enough to be observed in freeze-fracture electron microscopy (EM) of the membranes.[2] Thus thermodynamics of mixed helix–hydrocarbon chain (of membrane lipids) systems favors minimization of lipid–protein contact. Interestingly, this conclusion also indicates an unfavorable contribution to the insertion of a membrane protein into a membrane. This unfavorable entropy contribution is outweighed by the entropy advantage gained by removal of a hydrophobic protein surface from an aqueous medium.

B. Contacts between Lipid Headgroups and Integral Membrane Proteins

Now examine the polar headgroups of the lipids and their interactions with membrane proteins. Many of the polar headgroups are zwitterionic and carry a negative charge on the phosphate and a positive charge on the alcohol. Others carry a net negative charge. Furthermore, membrane proteins exhibit an interesting topology in their primary sequence. Clusters of charges are found at the interface between the hydrophobic and the polar regions of a membrane.

What would be expected to be the important free energy terms of the interaction between the membrane protein and the lipid headgroup? Favorable enthalpy terms may arise in some cases from electrostatic attraction between charged amino acid residues and oppositely charged phospholipids or portions of phospholipids. Thus positively charged amino acids, such as lysine or arginine, may favorably interact with negatively charged phosphates on phospholipids, or other negatively charged lipid headgroups. Likewise, negatively charged amino acids, like glutamate or aspartate, may interact favorably with positively charged portions of phospholipids, like the quaternary amines of PC and PE. Thus favorable enthalpic terms may occur describing the interaction of a limited number of phospholipids with membrane proteins at discrete sites.

What about the entropic terms of the interaction between phospholipid headgroups and membrane proteins? In contrast to the hydrocarbon chains, there may also be favorable entropic terms for this interaction. Phospholipid bilayer surfaces are extensively hydrated in biological membranes. This hydration is to some degree entropically unfavorable. It requires the ordering of water molecules around the phospholipid headgroups. When the cross-sectional area of the phospholipid headgroup is small compared to the cross-sectional area of the hydrocarbon chains, an increase in ordering of the water molecules must occur because of the greater space on the bilayer surface between the phospholipid headgroups that must be filled. Binding of a phospholipid headgroup to a membrane protein removes the necessity of ordering water molecules around the headgroup and of ordering water molecules in the binding site on the protein surface. This can therefore lead to a favorable entropy contribution to the interaction between phospholipid headgroups and membrane proteins (more pronounced for PE than for PC, for example).

These phospholipid headgroup–protein interactions can be described as

$$\Delta G_{tH} = \Delta H_{tH} + T\Delta S_{tH},$$

where G_{tH} represents the change in free energy of the lipid headgroups resulting from the transfer of a membrane protein from a aqueous phase into the membrane. There is likely a negative H_{tH} term, due to the interaction of oppositely charged moieties on the protein and on the lipid (unless dehydration is a problem). Furthermore there is a loss of entropy of binding of water molecules to the surface,

$$\Delta S_{tH} = \Delta S_h + \Delta S_w,$$

where ΔS_h refers to the ordering of the headgroup and ΔS_w refers to the ordering of water on the membrane surface. Binding of a phospholipid headgroup to a protein will reduce the motional freedom of the headgroup, so ΔS_h is negative. However, ΔS_w is likely larger and positive because of the reduction in ordered water at the membrane surface. Therefore,

$$\Delta S_{tH} \approx \Delta S_w > 0$$

and

$$\Delta G_{tH} < 0.$$

Taken together, the summation of the entropic and enthalpic terms describing the interaction between phospholipid headgroups and membrane proteins suggests that a favorable free energy of interaction might arise under some conditions. This binding energy may be an important element in protein insertion into membranes.

Additionally, in any binding reaction, a high concentration of the ligand will enhance binding. The membrane is a two-dimensional system within which the relative "concentration" of the phospholipids is high. The high concentration would drive the equilibrium toward interaction between lipids and proteins.

This theoretical discussion leads to an interesting balance in the thermodynamic terms describing the interaction of integral membrane proteins and lipids in membranes. Where there are binding sites on the protein for the lipid, as described above, the overall ΔG for the interaction between lipids and proteins can be expected to be favorable. However, the availability of binding sites for lipid headgroups on the protein is a crucial factor. If the binding sites were modified, such that ΔG_{tH} were no longer significantly favorable, then the overall ΔG describing the lipid–protein interactions would become less favorable relative to lipid–lipid interactions. This would be expected to enhance protein–protein interactions, as a partial substitution for the lipid–protein interaction. The important conclusion from the analysis is that modulation of lipid-headgroup–protein interactions may modulate the state of aggregation of membrane proteins in the membrane (which in turn may modulate membrane protein function).

C. Dynamics of Interactions between Lipids and Integral Membrane Proteins

The lifetime of a lipid on a membrane protein is crucial to developing an accurate model for any role lipid–protein interactions may play in membrane function. Can this also be modeled?

Lateral diffusion of the lipids and proteins can be modeled using Monte

Carlo techniques, which permits a time resolution of lipid–protein interactions. For the purposes of a simple calculation, hard disks of appropriate size (appropriate to the cross-sectional area of the referenced molecule) play the role of lipids and proteins in membranes. On providing these disk with the appropriate kinetic energy (such as $\frac{1}{2}$ kT) and allowing them to undergo random jumps of the appropriate magnitude (a few angstroms for each jump), one can directly observe lateral diffusion with a rate comparable to measured rates of lateral diffusion of lipids (B. A. Cornell and J. Middlehurst, unpublished results).

From this approach, an interesting conclusion emerges. The "escape" angle of a lipid from a position next to a protein is much more limited than when the lipid is next to other lipids. Imagine an infinitely large protein. To the lipid, the protein surface appears flat. The angle through which the lipid can escape from the protein surface is nearly half the angle available when a lipid is next to another lipid. Thus the likelihood of the lipid adjacent to the protein moving away from the protein and allowing another lipid to take its place is considerably reduced from that observed in the free diffusion of lipid in pure lipid bilayers. These calculations are dependent on the relative size of the lipids and the proteins and on their density of packing in the membrane. Qualitatively the lifetime of lipids in sites adjacent to a protein is long relative to the lifetime of lipids in sites in a pure lipid bilayer (Fig. 7.1). The number of lipids so restricted is just enough to cover the circumference of the protein that is cut by the surface of the lipid bilayer.

These calculations do not include any interactions between the lipid and the protein. Any favorable headgroup–protein interactions could increase the lifetime of lipids in sites adjacent to membrane proteins. Binding sites on the protein would increase the lifetime. "Sites" that trap lipid, thereby sterically reducing the escape rate, will also increase the lifetime of the lipid–protein complex. Under these conditions one could expect to see lifetimes of the lipid–protein complex similar to those of effector molecules modulating enzyme function through allosteric regulation. Thus there can be a subgroup of the lipids in contact with a membrane protein that has a significantly longer lifetime on the membrane protein. These are the kinds of interactions that would be expected to be specific for certain lipid headgroups and to be important to membrane protein function.

D. Contacts between Lipid Bilayer Surfaces and Integral Membrane Proteins

The lipid–protein interface may not just be along a plane normal to the membrane surface. Surface–surface interactions may occur in the plane

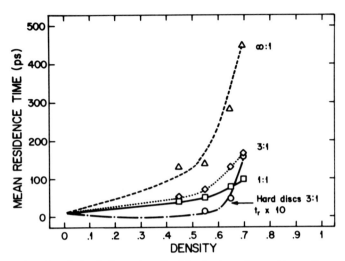

Fig. 7.1. Results of simulations, described in the text, for residence times of lipids (treated as modified hard disks, behaving as if their shape were modestly elastic) next to other molecules of similar or much larger size, as a function of density. The calculation extends to a density of 0.7, although a membrane would be better represented by a density greater than 0.7 and less than 0.8. At a ratio in sizes of 1 : 1, the lateral diffusion of lipids in a pure lipid bilayer is well represented. Lipid protein interactions are best modeled by a very large ratio, represented by the dashed line. Extrapolating to an appropriate density, one can estimate that the residence time of a lipid nest to a protein must be reduced from that experienced in lateral diffusion by more than an order of magnitude. This reduction will be dramatically enhanced by a rough protein surface, since the escape angle will be drastically reduced for some parts of the protein surface. Furthermore, some of the movements represented by this calculation will not represent an actual exchange of lipids in sites immediately adjacent to the protein with lipids removed from the protein surface. Therefore, the rate of escape of lipids out of the lipid–protein interface is predicted to be two or more orders of magnitude retarded relative to lateral diffusion of lipids in pure lipid bilayers in the liquid crystalline state. This extent of retardation is predicted in the absence of any specific interactions between the lipid and the protein. These calculations were carried out by B. A. Cornell and J. Middlehurst and were provided by Dr. B. A. Cornell.

of the membrane between the membrane surface and surfaces of the extramembranous portion of the membrane protein. Figure 7.2 shows schematically how this interaction might occur. If the surface of the protein in contact with the lipid bilayer surface is relatively hydrophobic, then the structure will be stabilized by lipid headgroups that are relatively hydrophobic (like PE). The relatively hydrophobic surfaces of the protein and the bilayer can approach each other closely because the exclusion of water from between the surfaces is thermodynamically favorable. Surface–surface interactions may involve a relatively large number of phos-

Fig. 7.2. Schematic representation of a mechanism for interaction between protein surfaces and membrane surfaces involving extramembranous portions of integral membrane proteins.

pholipids in a distinct lipid domain without requiring discrete binding sites on the protein.

E. Bilayer Free Volume and Integral Membrane Protein Function

The function of membrane proteins can also be affected by the properties of the "solvent" in which they exist: the lipid bilayer. Many proteins are known to undergo conformational changes as a necessary part of their functional cycle. A conformational change of an integral membrane protein will require a transient adjustment on the part of the lipid bilayer, as the detailed arrangement of the elements of the protein structure are altered in the bilayer. When one considers the protein as a whole, a protein conformational change may require a volume change for the protein within the lipid bilayer. This volume change of the protein demands a corresponding provision of volume by the bilayer.

As discussed in Chapter 4, lipid hydrocarbon chains exhibit kink formation and as a result transient packing defects appear in the bilayer. These packing defects are small volume elements that can facilitate the passive diffusion of small molecules across the membrane. They could also be recruited to the lipid–protein interface to permit the conformational changes required in the functional cycle of integral membrane proteins. The extent to which they were available to be recruited to the lipid–protein interface could then be an important factor in modulating membrane protein activity.

Even the introduction of a membrane protein into a lipid bilayer could require the recruitment of such volume elements to accommodate the "rough" surface of the protein as described at the beginning of this chapter. Litman and co-workers have developed a quantitative means to determine the extent of free volume in a bilayer available for the purposes being discussed here.[3] Measurements show that the presence of the protein reduces the free volume available in the membrane, as suggested by the considerations above. This subject will be explored more later in this chapter, as it provides the basis of an important means for modulation of integral membrane protein function.

F. Peptide–Lipid Interactions

The interaction of small peptides with membranes has been studied intensely.[4] Peptides that are hydrophobic will partition into the lipid bilayer. Basic peptides will bind to the surface of negatively charged bilayers (for example, bilayers containing PC). Even small peptides can alter lipid phase transitions and lipid conformation and even inhibit membrane fusion (see Chapter 9). Because these peptides have molecular size similar to that of the lipids, they do not influence lipid dynamics in the same manner as do integral membrane proteins.

II. CLASSES OF LIPID–PROTEIN INTERACTIONS IMPORTANT TO FUNCTION

For the purpose of obtaining a view on the structural aspects of lipid–protein interactions, two classes of lipid–protein interactions will be considered. One class of interactions involves lipids that do not bind directly to membrane proteins, a class that includes the majority of the membrane lipids. This means lipids that do not have long residence times on the protein (at the protein surface) as defined in Section I,C. In this class of interactions it is the bulk properties of the membrane lipids that are important to the function of the protein. For the purposes of this discussion, this class of lipid will be called interfacial lipid because all the lipid in this class is in relatively rapid communication with the integral membrane protein surface through exchange.

The second class of lipid–protein interactions that will be considered involves the lipid that binds to the protein. This is a lipid with a relatively long residence time on the protein (at the protein surface), as defined in Section I,C. Such lipid–protein interactions are usually structurally specific for particular lipids. These interactions mimic those found in solution protein chemistry in which modulators (effectors) regulate enzyme activity allosterically. For the discussion here this class of lipid will be called bound lipid.

A. Interfacial Lipid

The theoretical considerations earlier in this chapter suggest that there is a population of lipids in a membrane containing protein that is temporally confined at the interface between the membrane protein and the lipid bilayer. This lipid is in general in exchange with most other lipids in the membrane. The dynamics of all of these lipids are affected by the presence of the protein. Thus, all of these lipids will be grouped together and referred to as interfacial lipid. The number of lipid molecules in direct contact with the protein at any instant of time is well defined. This corresponds to the number of lipids required to close pack around the circumference of the portion of the protein exposed to the lipid bilayer at the polar–hydrophobic interface. Figure 7.3 shows schematically what this means. Of course, this two-dimensional picture is incomplete. There are two sides of the membrane, and the interfacial lipid must cover the circumference of the protein on both sides of the membrane, if the protein is a transmembrane protein. Furthermore, the circumference on one side need not be identical to the circumference on the other side of the membrane, since the circumference depends on the shape of the protein. If the protein is merely anchored, this latter consideration does not apply. Furthermore, if some of the protein surface is involved in protein–protein interactions, the amount of lipid–protein contact will be reduced. This will occur, for example, if the integral membrane protein forms dimers in the membrane.

How might one go about detecting the extent of lipid contact with protein? The size of the membrane protein relative to the size of a lipid causes a significant reduction in the site-to-site jump rate for those lipids in

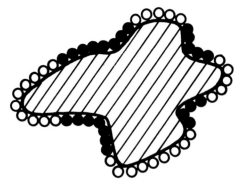

Fig. 7.3. Schematic representation of interfacial lipid. The lipids occupying invaginations of the protein surface may have longer residence times. The number of interfacial lipids is determined by the circumference of the cross section of the protein in the membrane.

sites adjacent to membrane proteins. However, diffusion is still relatively rapid. Therefore one must have a technique that can differentiate two populations between which lipids may be exchanging on the megahertz time scale. Electron-spin resonance (ESR) is eminently suitable for this question. Any lipid exchanging between two distinguishable lipid domains in a membrane at a rate slower than about 100 MHz will give rise to two distinguishable resonances in the ESR experiment. These membrane ESR experiments require spin labels that contain a stable free radical (Chapter 4). The synthetic spin labels must be artificially introduced into the membrane of interest.

An example of such an ESR experiment is presented in Fig. 7.4. Here a spin label, with the free radical on the hydrocarbon chain, is monitored in a membrane containing Ca^{2+}-ATPase from sarcoplasmic reticulum membranes.[5] The protein has been purified and reconstituted into a defined lipid medium. Several different lipid/protein mole ratios have been used. The ESR spectra from such samples show two distinguishable resonances. One resembles (but is not identical to) that obtained from the same spin label in lipid bilayers. The other resonance is characteristic of a motionally restricted environment. The relative population of the motionally restricted environment is directly proportional to the lipid/protein mole ratio. The higher the protein content, the greater the relative spectral intensity of the spectral component corresponding to the motionally restricted environment.

These data fit the theoretical expectations well. First, the presence of membrane proteins in the lipid bilayer does appear to induce the formation of two lipid domains, at least on the short time scale of the ESR experiment. This is one structural feature that was predicted.

Second, the population of lipids in the motionally restricted environment is proportional to the protein content of the membrane. This is what one expects if the lipid being detected in this experiment is indeed lipid in contact with protein surface. In particular, one can calculate the number of lipids necessary to fill the interfacial domain, given the mass of the protein, and estimating what percentage of that mass is in the membrane. Taking into account the state of aggregation of the membrane proteins, the number of lipids in the motionally restricted domain as detected by ESR spin labels is what would be required (using the best model building conventions) to cover the surface of the protein in the membrane.

Third, the exchange rate of lipids out of locations in contact with the protein surface is relatively rapid. Electron-spin-resonance measurements with spin-labeled lipids indicated that the frequency of exchange is in the range of 10 MHz.[6] This is significantly slower than free diffusion in the membrane, as was predicted by the discussion earlier in this chapter. As

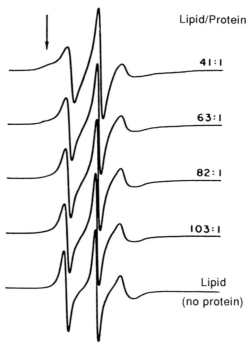

Lipid/Protein

41:1

63:1

82:1

103:1

Lipid
(no protein)

Fig. 7.4. ESR spin-labeled phospholipids included in a reconstitution of the Ca^{2+}-ATPase from rabbit muscle sarcoplasmic reticulum. The reconstitutions resulted in the lipid–protein ratios indicated on the right side of the figure. The visually dominant spin label spectrum represents relatively normal lipid bilayer. The arrow on the top left represents the leftmost portion of an ESR spectrum indicative of motionally restricted phospholipid. The remainder of the motionally restricted spectrum is visually obscured by the normal bilayer spectrum and must be analyzed by spectral simulation. As the lipid content of the reconstituted membrane increased, the motionally restricted spectrum diminished in relative magnitude. Reprinted with permission from J. R. Silvius, D. A. McMillen, N. D. Salley, P. C. Jost, and O. H. Griffin, *Biochemistry* **23** (1984): 538, American Chemical Society.

a result of this exchange and because the perturbing effects of the protein extend beyond the surface of the protein by several layers of lipid, all the lipid in biological membranes is perturbed relative to pure phospholipid bilayers.

Artifacts might interfere with interpretations of these data. The most important potential artifacts arise from the potential perturbing effect of the ESR spin labels. The spin labels are polar, as can be seen in the structures given in Chapter 4. This polar structure is being forced to reside in a hydrophobic region that could either encourage water penetration of the membrane or encourage the spin label to have some occupancy at sites near the membrane surface. A second problem is that these spin labels are rather large chemical structures relative to the size of the lipids on which they reside. Therefore, they can perturb the overall biolayer structure and thus primarily report the conditions of a perturbed microenvironment around the probe.

Calorimetry experiments also lead to the conclusion that there is a domain of interfacial lipids in membranes containing protein. As in the case of cholesterol, the presence of membrane proteins in lipid bilayers creates a domain of lipids that do not undergo the highly cooperative phase transition characteristic of pure lipid bilayers. The population of this domain is proportional to the protein content of the membrane. These measurements can be obtained only from membrane proteins reconstituted into lipid bilayers of saturated lipids. Saturated lipids often do not provide an environment conducive to membrane protein function. However, in several cases, the ESR and calorimetric measurements give quantitatively similar answers. In other cases, the calorimetric data suggest a more far-reaching effect of the protein, away from the protein surface. This is likely due to the much slower time scale of the calorimetry experiment, such that the effects of the protein are averaged over a larger number of lipids through exchange of lipids among lipid sites.

A good example of measuring the lipid in contact with protein surface is provided by rhodopsin reconstituted into phospholipid bilayers. The ESR spin-label results for rhodopsin indicate that about 24 lipids are temporally retarded in locations at the rhodopsin–lipid interface. Calculations suggest that about 24 lipids are required to cover the surface of the protein within the bilayer.

In the case of the calcium pump protein from sarcoplasmic reticulum, ESR measurements detect about 25 lipids in contact with the protein (per protein monomer).[5] Interestingly, X-ray diffraction measurements indicate that much of the mass of this protein lies outside the membrane.[7] Calculations suggest that about 35–40 lipids should be required to cover the surface of a monomer of the protein. The smaller number detected by ESR measurements corresponds to the number expected if the protein exists as a dimer in the membrane. In fact, radiation inactivation experiments demonstrate that the protein is a dimer (Fig. 7.5).[8]

In other experiments, the relative affinity for sites at the protein–lipid interface has been measured for a variety of lipids. An example is provided by the Na⁺,K⁺-ATPase, in which negatively charged lipids preferentially populate the lipid–protein interface in reconstituted membranes. Measurements of the relative affinity show that PS and stearic acid show relatively high affinity, and PE and PC show relatively low affinity for contact with the surface of the transmembrane portions of this integral membrane

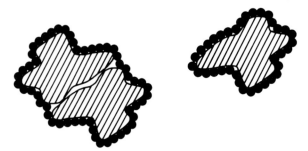

Fig. 7.5. Schematic representation of the effect of dimer formation on the lipid population of the lipid–protein interface.

protein. Electrostatic attraction plays only a partial role in determining this specificity.[9]

Not all integral membrane proteins show such relative specificity for negatively charged lipids. Rhodopsin reconstituted into phospholipid bilayers shows no strong preference for negatively charged lipids over PC or PE at the lipid–protein interface.[10]

B. Bound Phospholipids

The next question to consider is whether any lipids bind to membrane proteins. A bound lipid is a lipid with a long residence time on the protein (at the protein surface). The thermodynamic analysis suggested that the lipid headgroups provided the likely ligands for binding to the protein. Binding is likely to be specific for particular lipid headgroups structures.

To look for phospholipid binding, one needs a technique that is sensitive to a separation of lipid domains on a much longer time scale than that operative for ESR. One needs to see selectively domains that are in slow exchange with each other, separate from those that are merely retarded at the interface between lipid and protein due to the interfacial effects discussed above. Lipids that are bound to the protein will have residence times on the protein much in excess of the hundreds of nanoseconds associated with lipids that simply contact the protein surface. ^{31}P NMR of phospholipid headgroups in membranes provides a means to detect such lipid binding for phospholipids. ^{31}P NMR measures most directly phosphates of the phospholipid headgroups and is a nonperturbing technique. No labels need to be added since the ^{31}P nucleus is 100% naturally abundant. This is important because lipids bound to membrane proteins will not exchange effectively with labeled lipids to permit measurements of these domains with ESR (requiring spin-labeled lipids) or ^{2}H NMR (requiring ^{2}H-labeled lipids).[11] Thus the techniques of ESR and ^{2}H NMR cannot detect any truly protein-bound lipids under normal conditions. Intact biological membranes containing phospholipids can readily be studied in fully functional condition with ^{31}P NMR. Observation of separate spectral components in a ^{31}P NMR spectrum of a membrane implies an exchange rate between the domains that is slow compared to 1 kHz.

^{31}P NMR studies of some biological membranes show at least two domains of phospholipid headgroups, with quite different properties. One of the domains represents an environment similar (but not identical) to pure phospholipid bilayers. The other domain is motionally restricted. An example is provided by human erythrocyte glycophorin, as seen in Fig. 7.6. The ^{31}P NMR powder pattern reveals that the bound phospholipid is experiencing axial diffusion, corresponding to relatively strict rotational diffusion of a phospholipid–protein complex without accompanying headgroup motion typical of phospholipids in a bilayer. Figure 7.7 shows that a similar protein bound phospholipid component can be observed in a biological membrane, the sarcoplasmic reticulum from rabbit muscle.[12]

What is the rate characterizing the departure of a phospholipid headgroup as ligand from the protein binding site? In one system an upper limit to this rate has been measured. For the calcium pump protein of the sarcoplasmic reticulum, phospholipids leave binding sites on the protein at a rate ≤ 1 sec^{-1}.[13] In another system, the rate is much slower. Some of the PI molecules bound to glycophorin of the human erythrocyte mem-

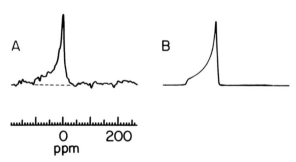

Fig. 7.6.   To determine the motional properties of phospholipids bound to a membrane protein in a lipid bilayer, glycophorin with 4 phospholipids tightly bound per protein was reconstituted in a digalactosyldiglyceride bilayer. The ^{31}P NMR spectrum of the phospholipid bound to glycophorin in this lipid bilayer is in (A), and the simulation in (B). These data show phospholipid that is strongly motionally restricted compared to pure phospholipid bilayers, is undergoing relatively slow axial rotation, and is in a different chemical environment bound to the protein than in a phospholipid bilayer as indicated by a change in the ^{31}P chemical-shift sensor.

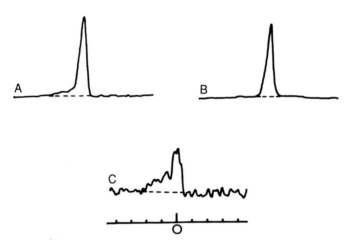

Fig. 7.7. (A) ^{31}P NMR of light sarcoplasmic reticulum and (B) total lipid extract of the same membrane. (C) Difference spectrum (A – B) showing a motionally restricted phospholipid component in this biological membrane.

brane apparently exhibit a vanishingly small rate of exchange.[14] The class of phospholipid headgroup binding sites characterized by such slow exchange would be expected to be important to membrane protein function.

What is the nature of the binding sites for phospholipids on membrane proteins? The answer to this question is yet to be determined. However, the primary structures of the membrane proteins provide some interesting clues. The primary sequences reveal several subdomains of hydrophobic stretches of amino acids that are capable of spanning the hydrophobic region of a membrane in the form of an α helix. Terminating these hydrophobic subdomains are regions in which clusters of charged amino acids are found. Because of the topology of membrane proteins, these clusters of charged residues lie in close proximity to the headgroups of the membrane lipids. The favorable nature of a putative interaction between a phospholipid phosphate (carrying a negative charge) and a lysine or arginine (carrying a positive charge) can be appreciated. It is just this kind of interaction that could be employed by the membrane system to decrease the off rate of a phospholipid from the lipid–protein interface.

Another kind of interaction that is possible between the membrane proteins and the lipid bilayer is the hydrophobic surface interaction described earlier in this chapter. This interaction would create a distinct domain in which the lipid headgroups are partially dehydrated compared to the normal lipid bilayer. The number of lipids inhabiting such a domain could be quite large, depending on the size of the extramembranous portion of the protein compared to the size of the intramembranous portion.

Phosphatidylethanolamine (PE), which alters surface hydration (see Chapter 4), stimulates the activity of the calcium pump from sarcoplasmic reticulum, from very low membrane PE content[15] to very high membrane PE content[16] (see Fig. 7.8). The large portion of this protein that protrudes in a mushroom shape from the transmembrane domain could have some contacts with the membrane surface. This could explain why PE has a general effect, far from any phase boundary, that does not show saturation with PE content.

So what is known about phospholipids that are bound as ligands to membrane proteins? First, they are distinguished from the interfacial lipid by their much slower exchange with phospholipid in the remainder of the membrane. Second, they are motionally restricted, compared to pure bilayer lipid, including phospholipid in the gel state. Third, they undergo rotational diffusion in the membrane, likely at a rate characteristic of protein rotational diffusion. Fourth, the binding sites can be specific for particular phospholipids, as distinguished by their headgroup structure. Fifth, the number of lipids bound to a membrane protein is less than the

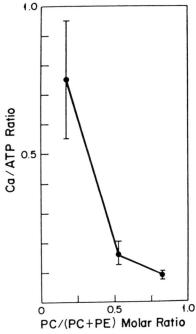

Fig. 7.8. Effect of membrane PE content on the activity of the calcium pump from sarcoplasmic reticulum. Reprinted with permission from K. Cheng, J. R. Lepock, S. W. Hui, and P. L. Yeagle, *J. Biol. Chem.* **261** (1986): 5081.

number of lipids in contact with the protein. Sixth, both phospholipids and cholesterol have been found as bound lipids in individual cases.

III. HOW DO LIPIDS AFFECT MEMBRANE PROTEIN FUNCTION?

Biological membranes contain a wide variety of lipid components. These lipids exhibit considerable variation in their behavior based on their structure. There must be specific roles for each of these membrane lipids in the function of a particular biological membrane. Some of the roles for modulating membrane protein function are beginning to be unraveled.

A. Interfacial Lipid

The hypothesis of free volume, described in Section I,E, has proven successful in explaining the effects of some lipid environments on membrane protein function. It is thus a model for the role of interfacial lipid in membrane activity. The most extreme example can be found in the gel phase of a membrane bilayer. In gel-state bilayers, the information in Chapter 4 indicates that the incidence of kinks (and thus packing defects) is reduced to a minimum. The free volume hypothesis would predict a loss in protein activity in the gel state due to the lack of volume elements in the bilayer for the necessary conformational change of the enzyme.

Membrane cholesterol provides a natural way to modulate free volume in the bilayer. Increases in membrane cholesterol lead to an ordering of the lipid hydrocarbon chains. This corresponds to a reduction in kink formation (and thus packing defects). The free volume hypothesis would then argue that protein conformational changes would be hindered by cholesterol due to a reduction in elements of free volume that can be recruited to the lipid–protein interface for required changes in volume occupied by a membrane protein as it undergoes a conformational change.

The equilibrium between the Meta I and Meta II states of rhodopsin was measured as a function of cholesterol and the dynamics of diphenylhexatriene was utilized to assess the extent of free volume available in the bilayer.[17] As can be seen in Fig 5.8, there is a direct correspondence between the free volume parameter and the equilibrium of rhodopsin between Meta I and Meta II. Since the Meta II state is required for activation of transducin (G_t) in the visual transduction pathway, activation of the target enzyme in this receptor-driven pathway would be expected to be inhibited by membrane cholesterol. In fact, plasma membrane cholesterol inhibits activation of the cGMP cascade by the rhodopsin in the rod outer segment plasma membrane.[18]

This same system was used to explore the role of unsaturation in the interfacial lipids. Unsaturation (*cis* as found in most natural lipids) introduces kinks and packing defects into the bilayer. This would be expected to increase free volume in the bilayer and lead to an activation of the rhodopsin, according to the free volume hypothesis. The Meta I–Meta II equilibrium favors the activated state in the presence of unsaturation. The stimulatory effects of the native rod outer segment disk lipids, which are rich in 22 : 6 fatty acids (nearly 50% of the total acyl chains), are dramatic. This suggests a fundamental role for polyunsaturation (*cis*) in biological membranes. Other physical effects of lipid unsaturation, such as increased cross-sectional area, perturbations to the order profile, or effects on the gel to liquid phase transition temperatures, provide no rationale for polyunsaturation in lipid hydrocarbon chains. However, the concept of free volume does provide a compelling rationale for polyunsaturation.

B. Bound Lipid

Some examples now exist of regulation of membrane protein function through binding of specific membrane lipids to membrane proteins. Some of these will be discussed briefly below. More examples are likely to emerge in the future as investigators directly look for such interactions. Purified membrane proteins may have specific phospholipids tightly bound, and thus show no apparent lipid specificity when reconstituted into bilayers of defined lipid content. The presence of such tightly bound lipids in a protein preparation can easily go unnoticed. One must explicitly assay for tightly bound lipids to know whether they are present.

An example of a specific phospholipid requirement to activate a membrane enzyme is provided by α-hydroxybutyrate dehydrogenase.[19] A series of elegant reconstitution experiments have demonstrated that phosphatidylcholine is required for proper activity of this enzyme. Detailed investigations using structural analogs revealed that the most important portion of the phosphatidylcholine molecule for activation of α-hydroxybutyrate dehydrogenase is the choline headgroup. Evidence suggests that the choline structure binds to the enzyme to promote activation of this protein.

The mitochondrial ADP–ATP exchange protein is responsible for exchange of ADP for ATP across the inner mitochondrial membrane for the purpose of delivering newly synthesized ATP to extramitochondrial regions of the cell and to provide new substrate for ATP synthesis within the mitochondria. Diphosphatidylglycerol (DPG) (cardiolipin) is important to support the activity of this protein. Six DPG molecules are tightly bound to the ADP–ATP exchange protein.[20] The activation by DPG is best explained by the effector model; DPG molecules bind to the protein and stabilize the active form. These tightly bound, slowly exchanging DPG molecules were detected using ^{31}P NMR, and were not observed in an ESR spin label study, in agreement with the theoretical analysis of the previous section in this chapter.

Human erythrocyte glycophorin can be isolated from the red blood cell membrane largely delipidated. However, unless stringent measures are taken, including extraction with organic phases, about five phospholipids remain tightly bound to the protein when it is isolated. These tightly bound phospholipids can be visualized using ^{31}P NMR as described earlier in this chapter. The dominant phospholipid bound to glycophorin is PI; three to four of those five bound phospholipids are PI, even though PI is a minor phospholipid component of the erythrocyte membrane.[21] Thus the binding of PI to the protein is specific; i.e., other phospholipids cannot compete effectively with PI for those phospholipid headgroup binding sites on the protein. These bound PI molecules play a functional role for glycophorin. Data suggest that the state of phosphorylation of the PI governs the affinity of the binding of band 4.1 to glycophorin. Band 4.1 also binds to the membrane skeleton via spectrin. Thus the state of phosphorylation of the specifically bound PI causes glycophorin to modulate the state of the membrane skeleton of the erythrocyte.[22, 23]

Experiments with mammalian Na^+,K^+-ATPase have demonstrated that cholesterol is required to activate this enzyme.[24] In the absence of cholesterol, this important plasma membrane enzyme exhibits little or no activity. Due to the structural specificity of the sterol activation of this enzyme (see Chapter 5), it has been suggested that cholesterol binds to this enzyme at a structurally specific site. To date, a physical measurement of such binding is not available.

Band 3, the anion transport protein of the human erythrocyte, binds cholesterol.[25, 26] Furthermore, cholesterol has been shown to inhibit the function of this protein.[27]

C. Bilayer Thickness

The long dimension (parallel to the bilayer normal) of the lipid bilayer is also important to lipid–protein interactions. The thickness of the hydrophobic portion of the lipid bilayer must match the thickness of the hydrophobic transmembrane portion of the protein. The transmembrane hydrophobic helices of the integral membrane proteins are of a defined length. If the hydrocarbon chains are too short, they will be unable to shield the full extent of the hydrophobic surface of the membrane protein from exposure to the aqueous phase. In cases of mismatch, some distortion at the lipid–protein interface must occur to satisfy the hydrophobic effect. This could lead to alterations in protein structure that might inhibit protein function. Such effects have been observed with the calcium pump protein from the sarcoplasmic reticulum. A bilayer width corresponding to a hydrocarbon chain with about 16–18 carbon atoms produced the maximum activity. Thicker or thinner bilayers apparently strained the protein conformation and inhibited the protein activity.[28]

IV. LIPID–PROTEIN INTERACTIONS AND LIPID STATE

When the lipid bilayer is in the gel state in reconstituted systems, many enzyme activities are inhibited. This result is not surprising, considering the rigid nature of the gel state (see above). This phenomenon has generated a number of studies of the effects of temperature on various membrane enzyme activities. Arrhenius plots have been extensively used to analyze the effects of temperature on these membrane functions. "Break points" have been noted in such plots by drawing straight lines through adjacent sets of data points and tabulating the temperatures at which the straight lines intersect. These break points have been ascribed to phase transitions or phase separations in the membranes. These are several cautionary notes one should take in evaluating such data. One is that the data sometime can be better represented by a curved line rather than as a collection of straight lines. A second potential problem is that sometimes independent measurements of the phase behavior of the membrane lipids are missing. In such cases, one does not know whether one is dealing with indirect effects due to the membrane lipids, or measuring thermal effects on the assay system or on the conformation of the protein involved. In one case, studies demonstrate that it is a protein structural transition, and not a membrane phase change, that leads to the "break" in the Arrhenius plot.[29] The best analysis of the present data would suggest that biological membranes function best when entirely in the liquid crystalline state, although they may retain function in certain membranes when as much as 90% of the membrane lipids are in the gel state.

V. RECONSTITUTION

Reconstitution is defined as follows. A membrane protein, with an assayable function, is isolated and purified from a biological membrane, and subsequently introduced into a membrane of defined lipid composition. If the function characteristic of this protein is regained in such a procedure, the protein is said to be reconstituted. Ideally this process results in a single protein in a defined lipid environment at a defined lipid/protein ratio, with which detailed biochemical and biophysical studies can be carried out.

It is fair to say that in the literature of membrane protein reconstitution, the reported results sometimes fail on one or more of three counts of this definition. First to adequately define the lipid/protein ratio, reconstituted membrane vesicles that are homogeneous with respect to density must be isolated. This isolation is usually done on a sucrose density gradient. Such an isolation procedure produces a product with a relatively homogeneous phospholipid/protein ratio. A second problem is the lack of definition of the lipid composition of the reconstituted system. What is put into the reconstituted system in terms of lipid content is not necessarily what one will get back in a reconstituted membrane. To define adequately the lipid composition of the reconstituted vesicles requires that the vesicles obtained from a sucrose density gradient be extracted with, for example, chloroform/methanol, and the lipid extract then run on a HPLC system for analysis of lipid composition. Finally, the most difficult problem in reconstitution is regaining full protein activity. To achieve this goal frequently requires numerous trial efforts, manipulating detergents, relative amounts of the various components, exogenous phospholipid composition, and detergent removal techniques until a combination that is successful is found.

The actual reconstitution procedures involve four steps. The first step is the isolation and purification of the protein of interest. The second is the recombination of the membrane protein with the desired lipid medium. The third step is the isolation of the reconstituted membrane on a sucrose density gradient. The fourth step is verifying the presence of function in the purified product.

A. Isolation of the Membrane Protein

Isolation of integral membrane proteins frequently involves the use of detergents to solubilize the membrane, followed by chromatography to isolate the particular membrane protein of interest from the other proteins of the source membrane. Detergent solubilization proceeds in at least three steps. The first step is the partitioning of the detergent into the biological membrane. Because of the hydrophobic nature of the interior of the membrane and the amphipathic nature of the detergent, at low concentrations the detergent preferentially partitions into the membrane. However, when the detergent molecules become as abundant as the lipid molecules in the membrane, the membrane becomes a less favorable site for the detergent. Then begins the second phase of detergent solubilization in which the detergent builds up in the aqueous phase as a monomer. There is a limitation to the solubility of the detergent in the aqueous phase. Eventually, at the CMC the detergent must begin to form micelles. As the detergent concentration is increased further, the population of micelles increases. During the third phase of the solubilization process, the membrane protein has the option to partition into the detergent micelle as well as the membrane. Thus one can expect mixed micelles of detergent and membrane protein to form, likely carrying considerable native membrane lipid also. (At high detergent concentrations, simpler micelles containing essentially only detergent and one membrane protein can form.) At this point, the membrane protein is in suspension in a detergent micelle, which allows for chromatography, including gel permeation chromatography, ion exchange chromatography, and affinity chromatography.[30] One must then determine when one reaches the point of a pure protein by suitable analytical techniques.

B. Recombination of Membrane Proteins with Exogenous Lipids

Two methods are commonly used to achieve the incorporation of solubilized membrane proteins into membranes of defined lipid composition. One is cosolubilization of the membrane protein with the lipids in detergent micelles. The detergent is then removed by dialysis or chromatography, or with absorbent beads as in the case of Triton or $C_{12}E_8$. This process of solubilization followed by detergent removal usually leads to a symmetrical distribution of membrane proteins, i.e., the membrane proteins face both ways in the reconstituted membrane (the proteins face one direction only in the native membrane).

The other common method for reconstitution of membrane proteins is insertion of the protein into performed membranes. In this procedure, detergent-solubilized membrane protein is diluted into a suspension of performed membranes such that the resulting detergent concentration is below the CMC of the detergent. In at least one case (rhodopsin) this leads to asymmetrical incorporation of the membrane protein; that is, all the rhodopsin molecules face the same way in the reconstituted membranes. This reconstituted produce more accurately reflects the topology of the native membrane than the first procedure described above.

C. Isolation and Characterization of the Reconstituted Membrane

The final important step is to isolate the reconstituted system on a sucrose gradient, hopefully as a single band. This isolated band is then analyzed and characterized. It needs to be analyzed for lipid/protein ratio and for lipid content. Finally the enzyme function is assayed. Only after all these steps can one really say that one has a reconstituted system for a particular membrane protein.

One of the interesting ways in which the concept of lipid–protein interactions enters the reconstitution literature is that proteins are not active in all detergents. Apparently only some detergents provide an environment sufficiently similar to the membrane to maintain activity. Furthermore, which detergent is best varies from protein to protein. Finally, some proteins denature irreversibly in detergents if completely delipidated. In such cases, some phospholipids must always be kept in the same micelle as the protein to maintain a protein conformation during purification that can be active upon reconstitution into a lipid bilayer.

VI. SUMMARY

Lipids and proteins are two of the major components of biological membranes. Three linked interactions involving these two components are important to the structure and function of membranes: lipid–lipid interactions, lipid–protein interactions, and protein–protein interactions. Any alteration in one of these pairwise interactions will necessarily lead to an alteration in another pairwise interaction. Therefore, for example, lipid–protein interactions can influence the extent of protein oligomerization in membranes.

On a structural level, the interactions between lipids and proteins can be divided into two classes. In one class, lipids encounter the protein surface through exchange among sites in the bilayer. These are termed interfacial lipids. A subpopulation of the interfacial lipids completely covers the surface of the integral membrane protein. These can be detected by techniques such as ESR of spin-labeled lipids.

In the other class, the lipids bind tightly and specifically to sites on the protein surface. These sites cannot be detected by any technique that requires the introduction of probes, such as ESR and ^{2}H NMR, because the probe molecules will not exchange readily with the tightly bound lipids. However, ^{31}P NMR offers a means in some cases to detect bound phospholipids because no probes need to be added for this measurement. These bound phospholipids take on the dynamics of the protein as a lipid-protein complex.

Interfacial lipids affect protein function through a response of the membrane protein to the bulk properties of the lipid bilayer. A mechanism for this regulation is offered by the hypothesis of free volume. If free volume elements can be recruited from the bilayer structure to the lipid-protein interface, then volume changes of the membrane protein within the membrane, required for conformational changes of the protein's functional cycle, can be accommodated. Unsaturation of lipid hydrocarbon chains can enhance provision of free volume elements and has been shown to support greater protein function. Cholesterol can reduce the prevalence of free volume elements within the bilayer and consequently inhibit some protein function.

Bound lipids affect protein function through allosteric mechanisms, binding to sites on the protein and altering the conformation and function of the membrane protein. In testing for this mechanism, it should be noted that it can be difficult to totally delipidate membrane proteins. If membrane proteins specifically and tightly bind particular phospholipid, like cardiolipin or PI, the influence of added lipids in a reconstitution experiment may not be apparent because crucial lipids may remain bound to the protein of interest throughout the reconstitution process.

Reconstitution involves the isolation of pure membrane proteins, recombination of the membrane protein with membrane lipids, and purification of a homogeneous product with full function for study.

REFERENCES

1. Wang, J. and A. Pullman, "Do helices in membranes prefer to form bundles or stay dispersed in the lipid phase?" *Biochim. Biophys. Acta* **1070** (1991): 493–496.
2. A. Y., Romans, P. L. Yeagle, S. E. O'Conner, and C. M. Grisham, "Interactions between glycophorin and phospholipids in recombined systems," *J. Supramol. Struct.* **10** (1979): 241–251.
3. Straume, M., and B. J. Litman, "Equilibrium and dynamic bilayer structural properties of unsaturated acyl chain phosphatidylcholine–cholesterol–rhodopsin recombinant vesicles and rod outer segment disk membranes as determined from higher order analysis of fluorescence anisotropy decay," *Biochemistry* **27** (1988): 7723–7733.
4. Epand, R. M., and R. F. Epand, "Lipid–peptide interactions," In *Structure of Biological Membranes* (P. L. Yeagle, ed.) (Boca Raton: CRC Press, 1992), 573–602.
5. Silvius, J. R., D. A. McMillen, N. D. Salley, P. C. Jost, and O. H. Griffin, "Competition between cholesterol and phosphatidycholine for the hydrophobic surface of sarcoplasmic reticulum Ca^{2+}-ATPase," *Biochemistry* **23** (1984): 538–547.
6. Peelen, S. J., J. C. Sanders, M. A. Hemminga, and D. Marsh, "Stoichiometry, selectivity, and exchange dynamics of lipid–protein interaction with bacteriophage M13 coat protein studied by spin label electron spin resonance," *Biochemistry* **31** (1992): 2670–2677.
7. Herbette, L., P. DeFoor, S. Fleischer, D. Pascolini, A. Scarpa, and J. K. Blasie, "The separate profile structures of the functional calcium pump protein and the phospholipid bilayer within isolated sarcoplasmic reticulum membranes determined by X-ray and neutron diffraction," *Biochim. Biophys. Acta* **817** (1985): 103–122.

8. Hymel, L., A. Maurer, C. Berenski, C. Y. Jung, and S. Fleischer, "Target size of calcium pump protein from skeletal muscle sarcoplasmic reticulum," *J. Biol. Chem.* **259** (1984): 4890–4895.

9. Esmann, M., and D. Marsh, "Spin-label studies on the origin of the specificity of lipid-protein interactions in NaKATPase membranes from squalus acanthias," *Biochemistry* **24** (1985): 3572–3578.

10. Watts, A., I. D. Volovski, and D. Marsh, "Rhodopsin–lipid associations in bovine rod outer segment membranes: Identification of immobilized lipid by spin-labels," *Biochemistry* **18** (1979): 5006–5013.

11. Horvath, L. I., M. Drees, K. Beyer, M. Klingenberg, and D. Marsh, "Lipid–protein interactions in ADP–ATP Carrier/egg phosphatidylcholine recombinants studied by spin-label ESR spectroscopy," *Biochemistry* **29** (1990): 10664–10669.

12. Selinsky, B. S., and P. L. Yeagle, "Effects of potassium on lipid–protein interactions in light sarcoplasmics reticulum," *Biochemistry* **29** (1990): 415–421.

13. Selinsky, B. S., and P. L. Yeagle, "Phospholipid exchange between restricted and non-restricted domains in sarcoplasmic reticulum," *Biochim. Biophys. Acta* **813** (1985): 33–40.

14. Yeagle, P. L., and D. Kelsey, "Phosphorus NMR studies of lipid–protein interactions: Human erythrocyte glycophorin and phospholipids," *Biochemistry* **28** (1989): 2210–2215.

15. Hidalgo, C., D. A. Petrucci, and C. Vergara, "Uncoupling of calcium transport in sarcoplasmic reticulum as a result of labeling lipid amino groups and inhibition of Ca-ATPase activity by modification of lysine residues of the Ca-ATPase polypeptide," *J. Biol. Chem.* **257** (1982): 208–216.

16. Cheng, K., J. R. Lepock, S. W. Hui, and P. L. Yeagle, "The role of cholesterol in the activity of reconstituted Ca-ATPase vesicles containing unsaturated phosphatidylehthanolamine," *J. Biol. Chem.* **261** (1986): 5081–5087.

17. Mitchell, D., M. Straume, J. L. Miller, and B. J. Litman, "Modulation of metarhodopsin formation by cholesterol-induced ordering of bilayers," *Biochemistry* **29** (1990): 9143–9149.

18. Boesze-Battaglia, K., and A. Albert, "Cholesterol modulation of photoreceptor function in bovine rod outer segments," *J. Biol. Chem.* **265** (1990): 20727–20730.

19. Isaacson, Y. A., *et al.*, "The structural specificity of lecithin for activation of purified D-β-hydroxybutyrate apodehydrogenase," *J. Biol. Chem.* **254** (1979): 117–126.

20. Beyer, K., and M. Klingenberg, "ADP/ATP carrier protein from beef heart mitochondria has high amounts of tightly bound cardiolipin, as revealed by ^{31}P NMR," *Biochemistry* **24** (1985): 3821–3826.

21. Buckley, J. T., "Coisolation of glycophorin A and polyphosphoinositides from human erythrocyte membranes," *Can. J. Biochem.* **56** (1978): 349–351.

22. Anderson, R. A., and V. T. Marchesi, "Associations between glycophorin and protein 4.1 are modulated by polyphosphoinositides: A mechanism for membrane skeleton regulation," *Nature (London)* **318** (1985): 295–298.

23. Anderson, R. A., and R. E. Lovrien, "Glycophorin is linked by band 4.1 protein to the human erythrocyte membrane skeleton," *Nature (London)* **307** (1984): 655–658.

24. Yeagle, P. L., D. Rice, and J. Young, "Effects of cholesterol on (Na,K)-ATPase ATP hydrolyzing activity in bovine kidney," *Biochemistry* **27** (1988): 6449–6452.

25. Klappauf, E., and D. Schubert, "Band 3-protein from human erythrocyte membrane strongly interacts with cholesterols," *FEBS Lett.* **80** (1977): 423–425.

26. Schubert, D., and K. Boss, "Band 3 protein–cholesterol interactions in erythrocyte membranes," *FEBS Lett.* **150** (1982): 4–8.

27. Grunze, M., B. Forst, and B. Deuticke, "Duel effect of membrane cholesterol on simple and mediated transport process in human erythrocytes," *Biochim. Biophys. Acta* **600** (1980): 860–868.
28. Caffrey, M., and F. W. Feigneson, "Fluorescence quenching in model membranes. 3. Relationship between calcium adenosinetriphosphatase enzyme activity and the affinity of the protein for phosphatidycholines with different acyl chain characteristics," *Biochemistry* **20** (1981): 1949–1961.
29. Dean, W. L., and C. P. Suarez, "Interactions between sarcoplasmic reticulum calcium adenosinetriphosphatase and nonionic detergents," *Biochemistry* **20** (1981): 1743–1747.
30. Stubbs, G. W., H. G. Smith, and B. J. Litman, "Alkyl glucosides as effective solubilizing agents for bovine rhodopsin. A comparison with several commonly used detergents," *Biochim. Biophys. Acta* **426** (1976): 46–56.

8

Transport

One of the natural consequences of the structure of the membranes of cells is that a cell membrane separates two aqueous domains and controls communication between them. The hydrophobic barrier of the membrane largely inhibits the passage through the lipid bilayer of charged or highly polar species. Each domain exhibits a different composition. This results in a difference in chemical potential across the membrane, inducing a stress on the membrane. Such differences in composition may be crucial to cellular function.

A difference in chemical potential for a solute between a compartment on one side of the membrane and a compartment on the opposite side of the membrane will lead to pressure for the diffusion of the solute down the gradient in chemical potential. This diffusion corresponds to transport of solute across the membrane. Transport offers a means to reduce the transmembrane difference in chemical potential. This permits the system to approach equilibrium.

In the biology of cells, equilibrium frequently means death because gradients of solutes are often required for a cell to function. For example, sodium concentration is generally low inside a cell and high outside, as is calcium. Equalizing the concentrations of these cations outside and inside the cell (which is the effect of eliminating the difference of the chemical potential of the solute across the plasma membrane) would disrupt crucial processes inside the cell and lead to the demise of that cell. To prevent this catastrophe, the cell engages in an energy-dependent process to transport these solutes out of the cell.

There are two general classes of membrane transport. The first is passive, or in some cases facilitated, diffusion in which the solute moves from a region of relatively high chemical potential to a region of relatively low chemical potential (toward thermodynamic equilibrium). No cellular energy, such as that in the form of ATP, is required for this diffusion.

The energy for the transport process comes from the concentration (or activity) gradient, from one side of the membrane to the other.

The second form of transport is active transport. Here "high energy" compounds, such as ATP, are hydrolyzed. The free energy released by this hydrolysis is employed to pump solutes across the membrane, usually against a concentration gradient. Alternatively, the energy stored in the transmembrane gradient of another species can be used to energize active transport.

I. PASSIVE DIFFUSION

The simplest form of transport, passive diffusion, refers to the movement of a species by random processes through the lipid bilayer portion of a membrane, independent of any metabolic energy.[1] Net flux of the transported species occurs only when there is a difference in chemical potential of the species on one side of the membrane compared to the other side. Usually this chemical potential difference is the result of a difference in concentration (or activity) of the species on one side versus the other. However, if the transported species is charged, the influence of a transmembrane electrical potential on the diffusion of the charged species can be considerable (Fig. 8.1).

The difference in chemical potential is directly related to the difference in concentration or the difference in activity, of the species on the two sides of the membrane. This is expressed in

$$\Delta \mu_{i \to j} \sim \ln\left(\frac{aj}{ai}\right)$$

where ai and aj are the activities of the species in question on either side of the membrane. The greater the difference in activity of the species on

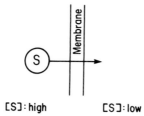

Fig. 8.1. Schematic representation of passive diffusion.

either side of the membrane, the greater the difference in chemical potential of the species from one side of the membrane to the other. Therefore, the greater the activity gradient across the membrane, the greater the driving force for passive diffusion of the species in question across the membrane.

The flux of an uncharged species across a membrane is responsive to this activity gradient. The flux, or the number of molecules passing through the membrane per unit time, can be approximated under these conditions as the product of the activity gradient and the diffusion coefficient,

$$J_x = -DK\frac{da}{dx}$$

where J_x is the flux, D is the diffusion coefficient, and K is the partition coefficient of the diffusing species between membrane and water. As mentioned above, there can be an influence of a transmembrane electrical potential on the flux if the species is charged. This can be represented as

$$J_x = -DK\left(\frac{da}{dx} + \frac{aqF}{RT}\frac{d\chi}{dx}\right)$$

where q is the ionic valence, F is the Faraday constant, and χ is the electrical potential.

The activity gradient refers to the activities of the polar species in the aqueous domains on either side of the membrane. Nonpolar species can be expected to partition largely into the hydrophobic interior of the membrane, due to the hydrophobic effect, and to exhibit a negligible concentration in solution.

As a polar species transits a membrane, the most obvious barrier to its transport is the hydrophobic interior of the membrane. It is energetically unfavorable for the polar solute to inhabit the hydrophobic interior due to the hydrophobic effect. It has been estimated that the energy cost of introducing a small ion into the hydrophobic interior of the membrane is on the order of tens of kilocalories per mole.

How does a polar solute transit the hydrophobic membrane interior? One likely mechanism involves the kinks that hydrocarbon chains form under the influence of thermal motion. These kinks travel up and down the chains. These kinks correspond to defects in the bilayer structure that can be transiently occupied by solutes. Thus the kinks can provide a pathway of small volume elements through which the solute can cross the membrane. This mechanism provides a good explanation for the rapid transit of water across a membrane, for example.

One consequence of this mechanism is that the larger the solute, the less favorable the fit between the solute and the volume elements just defined. Thus the larger the polar species, the less effective such diffusion will be. For polar species of the same family, this expectation is fulfilled. For example, the permeability of glucose is less than the permeability of water because of the greater size of the glucose. Species carrying a formal charge are less capable of entering the bilayer than polar, uncharged species (because of the hydrophobic effect), and their permeability is consequently less. For example, the permeability of a lipid bilayer to sodium ion is less than the permeability of that same lipid bilayer to glucose.

The other consequence of this mechanism for polar solute transport across a lipid bilayer is that anything that enhances the occurrence of defects in the bilayer will enhance passive diffusion of polar solutes across the membrane. A lipid bilayer at the phase transition temperature experiences an increase in the incidence of defects in its structure (at the interface between solid and fluid phase) and a corresponding increase in permeability. An increase in unsaturation also leads to an increase in the incidence of defects as well as an increase in membrane permeability. Even the reconstitution of membrane proteins into a lipid bilayer can increase the membrane permeability quite apart from any channels the protein might form by itself. Such an enhanced permeability may be due to packing defects at the lipid–protein interface due to transient mismatches between the rough protein surface and the lipid hydrocarbon chains.

Transit of the hydrophobic interior of the lipid bilayer is not the only barrier to passive diffusion through a membrane. Before encountering the membrane interior, a solute must get through an interfacial region with considerably different properties than the bulk solution. This has been referred to as an unstirred layer. The surface of the membrane is defined by the interface between the lipids and the aqueous medium surrounding the membrane. A number of water molecules are ordered next to the membrane surface, and some of these water molecules are bound to the lipid headgroups. Furthermore, the charges on the lipid headgroups create a double layer of oppositely charged species. The membrane surface generally has a negative charge, and opposing that is a layer of positive charge consisting of counterions from solution, such as sodium ions (Fig. 8.2). Furthermore there is a hydrodynamic effect due to the membrane surface. The water molecules near the surface, even those that are not bound, tend to experience a drag opposing their movement. Therefore, the surface of the membrane is a partially ordered array of solvent and solute molecules that extends for some distance from the surface. This rather special region presents a barrier through which a solute molecule must pass before encountering the membrane lipid bilayer.

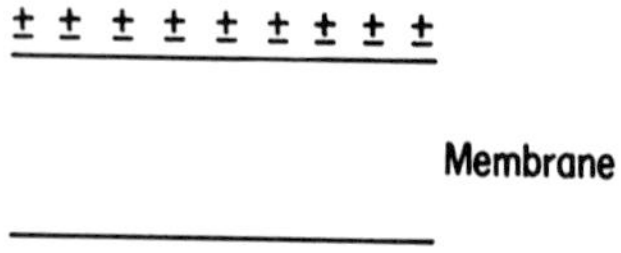

<pre>± ± ± ± ± ± ± ± ±</pre>

Membrane

Fig. 8.2. Schematic representation of the organization of charges on the membrane surface.

One other barrier to transport across the membrane is presented by the hydration of the species to be transported. Generally this hydration shell must be stripped away before the solute can enter the membrane lipid bilayer. Therefore, the energy of dehydration of the solute must be considered a barrier to the transport of that solute. Despite these barriers, solutes do manage to get across membranes. Table 3.3 summarizes some of the permeability properties of lipid bilayers.

So far the discussion has focused on passive diffusion of polar, nonionizable solutes across lipid bilayers. Ionizable solutes sometimes have access to another important mechanism for transport across a lipid bilayer. For example, carboxylic acids can exist in two distinctly different forms. In one form, the carboxyl is ionized and the molecule carries a net negative charge. This charged species does not effectively partition into the hydrophobic interior of the membrane. However, the protonated molecule is reasonably nonpolar (except for the carboxyl, the molecule is frequently essentially hydrocarbon), and its partition coefficient with respect to the hydrophobic interior is more favorable than that for the ionized species.

Therefore one mechanism for transport of ionizable organic acids across a membrane exploits the ability of the molecule to exist in two forms. The portion of the molecules that is in the protonated form (governed by the pK_a of the acid and the pH of the solution) readily diffuses through the membrane. Even if there is only a very small percentage in the protonated form, depletion of that form by passage through the membrane is compensated by protonation of more solute molecules on the side with the greater concentration. Thus a net movement of solute through the membrane is achieved by funneling solute through the membrane-permeable form. The more the pH of the medium and pK_a of the titratable group on the solute favor the permeable form, the greater will be the rate of transport of the species in question (activity gradients across the membrane being equal).

II. FACILITATED DIFFUSION

One of the great barriers to passage of polar solutes across the membrane is the incompatibility of polar molecules with the hydrophobic interior of the membrane, governed by the hydrophobic effect. What if a vehicle could be designed to mask the hydrophilic nature of the polar species, thereby making it compatible with the membrane interior? This vehicle would considerably lower the energy barrier to transport across the membrane and likely increase the rate of transport. Such an effect is referred to as facilitated diffusion.

Many examples of facilitated diffusion fall into two general categories. One category is characterized by the term carrier and the other category is characterized by the term channel.

A. Carriers

The term carrier describes a species that relatively specifically binds a solute to be transported and renders that solute less polar. It does so by structurally presenting a compatible polar interior as a binding site for the solute and exhibiting a hydrophobic exterior with which to transit the hydrophobic portion of the membrane. As the term carrier implies, the complex of the polar species and the carrier may diffuse as a unit across the membrane. This process therefore implies association of the carrier and solute on one side of the membrane to form a complex, followed by diffusion and dissociation of the complex on the other side of the membrane.

The carrier-mediated process is saturable since the carrier binds the solute. Because it involves binding of the solute, this process can exhibit specificity for solute structure. It can also, in some cases, be inhibitable by structural analogs to the transported solutes.

Because carrier-mediated facilitated diffusion involves transmembrane diffusion of the complex, the state of the membrane can be expected to affect the transport process. This process is inhibited by a transition of the phospholipid bilayer to a gel state through which diffusion is difficult.[2]

As in passive diffusion, carrier-mediated transport can be driven either by a gradient in concentration of the chemical species from one side of the membrane to the other or, if the species is charged, by an electrical potential gradient across the membrane. In all cases the transport is down the gradient or in the direction of thermodynamic equilibrium. Carrier-mediated transport is not capable of moving solutes against a concentration gradient.

One of the most characteristic features of carrier-mediated transport is the flux. Turnover numbers are typically in the range 100–3000 sec^{-1}. This rate is determined by the transmembrane diffusion rates of the carrier and the off rate of the transported species from the carrier. If the flux corresponds to a turnover number much greater than that, carrier-mediated transport is not likely. Instead, one must suspect the role of a channel as will be discussed shortly.

1. VALINOMYCIN

Antibiotics provide examples of carrier-mediated transport. Valinomycin is such an antibiotic. Valinomycin is an ionophore that makes membranes permeable to cations, particularly potassium.[3] Ionophores such as this are sometimes found to be synthesized by bacteria and fungi. The ability of these ionophores to make membranes permeable to important species can have profound effects on biological function.

Valinomycin is a cyclic dodecapeptide, although not all its substituents are, strictly speaking, amino acids. Its structure appears in Fig. 8.3. A curious feature of its structure makes this cyclic molecule particularly suitable for its transport function. All the carbonyls of the amide and ester bonds in the molecule face the inside of the cyclic structure. The carbonyl

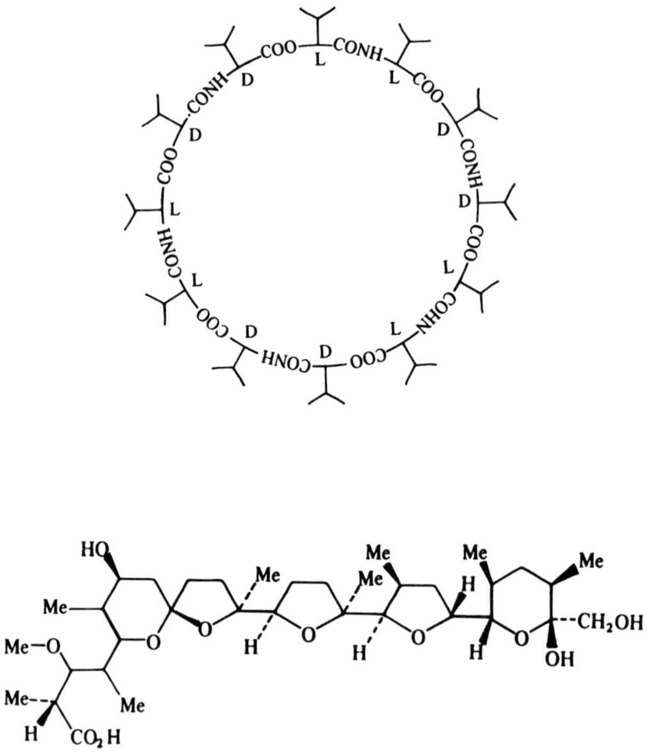

Fig. 8.3. Chemical structures of valinomycin (top) and monensin (bottom).

is a polar structure. In fact, because of the electronegativity difference between the oxygen and the carbon, the oxygen carries a partial negative charge. Thus these carbonyls are well suited to function as ligands for a cation.

The interior of valinomycin is where the cation, usually K^+, binds.[4] The size of the ring made by all the carbonyls nearly perfectly matches the ionic radius of the K^+. Thus there is a good fit and the dehydrated K^+ ion binds effectively and specifically. Normally, valinomycin exhibits specificity for K^+ over Na^+ of more than an order of magnitude in relative affinity. The more favored binding of K^+ can be understood by examining the ionic radii of these two cations. The ionic radius of Na^+ is roughly 30% smaller than that for K^+. Therefore, the cation binding site in valinomycin is too large for an effective fit for Na^+. In fact in the crystal structure of the sodium complex with valinomycin, the sodium ion was not found in the potassium binding site.[5] It is from this fact that valinomycin carrier-mediated transport attains its specificity for K^+.

The mechanism for transport of K^+ by valinomycin (Fig. 8.4) consists of three steps. The first is association of the cation and the ionophore. The binding of K^+ triggers a conformational change in the ionophore. The ring closes in and arranges itself such that the methyl groups dominate the exterior. This creates a hydrophobic exterior, with the cation tucked

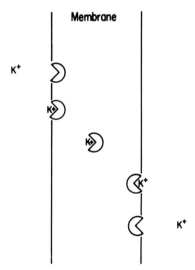

Fig. 8.4. Schematic representation of potassium transport by valinomycin by a carrier mechanism.

into a polar pocket in the interior. This conformation allows the second step of the transport to proceed, which is translocation across the membrane. Complexation of the cation with the ionophore makes effectively a much less polar species out of the cation. The valinomycin–K^+ complex is more soluble in the hydrophobic interior of the membrane than K^+. The increase in solubility permits an increase in cation permeability for the membrane. Translocation of the complex leads to the last step of the process, which is dissociation of the complex on the side of the membrane trans to the site of complex formation. Unless the facilitated diffusion of potassium is dominated by a transmembrane electrical potential, there is also a counter-ion requirement in this process.

2. MONENSIN

A process similar to that for valinomycin obtains for monensin. The structure of this ionophore appears in Fig. 8.3. It is a linear molecule, in contrast to the cyclic structure observed for valinomycin. However, on binding a cation, this molecule folds into a circular structure with much the same properties as valinomycin. It wraps its hydrophilic carbonyl ligands around the cation and presents methyl groups to the outside. This once again creates a hydrophobic surface surrounding a cation, increasing its solubility in the hydrophobic interior of the membrane. The consequence, as in the case of valinomycin, is an increased permeability to cations.

3. CARRIERS IN CELL MEMBRANES?

In general, one would not expect membrane proteins to function as carriers as defined here. Proteins usually contain highly polar amino acids, as well as hydrophobic amino acids. Translocation of a protein–cation complex would require submersion of the polar amino acids in the hydrophobic interior of the membrane. Thermodynamically, this is even less favorable than introduction of the bare cation into the hydrophobic membrane interior.

An interesting experiment was performed to rule out a rotating carrier model for transport of calcium by the sarcoplasmic reticulum.[6] Antibodies to the calcium pump were obtained that bound to the protein but did not significantly inhibit calcium transport. An antibody is too large to transport through the membrane. Therefore this experiment ruled out a rotating carrier model for this transport process.

B. Channels Formed by Antibiotics

Channels have several characteristics in common with carriers. Transport across membranes by channels is driven by a difference in chemical potential between the two sides of the membrane. Transport is therefore usually down a concentration gradient. A channel can also exhibit specificity with respect to size, for example, as is the case for carriers.

However, in other characteristics a channel is different from a carrier. A channel forms, as its name implies, a hole through the membrane of suitable size and polarity to transport solutes. However, transport does not occur by diffusion of an ionophore–cation complex across the membrane. Thus channel-mediated transport is less sensitive to the state of the membrane than is carrier-mediated transport. More than one monomer of the ionophore may be involved in a channel formation.

The key functional difference between a carrier and a channel is that channels are capable of supporting much greater fluxes of the transported species than are carriers.[7]

1. GRAMICIDIN

The first example of a channel to be considered is the channel formed by gramicidin A. The crystal structure of gramicidin appears in Fig. 8.5.[9] In a hydrophobic media, gramicidin A forms a helical structure.[8] It is estimated that a little more than six residues are required to form a turn of this helix. This leaves a pore in the middle of the helix, in contrast to the α helix. This pore is about 4 Å in diameter. Note that all the side chains of the constituents of gramicidin are hydrophobic. The helical conformation will put those hydrophobic structures on the outside of the

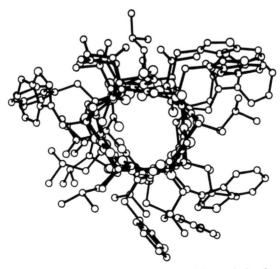

Fig. 8.5. Crystal structure of gramicidin. Reprinted with permission from D. A. Langs, *Science* **241** (1988): 188.

helix, thereby making the hydrophobic helix compatible with the hydrophobic interior of the membrane.

The kinetics of the transport mediated by gramicidin A are consistent with the formation of a dimer in the membrane. Model building shows that a dimer of gramicidin in the helical conformation described above would form a channel 25 to 30 Å in length. This channel is sufficiently long to span completely the hydrophobic region of a typical phospholipid bilayer. Thus channel formation by gramicidin would appear to result from aggregation of two gramicidin monomers in the membrane that form helices in tandem, spanning the bilayer. The gramicidin dimer transiently creates a pore of particular size in the membrane, and, as long as that pore is in existence, any concentration gradient of ions that are capable of entering the gramicidin pore will be dissipated. The selectivity of ions that will be transported by a gramicidin channel is $H^+ > NH_4^+ > Cs^+ > Rb^+ > K^+ > Na^+ > Li^+$. This selectivity is determined by both size and hydration energies. The ion cannot be too large for the size of the pore, or it will not enter that pore and be transported. Furthermore, to enter the pore, the ion must be stripped of its bound water molecules. The ease with which this dehydration event can take place is therefore a determining factor in transport by gramicidin. Li^+, the most strongly hydrated of the series, has the most difficulty in entering the channel. Protons head the series because they are so small. Conduction across the channel can be envisioned as sequential binding of the cation to a series of "sites" along the length of the gramicidin channel.

Individual channel formation by gramicidin is controlled by random collision through lateral diffusion of gramicidin molecules in the plane of the membrane. Consequently the opening and closing of gramicidin channels in the membrane are random.

The conduction, or transport, of ions by gramicidin is a quantized event. A single channel is either open or closed. When it is open, it contributes a fixed increment to the overall membrane conductance of ions. When it is closed, it makes no contribution to the conductance. Therefore, electrical measurements of the conductance of ions across a membrane by gramicidin show discrete steps in the conductance, as shown in Fig. 8.6. The conductance is a fluctuating phenomenon for which a noise analysis is appropriate. The total conductance of the membrane, as determined by gramicidin, is dependent on two factors: the single channel conductance and the length of time a typical channel is open. This latter factor is in turn dependent on the temperature, the membrane composition and resulting thickness, and the concentration of the gramicidin in the membrane.

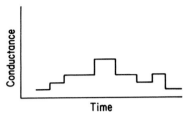

Fig. 8.6. Gated nature of the conductance of a set of channels randomly opening and closing.

2. AMPHOTERICIN

Amphotericin is polyene antibiotic that can increase the permeability of membranes. The structure of this molecule appears in Fig. 8.7. Amphotericin is capable of forming channels in membranes, but the channel-forming event is more complex than in the case of gramicidin. These channels are voltage gated. Substantially more than two molecules are required to form a channel. Model building shows that six or eight amphotericin molecules could aggregate to form a channel 7 Å in diameter and about 15–20 Å long. Two of these complexes end to end could then form a channel across the membrane. The structure of amphotericin shows that it could form an amphipathic structure, that is, a structure in which one side of the molecule will be hydrophobic and the other side will be hydrophilic. Orientation of the hydrophobic side of the amphotericin to the outside of the channel (i.e., facing the hydrophobic hydrocarbon portion of the membrane) leaves a hydrophilic interior to the channel. This structure is then well organized for transport of ions across membranes.

An interesting additional feature of this molecule is that it binds preferentially to membranes containing sterol, such as cholesterol. Amphotericin does not increase the conductance of membranes in the absence of sterols. The amphotericin–sterol interaction may be a preliminary interaction necessary to the formation of channels.

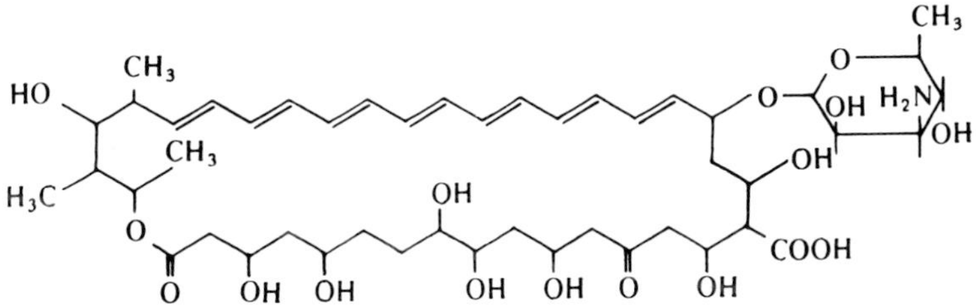

Fig. 8.7. Structure of amphotericin.

C. Protein Channels

Gram-negative bacteria have three layers determining the outside of the cell. The innermost layer is the plasma membrane of the cell, which resembles the inner membrane of mitochondria from eukaryotic cells. The plasma membrane exhibits the permeability barrier characteristic, in general, of most biological membranes. The next layer is the cell wall, a complex, highly crosslinked peptidoglycan structure, held together by covalent bonds. The cell wall does not present a serious permeability barrier to small solutes, in contrast to the plasma membrane. The outer membrane is constructed of lipids and proteins without the extensive covalent crosslinking of the cell wall. In this sense, the outer membrane more resembles the structure of the inner membrane than the cell wall. However, unlike the inner membrane, the outer membrane is permeable to a variety of small solutes. The outer membrane appears to function as a molecular sieve, letting through small hydrophilic solutes, but excluding large ones. This permeability characteristic is conferred on the outer membrane by a class of proteins called porins (these have also been referred to as matrix proteins).

The porins are a class of transmembrane proteins that form channels in the outer membranes of gram-negative bacteria.[10] They do so by aggregating into multimers in the membrane in such a way to leave an aqueous transmembrane channel in the middle of the aggregate. These channels, whose size is defined by the porin proteins forming the channel, provide the permeability characteristics associated with the outer membrane.

The size of the channels formed is different depending upon the particular bacterial porin. In the case of *Escherichia coli*, hydrophilic molecules as large as molecular weight 600 can penetrate the outer membrane. In *Salmonella typhimurium*, the exclusion limit is similar, about 700 Da. For *Pseudomonas aeruginosa*, molecules of molecular weight 6000 can penetrate the porin channels.

Another way to measure the size of the channels is to measure the conductance supported by ions. In this way, the diameter of the porin channels from the above bacterial species can be estimated. Table 8.1 presents the results. As you can see, the diameter so estimated for the *E. coli* and *S. typhimurium* species are similar, in line with their similar molecular weight cutoffs for solute permeability. Also, as expected, the diameter of the porin channels for *P. aeruginosa* is substantially larger, reflecting the much larger molecular weight cutoff for the passage of hydrophilic solutes through its porin channels.

All indications are that these porin channels are passive aqueous channels in the outer membrane. No energy is required for transport. Therefore

TABLE 8.1
Comparison of the Pores Formed by
Porins of Gram-Negative Bacteria[a]

Pore	d (nm)
E. coli	
OmpF (Ia)	1.3
OmpC (Ib)	1.3
PhoE (Ic)	1.2
LamB	1.5
S. typhimurium	
M 38,000	1.4
M 39,000	1.4
M 40,000	1.4
P. aeruginosa	
F	2.2

[a] From R. Benz, *Curr. Topics Membr. Transp.* **21** (1984): 199–219.

passage of solutes is strictly according to their respective chemical potential differences across the outer membrane. Furthermore, little selectivity is noted, other than according to size, as determined by the diameter of the channel. One exception to this is a channel for anions in the outer membrane for *P. aeruginosa,* in which anions larger than chloride are inhibited from transport.

The conductance of these channels formed by porin behave in a manner similar to that of the antibiotics described above. That is, an individual channel is either open or closed, and when it is open it exhibits a fixed conductance. The total conductance exhibited by the membrane is then determined by the sum of the number of channels open at any given instant of time. Since the number open will fluctuate, the conductance will change in stepwise manner, as channels open or close.

The secondary structures of the porins show an interesting deviation from the pattern observed for other integral membrane proteins. For a polypeptide chain that penetrates the hydrophobic interior of a membrane, an important requirement is that the polarity of the amide bond be masked. The formation of α-helices in which all the amide carbonyls are hydrogen-bonded to amide nitrogens reduces the effective polarity of those structures as sensed by the environment. Another structure well known from water soluble proteins is the β sheet. In the β sheet, which comes in several subcategories, the amide bonds are involved in hydrogen-bonding to adjacent strands of polypeptide chain, rather than to residues three or four residues away in the primary sequence. The structures formed are more like sheets than the cylindrical structure of the α helix.

The β-sheet structure may be important for the transmembrane porins rather than the α helix.[11] This structure may be particularly suited to the formation of a channel by a membrane protein or by oligomers of a membrane protein.

D. Band 3

An example of an integral membrane protein facilitating diffusion of specific species is band 3, the anion transport system of the erythrocyte.[12] This transmembrane protein, one of the most abundant proteins in the erythrocyte membrane, has a molecular weight of about 95,000. It is glycosylated to a limited extent. Band 3 appears to exhibit some poly-dispersity when run on SDS–polyacrylamide gels, since it appears as a relatively broad band on Coomassie-stained gels of erythrocyte membranes.

Functionally, this protein is responsible for the exchange of bicarbonate and chloride across the erythrocyte plasma membrane.[13] Carbon dioxide is passed from oxidizing tissues to the blood to be removed from the body. It is essential that it be removed, but carbon dioxide has a limited solubility in aqueous media. The erythrocyte contains a means to solve this problem. Carbonate dehydratase in the red cell catalyzes the rapid conversion of carbon dioxide to bicarbonate ion inside the red cell. Thus carbon dioxide can diffuse inside the red cell and be rapidly converted to the highly water-soluble bicarbonate ion. Then to prevent the excessive accumulation of bicarbonate ion on the inside of the erythrocyte, bicarbonate is transported outside the cell in exchange for an abundant anion, the chloride ion. This whole process is fast, to allow for rapid uptake by blood of the carbon dioxide produced by the tissues. The reverse process must then occur rapidly in the lungs to allow for the expulsion of carbon dioxide as a waste product from the organism.

The transport process facilitated by band 3 is electrically neutral. One negatively charged bicarbonate ion is exchanged for one negatively charged chloride ion. No energy is required for the process except the response of the transport system to a chemical potential gradient of, for example, bicarbonate.

Anion transport by band 3 uses discrete sites on the protein. It would appear that there are specific anion binding sites for which other anions can compete. Because of the existence of discrete sites, anion transport exhibits the phenomenon of saturation. The reaction mechanism of anion transport is consistent with a classic "ping-pong" mechanism. Transport is specifically inhibited most characteristically by 4,4'-diisothiocyano-2,2'-stilbenesulfonic acid (DIDS).[14] This inhibitor can bind tightly and irreversibly to band 3; DIDS binding has been used in the identification of band 3 on SDS–polyacrylamide gel electrophoresis as the membrane protein responsible for anion transport.

Much of the structure of band 3 in the membrane has been worked out through extensive proteolysis and labeling studies as well as sequence analysis.[15, 16] Apparently 12–14 segments of the protein traverse the membrane in an α-helical conformation[17] The carboxyl terminus and the amino terminus are found on the cytoplasmic face of the plasma membrane. Band 3 exists as a dimer of dimers, or tetramer, in the membrane. Band 3 forms a hydrophilic channel in the membrane for passage of anions. It may be that the transmembrane α-helices are organized to form this channel. Amphipathic helices, in which one side is hydrophobic to interact with the lipid hydrocarbon chains, and the other side is hydrophilic, would be an excellent means to create an aqueous channel. It might also be that the dimer of the protein is required to form the channel. This issue has not yet been clarified.

Band 3 is attached to the membrane skeleton, which is built with spectrin and actin.[18] The attachment is via band 2.1, or ankyrin, which has binding sites for band 3 and for spectrin. A fraction of band 3 is immobile, presumably the fraction that is involved in binding to ankyrin and the membrane skeleton.[19]

The activity of band 3 is affected by its lipid environment in the membrane. For example, cholesterol in the membrane will inhibit the ability of band 3 to transport anions. It has been suggested that this may be due to the binding of cholesterol to band 3 in the membrane.[20]

E. Chloride Channel

Another example of an anion channel is the chloride channel. A chloride channel has been implicated in the genetic disease cystic fibrosis.[21] Unlike the anion channel above, it appears that the choride channel of cystic fibrosis is not sensitive to DIDS. The polypeptide is predicted to be 168 kDa from it amino acid sequence deduced from the nucleotide sequence of the gene.[22] A complex transmembrane domain is suggested by analysis of the sequence, as in the case of band 3, with as many as 12 transmembrane α-helices predicted. Expression of the gene product in Chinese hamster ovary (CHO) cells led to chloride conductance, which supported the contention that the defect in cystic fibrosis involves, at least in part, a chloride transport protein.[23]

E. Glucose Transporter

Glucose uptake is one of the most important transport functions of the plasma membrane of many cells because glucose is a primary energy source through the reactions of glycolysis. Glucose transport is mediated by a membrane protein in the plasma membrane. The process exhibits properties characteristic of facilitated diffusion. Glucose transport can be inhibited by structural analogs, is stereospecific for D-glucose, and is saturable.[24]

This transport protein is often referred to as the glucose carrier. However, as indicated earlier, proteins cannot be expected to operate as a carrier in the same manner as valinomycin, shuttling back and forth across the membrane. In fact, since the glucose carrier in the red cell is of molecular weight about 55,000, it is much larger than the cross section of the membrane it inhabits.

Glucose transprt can take place totally by facilitated diffusion, in which case the free glucose concentrations on both sides of the membrane are at equilibrium with each other. This system is operative for the red cell and other nonepithelial tissues. (For epithelial tissues, glucose uptake can be active by coupling with the sodium gradient, as discussed at the end of this chapter. The Na^+, K^+-ATPase creates a sodium gradient, from the outside to the inside of the cell. Coupling glucose transport with the inward flux of sodium therefore allows glucose to be accumulated inside the cell against a concentration gradient. The protein involved in this process is distinct from the glucose transporter protein involved in facilitated diffusion.)

Since much is known about the glucose transporter of human erythrocytes, that system will be used as an example. However, the protein involved in this system appears to have much in common with the protein responsible for glucose transport in some other cell types.

The human erythrocyte glucose transporter shows up as band 4.5 on SDS–polyacrylamide gel electrophoresis, stained with Coommassie, of the erythrocyte plasma membrane. It is a prominent band on the gel and is a major polypeptide of the red cell membrane. Approximately 500,000 copies are found in the cell. It is a glycoprotein. Sequence analysis indicates a protein with as many as 12 transmembrane segments.[25] Evidence suggests that these transmembrane segments are α helices.[26] The effective molecular weight of the active unit for glucose transport in the red cell membrane was determined by radiation inactivation (target size analysis) to be 185,000. This may imply that three or four monomers of the transporter are aggregated into a unit in the native membrane.[27] It is not clear whether the protein must form an oligomer for full activity, or whether the monomer can be fully active.

A specific inhibitor of glucose transport by this transport protein, cytocholasin B, has proven to be helpful in exploring the properties of this system. Binding of this molecule to the glucose transport protein inhibits the transport of glucose. Binding appears to be specific for the glucose transport protein. There is one cytocholasin B binding site per monomer of the protein and the site is found on the cytoplasmic surface of the plasma membrane.

The mechanism of transport by this system involves the alternate exposure of the glucose binding site to the extracellular surface of the plasma membrane of the cell and to the cytoplasmic surface of the plasma membrane. As in the case of virtually all the transport systems to be explored here, the relationship between the detailed molecular mechanism of transport and the structure, and perhaps conformational changes of the transport protein, is unknown. Much remains to be learned in this area.

Investigation of the regulation of glucose transport by insulin has revealed an interesting mechanism for control of glucose transport.[28] In adipose tissue, for example, stimulation by insulin produces a two- to fivefold increase in glucose uptake. This stimulation occurs in a matter of seconds and results in an increase in V_{max}, without an increase in K_m associated with the transport. One mechanism suggested by this result is an increase in the number of transport proteins, without an increase in the activity of individual proteins. However, new protein synthesis is not required for this apparent increase in transport proteins.

In response to observations of this sort, it has been proposed that insulin stimulation of glucose transport operates via a recruitment mechanism. This hypothesis states that insulin stimulation causes an increase in glucose transporters in the plasma membrane by recruiting them from a preexisting intracellular membrane pool. Studies with cytocholasin B binding have produced data that are consistent with this hypothesis. The plasma membrane is enriched with transport proteins at the same time a corresponding decrease in glucose carriers is observed in intracellular membranes. Recruitment of transporters offers a viable mechanism for the short-term response to insulin.

III. ACTIVE TRANSPORT

In both passive diffusion and facilitated diffusion the primary driving force for transport across the membrane is the difference in chemical potential for the species in question from one side of the membrane to another. If the chemical potential of the species is equal on both sides of the membrane, one cannot expect to see any net transport (although exchange of the species across the membrane, without net transport, will still occur). Biologically, this process works well when the transported species is being changed or consumed preferentially on one side of the membrane. Such a constant depletion of the concentration of free solute on one side of the membrane will lead to a continual flux of that solute across the membrane toward the side where the solute is being depleted.

In contrast, there are cases where a concentration gradient of a solute is maintained across a membrane. One example is the difference in sodium ion concentration between the inside and the outside of most cells. Generally, sodium ion is at a much lower concentration inside a cell than outside that cell. This difference in concentration represents a difference in chemical potential that cannot be maintained in the face of passive or facilitated diffusion across the membrane.

In the cytoplasm, the free calcium concentration is quite low, in the micromolar range, or even less. Outside the cell, calcium concentration is several orders of magnitude higher, depending on the medium. Even in the lumen of organelles inside the cell, the calcium concentration can be much higher than that in the cytoplasm. For example, in the lumen of the sarcoplasmic reticulum, the calcium concentration can be in the millimolar range.

Thermodynamics requires that these gradients be dissipated if possible. Even with the general low permeability of membranes to ions, passive diffusion of solutes through the membrane does take place. How then does the cell maintain a difference in concentration of ions across a membrane of three orders of magnitude or even more?

The answer to this question lies in the activity of the membrane pumps. These pumps are enzymes that perform active transport in those membranes that support concentration gradients. That is, these enzymes utilize energy to pump ions against a concentration gradient. In the absence of an energy source, frequently ATP, these pumps are incapable of catalyzing active transport.

It is a calcium ATPase that catalyzes the transport of calcium into the lumen of the sarcoplasmic reticulum in muscle against a concentration gradient of calcium. It is another a calcium ATPase that catalyzes the transport of calcium across the plasma membrane to maintain the low level of calcium in the cytoplasm of the cell. It is the Na^+,K^+-ATPase in the plasma membrane that pumps sodium out of the cell (and potassium into the cell), thereby maintaining a substantial sodium gradient across the plasma membrane. All of these pumps are integral transmembrane proteins that function as enzymes, catalyzing the simultaneous hydrolysis of ATP and ion transport across the membrane.

So what is the key to this process of active transport? How can materials be pumped against a concentration gradient? The concentration gradient corresponds to a difference in chemical potential across the membrane.

This chemical potential difference corresponds to an unfavorable change in free energy for transport against a concentration gradient. The free energy cost of transporting ions against that concentration gradient can be calculated by

$$\Delta G = RT \ln\frac{a_l}{a_c} + nF^*,$$

where a_l is the activity of the calcium inside the lumen of the sarcoplasmic reticulum, a_c is the activity of calcium in the cytoplasm, n is 2 corresponding to the charge on calcium, and F^* is the membrane potential. Although the exact value of the ΔG depends on the membrane potential, the difference in concentration of calcium across the sarcoplasmic reticulum corresponds to several kilocalories per mole of unfavorable free energy change for transport of calcium against that concentration gradient. Clearly such transport must be a highly unfavorable event.

Yet such transport occurs. The secret for the above-mentioned examples of active transport lies in the ability of these ion pumps to catalyze the hydrolysis of ATP and make use of some of the energy that is released; ATP is a high-energy compound, due to the large favorable free energy change that occurs on the hydrolysis of ATP. Depending on the conditions this can amount to perhaps 13 kcal/mol (it can vary considerably, depending on the environment of the reaction).

Now consider the following equations:

ATP $\rightarrow$ ADP + P$_i$		$\Delta G = -13$ kcal/mol
$2Ca^{2+}_c \rightarrow Ca^{2+}_l$		$\Delta G = +11$ kcal/mol
ATP + $2Ca^{2+}_c \rightarrow$ ADP + P$_i$ + Ca^{2+}_l		$\Delta G = -2$ kcal/mol

These reactions are meant to convey that transport of calcium from the cytoplasm to the lumen cannot occur without hydrolysis of ATP and hydrolysis of ATP cannot occur without transport of calcium from the cytoplasm to the lumen. This is referred to as coupling of the reactions.[29] The coupling is carried out by the mechanism of catalysis by the pump protein in the membrane. An unfavorable reaction is coupled to a favorable reaction to allow the unfavorable reaction to go forward. The favorable reaction in membrane transport can also be the flow of a solute down its own concentration gradient, which can be used to transport another solute against its concentration gradient. This is a hypothetical example of what can happen in active transport. The values for ΔG are for illustration only.

A. Calcium Pump Protein of the Sarcoplasmic Reticulum

In order to gain a better understanding of the active transport process, consider the Ca^{2+}-stimulated Mg^{2+}-ATPase, or calcium pump protein of the muscle sarcoplasmic reticulum. This is the enzyme responsible for the sequestration of calcium in the lumen of the sarcoplasmic reticulum against a considerable concentration gradient. The Ca^{2+}-ATPase both pumps calcium across the membrane and hydrolyzes ATP. Therefore it is a good example of the coupling process introduced above. The question is how does the transport process work?

The calcium pump protein of the sarcoplasmic reticulum is a single polypeptide with approximately 40% of its mass buried in the hydrophobic interior of the membrane. The remainder of the mass of the protein lies outside the membrane, asymmetrically distributed toward the cytoplasmic side of the sarcoplasmic reticulum. Its molecular weight is about 110,000. A variety of evidence suggests that in the plane of the membrane, the protein can exist in several states of aggregation, in particular a dimer.[30] Depending upon the conditions, however, monomers, dimers, or tetramers may be observed. Unfortunately, since all the conditions under which the state of aggregation in the membrane has been measured are in some sense artificial, the true state of aggregation in the native membrane is not known. However, although it is not known whether a monomer of the protein can be active in calcium transport, it has been observed that a monomer of the protein can be active in ATP hydrolysis. Furthermore, dimers of the calcium pump protein are capable of calcium transport in the native membrane.

The calcium pump protein is the major integral membrane protein of the sarcoplasmic reticulum. However, depending in part on the source of the sarcoplasmic reticulum, various amounts of other proteins are present. These include glycoproteins and water-soluble calcium binding proteins. Although these other proteins are likely to be important in the regulation of sarcoplasmic reticulum function, reconstitution studies have shown that the calcium pump protein by itself is sufficient to exhibit calcium transport coupled to ATP hydrolysis.

Adenosine triphosphate hydrolysis by the Ca^{2+}-ATPase requires 30 phospholipids in a complex with the enzyme for maximal activity as determined by delipidation experiments.[31] Some detergents such as $C_{12}E_8$ are also capable of supporting ATP hydrolysis activity.[32] Calcium transport requires a closed membrane system.

The coupling of ATP hydrolysis to calcium transport has been idealized to be two calcium ions transported per ATP hydrolyzed. In practice this ratio is rarely seen experimentally, when the calcium transport and ATP hydrolysis are measured under identical conditions. The ratio may be variable, depending on the environment of the membrane. Available experimental evidence suggests that the latter is likely. Furthermore it is conceivable that the *in vivo* ratio is modulated by the state of the muscle.

However, such suggestions are purely hypothetical at present, since the *in vivo* function is not readily measurable.

Each of the steps in the kinetic mechanism of the calcium pump is reversible.[33] Therefore it should be possible, under appropriate conditions, to reverse the pump. This calls for the calcium ions to flow in the direction opposite to what is normal, and consequently to synthesize ATP. The experimental observation of this phenomenon lent considerable support to the kinetic scheme that has been developed.

Scheme 8.1 represents the most simple of kinetic schemes to describe the action of the calcium pump protein.

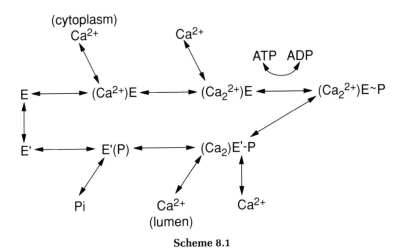

Scheme 8.1

In words, Scheme 8.1 can be described as follows. Consider first an unbranched cycle. Step one is the rapid binding of one calcium to a high-affinity binding site on the cytoplasmic side of the sarcoplasmic reticulum (on the dephosphorylated enzyme). This facilitates the binding of a second calcium to another calcium binding site on the protein. Then ATP binds to the calcium pump protein at the active site for ATP hydrolysis, leading to release of ADP and phosphorylation of the enzyme.

At this point, the reaction scheme describes in a single step what must be a multifaceted process.[34] The enzyme changes conformation, going from the E to the E′ state. Apparently simultaneously, the calcium binding sites are exposed to the inside or lumen of the sarcoplasmic reticulum. The calcium binding sites concurrently become low-affinity calcium sites. Sequentially, these calciums are then released into the lumen. The phosphorylated enzyme is then hydrolyzed, and the cycle is completed by the calcium pump protein returning to the E state.

This unbranched reaction cycle calls for strict stoichiometry of two calciums pumped per ATP hydrolyzed. However, under some experimental conditions other pump stoichiometries (with less than two calciums pumped per ATP hydrolyzed) have been observed. One way to account for those latter observations is to allow some branching in the reaction mechanism.[35] Such branching is represented in Scheme 8.2.

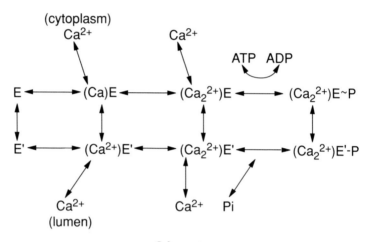

Scheme 8.2

Scheme 8.2 allows for calcium efflux as well as calcium pumping. The calcium pumping can appear to occur at different calcium/ATP ratios, when one is measuring the overall reaction.

These are not the only ways to represent the reactions of the calcium pump protein, but they allow a discussion of the fundamental mysteries of this ion pump and thereby shed some light on other, less well studied, ion pumps. Kinetically, these steps have been well characterized. The steps involving calcium binding to high-affinity sites can be understood in the same terms as ion binding to other high-affinity sites on proteins. Likewise calcium release from low-affinity sites, i.e., low enough affinity that even in the high calcium concentration of the lumen of the sarcoplasmic reticulum these sites are not saturated, can be understood in terms of ion binding to low-affinity sites on other proteins. Furthermore, ATP hydrolysis is a fairly common biochemical reaction that has been studied extensively in other enzyme systems.

What is not understood is how a thermodynamically unfavorable reaction (transport of calcium) is coupled to a thermodynamically favorable reaction (hydrolysis of ATP) to permit the unfavorable reaction to proceed. Presumably there are some movements of parts of the protein that change the exposure of the calcium binding sites from one side of the membrane to the other, and change the calcium affinity. These changes do not involve the rotation of the whole protein. Just what they do involve remains to be elucidated.

B. Na$^+$,K$^+$-ATPase

Having examined the case of Ca^{2+}-ATPase, the description of the sodium, potassium ATPase becomes much easier.[36] Structurally Na$^+$,K$^+$-ATPase is more complicated than Ca^{2+}-ATPase. It consists of two nonidentical subunits.[37] One has a molecular weight of about 110,000 and the other is about half that size. The larger, or α subunit, is the catalytic subunit because it contains the active site for ATP hydrolysis. As noted in Chapter 6, the sequence of this subunit has some homology to Ca^{2+}-ATPase.[38] The β subunit is a glycoprotein whose function is currently unknown but is required for activity.

The Na$^+$,K$^+$-ATPase is responsible for the pumping of sodium out of the cell and potassium into the cell, both against concentration gradients.[39] Because of the roles the gradients of sodium and potassium play, this is one of the most important enzymes in the plasma membrane of the cell. As in the case of calcium ATPase, Na$^+$,K$^+$-ATPase hydrolyzes ATP to make the pumping of sodium and potassium against their respective concentration gradients energetically feasible. In the red cell, three sodium ions and two potassium ions are pumped for every ATP hydrolyzed.

The activity of Na$^+$,K$^+$-ATPase is readily identified because of the availability of specific inhibitors of the enzyme. Most prominent among these is the cardiac glycoside ouabain. Ouabain binds specifically and tightly to the enzyme inhibiting ATP hydrolysis and consequently ion pumping. Ouabain binding is sufficiently specific that it can be used to count the number of Na$^+$,K$^+$-ATPase molecules on the cell surface. Furthermore, the activity of Na$^+$,K$^+$-ATPase is usually measured as that portion of the total ATPase activity (in the presence of sodium and potassium) that is inhibited by ouabain.

The structure of the protein will be considered a little further before going on to a discussion of the activity of the pump.[40] The enzyme is a transmembrane protein. The ATP binding site is on the cytoplasmic surface of the plasma membrane, and the ouabain binding site is on the extracellular surface of the plasma membrane. Some of the mass of the protein protrudes from the surface of the membrane on both sides of the plasma membrane. The α subunit is exposed on both sides of the membrane with the greatest mass protruding on the cytoplasmic side. The glycoprotein subunit is exposed on the extracellular surface. About 30% of the total mass of the enzyme is buried in the membrane. As in the case of the ATPase, it is not clear whether the enzyme is active in transport as a monomer (of two subunits) or a dimer (of four subunits) in the native membrane. More details of the structure can be found in Chapter 6.

Phospholipids, or, in some special cases, detergents, are required to sustain ATPase activity. (Of course, transport can be observed only in closed cellular or vesicular systems.) Acidic phospholipids are particularly effective at activating Na^+,K^+-ATPase. Studies on the relative affinity of the enzyme for lipids using ESR spin labels suggest that acidic lipids are preferentially found in the lipid–protein interface for this protein.[41]

The protein is capable of a complex set of conformational changes, depending on the composition of the media. These have largely been elucidated by fluorescence measurements. In general, conformational changes observed for this enzyme are consistent with small changes in the orientation of particular segments of the protein during the ion pumping process. The details of those conformational changes are not necessary to pump ions and that rotation of the protein from one side of the membrane to the other is not a part of the transport function.

The reaction mechanism of Na^+,K^+-ATPase can be described in a manner analogous to that of Ca^{2+}-ATPase. The following model is a partial description of the chemical reactions involved in sodium and potassium transport by this enzyme. Scheme 8.3 derives from what is referred to as the Albers–Post model.

Scheme 8.3 can be described as follows. There are two enzyme conformations, E1 and E2. E1 will bind ATP, Mg^{2+}, and Na^+ from the inside,

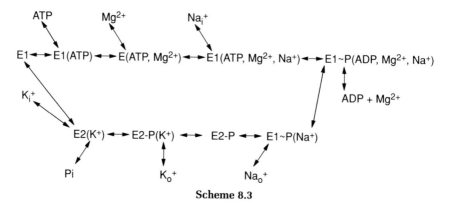

Scheme 8.3

in that order. The sodium binds to a high-affinity site available only from the inside. The binding of the sodium stimulates the enzyme to hydrolyze the ATP, forming a phosphorylated enzyme intermediate. This then exposes the Na^+ to the outside surface and changes the Na^+ binding to a low-affinity type. The Na^+ is then released to the outside and E2 is formed. Also exposed to the outside surface by the E2 phosphorylated enzyme is a potassium binding site. When potassium binds, the phosphoenzyme is hydrolyzed. This stimulates the enzyme to expose the potassium binding site to the inside surface of the membrane and change to a lower affinity binding site. Release of the potassium to the inside of the cell then follows. Inside and outside in this mechanism refer to the inside and the outside of the cell plasma membrane in which the Na^+,K^+-ATPase resides. This scheme calls for three sodiums to be pumped for two potassium ions pumped, although that stiochiometry is not explicitly written in the above scheme.

There are two important steps with regard to enzyme conformation in this mechanism. These steps roughly correspond to the one equivalent step in the Ca^{2+}-ATPase mechanism during which the calcium ion binding site is changed in exposure from one side of the membrane to another. For Na^+,K^+-ATPase, one of these steps is the change from E1 to E2 after phosphorylation. During this step, the sodium binding site is changed from a cytoplasmic exposure to an extracellular exposure with a corresponding change in binding affinity. Spectroscopic studies suggest that conformational changes that may be associated with this step occur, but the details remain unknown. Likewise conformational changes must occur for the second important step in which the phosphorylated E2 enzyme is dephosphorylated by the binding of potassium and the potassium binding site is changed in exposure from the outside to the inside (with a corresponding change in affinity). Thus, as in the case of Ca^{2+}-ATPase, much remains to be understood concerning the molecular mechanism of Na^+,K^+-ATPase.

It is worth adding here that other activities can be expressed by Na^+,K^+-ATPase, depending on the environment. The enzyme is capable of the exchange of sodium ion across the membrane, as well as the exchange between ATP and ADP. Net sodium efflux and exchange across the membrane of potassium are also observed for Na^+,K^+-ATPase under certain conditions.

C. Cotransport of Glucose and Sodium

With the currents of ion movements in and out of the cell under the influence of membrane ion pumps as just described, the resulting ion gradients are used for other processes. One example is found in the co-transport of glucose and sodium in the intestinal brush border membrane. Glucose is transported across the brush border membrane from the lumen of the intestine and through the epithelial cell layer against a concentration gradient. As in some of the other transport proteins, this transporter exhibits about 12 transmembrane segements in its amino acid sequence.[42]

Glucose transport in this system is accomplished through the use of the sodium gradient established by Na^+,K^+-ATPase. The transport system requires one sodium and one glucose to be transported simultaneously. In so doing, the transport system allows sodium to flow down its concentration gradient. This is another example of coupling of two reactions. The favorable reaction of sodium transport with its concentration gradient is coupled with the unfavorable transport of glucose against its concentration gradient. The overall reaction is then thermodynamically favorable up to the observed intracellular concentration of glucose. Higher intracellular glucose levels cannot be reached because the free energy in the sucrose gradient would exceed the free energy available to the transport system from the sodium gradient.

The requirement of the chemical potential difference of sodium across the membrane in this process classifies the symport (obligatory transport in the same direction) of glucose and sodium as an active (energy-requiring) transport process. A symport is not the only means to use energy stored in a concentration gradient. An antiport system transports substrates in opposite directions, utilizing the energy stored in the concentration gradient of one of the species to transport another species against its concentration gradient.

D. Proton Pumps

Some membrane ATPases simply pump H^+.[43] Examples include proton pumps in coated vesicles that lead to acidification of the lumen of the vesicle after vesicle formation,[44] and acid secretion (out of the cell).[45, 46] Kinetically and structurally these H^+-ATPases resemble other pumps already discussed.[47]

E. The Phosphotransferase System of Bacteria

Some bacteria, including *E. coli,* take up carbohydrate, such as glucose, through a concurrent transport and phosphorylation scheme.[48] This differs from glucose uptake in mammalian systems, where glucose phosphorylation by hexokinase occurs subsequent to transport. The phosphotransferase (PTS) system does not lead to the accumulation of any unphosphorylated substrate in the cell. The phosphorylated product is metabolized directly.

The PTS system consists of both integral and peripheral (cytoplasmic) membrane proteins. Figure 8.8 schematically represents the PTS-catalyzed transport for glucose. The system functions via a series of phosphate transfers beginning with phosphoenolpyruvate, an intermediate in glycolysis. The phosphate is transferred to enzyme I (EI); EI is a soluble enzyme of 60–70 kDa. In the active form it is a dimer. The phosphorylated form of EI (EI-P) can transfer a phosphate to a heat-stable protein (HPr). Both these protein phosphate donors have a high phosphate transfer potential, substantially higher than that of ATP.

The phosphoprotein HPr-P can transfer phosphate either directly to enzyme II (EII) or via an intermediate, enzyme III (EIII). In the latter case, this represents the substrate-specific part of the PTS system. EIII is specific for particular EII which in turn are specific for carbohydrates (EII for glucose is distinct from EII for mannitol, for example). Enzyme III is about 20 kDa in some bacteria and forms oligomers.

Enzyme II is an integral tansmembrane protein. In the case of *S. typhimurium,* the mannitol-specific protein is about 68 kDa. Enzyme II is exposed to only a modest extent on the extracellular surface, but a large part of the protein mass is exposed on the cytoplasmic surface; EII is apparently responsible for the concerted reactions that concurrently lead to substrate phosphorylation and substrate transport.

The PTS system is essential for provision of carbohydrate for glycolysis. The PTS system also functions in chemotaxis for carbohydrate. When active, the PTS system appears capable of modifying random bacterial motion to produce net movement in the direction of the carbohydrate gradient in the medium through control of the flagella. This is separate from chemotaxis mediated by protein methylation.

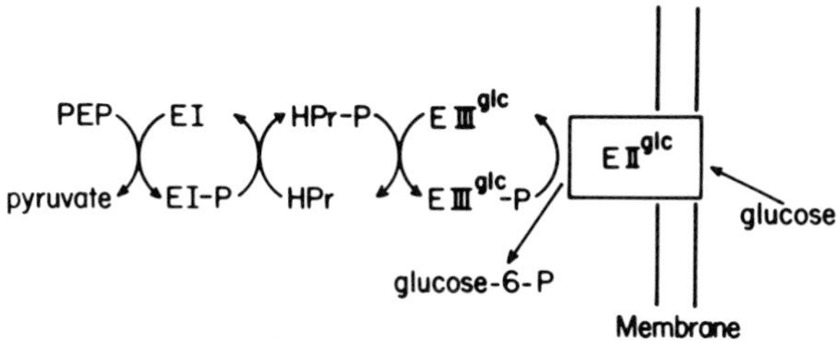

Fig. 8.8.   Schematic representation of the phosphotransferase system of bacteria.

IV. SUMMARY

Transport of solutes across membranes can occur via three classes of mechanisms. One is passive transport. No energy is required, since the solute achieves net flux across a membrane only by moving down a concentration gradient. Passive diffusion may occur via structural discontinuities in the lipid bilayer. Glucose transit of membranes is adequately described by this mechanism.

Facilitated transport takes place with the assistance of carriers or pores. Carriers bind the solute, rendering it more lipid soluble and carry it across the membrane to be released on the other side. Valinomycin is such a carrier. Channels form pores in membranes through which small solutes may pass. Porins from bacteria are representative of channels in membranes. Much greater fluxes of solutes can be achieved with channels than with carriers.

Active transport occurs by a coupling a thermodynamically favorable reaction with a transport reaction that is unfavorable, i.e., transport against a concentration gradient. This can be achieved by a membrane bound enzyme that catalyzes the hydrolysis of ATP and couples that favorable reaction to transport of ions against a concentration gradient. Coupling means that, on the pump protein, the hydrolysis of ATP cannot occur without transport, and vice versa. The mechanism for this energy transduction is as yet unknown. Energy for transport can also be extracted from concentration gradients of other chemical species, such as sodium. Active transport can further be supported by chemical modification of the transported species, thus keeping the concentration of the free species low.

REFERENCES

1. Ohki, S., and R. A. Spangler, "Passive and facilitated transport," In *The Structure of Biological Membranes* (P. L. Yeagle, ed.) (Boca Raton: CRC Press, 1992).
2. Krasne, S., G. Eisenman, and G. Szabo, "Freezing and melting of lipid bilayers and the mode of action of nonactin, valinomycin, and gramicidin," *Science* (1971): 412–415.
3. Tosteson, D. C., P. Cook, T. E. Andreoli, and M. Tieffenberg, "The effect of valinomycin on potassium and sodium permeability of H.K. and L.K. sheep red cells," *J. Gen. Physiol.* **50**, (1967): 2513.
4. Pinkerton, M., M. K. Steinrauf, and P. Dawkins, "The molecular structure and some transport properties of valinomycin," *Biochem. Biophys. Res. Commun.* **35**, (1969): 512.
5. Steinrauf, L. K., J. A. Hamilton, and M. N. Sabesan, "Crystal structure of alinomycin–sodium picrate: Anion effect on valinomycin-cation complexes," *J. Am. Chem. Soc.* **104** (1982): 4085.
6. Dutton, A., E. D. Rees, and S. J. Singer, "An experiment eliminating the rotating carrier mechanisms for the active transport of Ca ion in sarcoplasmic reticulum membranes," *Proc. Natl. Acad. Sci. U.S.A.* **73** (1976): 1532–1536.

7. Hladky, S. B., and D. A. Haydon, "Ion transfer across lipid membranes in the presence of gramicidin A," *Biochim. Biophys. Acta* **274** (1972): 294.

8. Urry, D. W., "The gramicidin A channel: A proposed pi helix," *Proc. Natl. Acad. Sci. U.S.A.* **68** (1971): 672.

9. Langs, D. A., "Three dimensional structure at 0.86 Å of the uncomplexed form of the transmembrane ion channel peptide gramicidin A," *Science* **241** (1988): 188.

10. Benz, R., "Structure and selectivity of porin channels," *Curr. Topics Membr. Transp.* **21** (1984): 199–219.

11. Kleffel, B., R. M. Garavito, W. Baumeister, and J. P. Rosenbusch, "Secondary structure of a channel-forming protein: Porin from *E. coli* outer membranes," *EMBO J.* **4** (1985): 1589–1592.

12. Jennings, M. L., "Structure and function of the red blood cell anion transport protein," *Annu. Rev. Biophys. Chem.* **18** (1989): 397.

13. Roughton, F. J. W., "Recent work on carbon dioxide transport by the blood," *Physiol. Rev.* **15** (1935): 241.

14. Cabantchik, Z. I., and A. Rothstein, "Membrane proteins related to anion permeability of human red blood cells. I.Localization of disulfonic stilbene binding sites involved in permeation," *J. Membr. Biol.* **15** (1974): 207.

15. Tanner, M. J. A., P. G. Martin, and S. High, "The complete amino acid sequence of the human erythrocyte membrane anion transport protein deduced from the cDNA sequence," *Biochem. J.* **256** (1988): 703–712.

16. Rothstein, A., "The functional architecture of band 3, the anion transport of the red cell membrane," *Can. J. Biochem. Cell Biol.* **62** (1984): 1198–1204.

17. Lux, S. E., K. M. John, R. R. Kopito, and H. F. Lodish, "Cloning and characterization of band 3, the human erythrocyte anion-exchange protein," *Proc. Natl. Acad. Sci. U.S.A.* **86** (1989): 9089.

18. Bennett, V., "The membrane skeleton of hyman erythrocytes and its implication for more complex cells," *Annu. Rev. Biochem.* **54** (1985): 273–304.

19. Cherry, R. J., "Rotational and lateral diffusion of membrane proteins," *Biochim. Biophys. Acta* **559** (1979): 289–327.

20. Schubert, D., and K. Boss, "Band 3 protein–cholesterol interactions in erythrocyte membranes," *FEBS Lett.* **150** (1982): 4–8.

21. Tabcharani, J. A., W. Low, D. J. Elie, and W. Hanrahan, "Low-conductance chloride channel activated by cAMP in the epithelial cell line T84," *FEBS Lett.* **270** (1990): 157–164.

22. Riordan, J. R., J. M. Rommens, B. Kerem, N. Alon, R. Rozmahel, Z. Grzelczak, J. Zielenski, S. Lok, N. Plavsic, J. L. Chou, M. L. Drumm, M. C. Ianuzzi, F. S. Collins, and L. C. Tsui, "Identification of the cystic fibrosis gene: Cloning and characterization of the complimentary DNA," *Science* **245** (1989): 1066–1073.

23. Tabcharani, J. A., X-b. Chang, J. R. Riordon, and J. W. Hanrahan, *Nature (London)* **352** (1991): 628–631.

24. LeFevre, P. G., and J. K. Marshall, "The attaachment of phloretin and analogues to human erythrocytes in connection with inhibition of sugar transport," *J. Biol. Chem.* **234** (1959): 3022.

25. Mueckler, M. C. Caruso, S. A. Baldwin, M. Pacino, I. Blench, H. R. Morris, W. J. Allard, G. E. Lienhard, and H. F. Lodish, "Sequence and structure of human glucose transporter," *Science* **229** (1985): 941–945.

26. Chin, J. J., E. K. Y. Jung, V. Chen, and C. Y. Jung, "Structural basis of human erythrocyte glucose transporter function in proteoliposome vesicles: Circular dichrosim measurements," *Proc. Natl. Acad. Sci. U.S.A.* **84** (1987): 4113–4116.

27. Hebert, D. N., and A. Carruthers, "Cholate-solubilized erythrocyte glucose transporters exist as a mixture of homodimers and homotetramers," *Biochemistry* **30** (1991): 4654–4658.

28. Simpson, I. A., and S. W. Cushman, "Hormonal regulation of mammalian glucose transport," *Annu. Rev. Biochem.* **55** (1986): 1059–1089.

29. Jencks, W. P., "What is a coupled vectorial process?" *Curr. Top. Membr. Transp.* **19** (1983): 1–19.

30. Hymel, L., A. Maurer, C. Berenski, C. Y. Jung, and S. Fleischer, "Target size of calcium pump protein from skeletal muscle sarcoplasmic reticulum," *J. Biol. Chem.* **259** (1984): 4890–4895.

31. Warren, G. B., P. A. Toon, N. J. M. Birdsall, A. G. Lee, and J. C. Metcalfe, "Reconstitution of a calcium pump using defined membrane components," *Proc. Natl. Acad. Sci. U.S.A.* **71** (1974): 622–626.

32. Dean, W. L., and C. P. Suarez, "Interactions between sarcoplasmic reticulum calcium adenosinetriphosphatase and nonionic detergents," *Biochemistry* **20** (1981): 1743–1747.

33. Inesi, G., "Mechanisms of calcium transport," *Annu. Rev. Physiol.* **47** (1985): 573–601.

34. Tanford, C., "The sarcoplasmic reticulum pump: Localization of free energy transfer to discrete steps of the reaction cycle," *FEBS Lett.* **166** (1984): 1–7.

35. Gafni, A., and P. D. Boyer, "Modulation of stiochiometry of the sarcoplasmic reticulum calcium pump may enhance thermodynamic efficiency," *Proc. Natl. Acad. Sci. U.S.A.* **82** (1985): 98–101.

36. Skou, J. C., "The identification of the sodium pump as the membrane-bound Na,K-ATPase," *Biochim. Biophys. Acta* **1000** (1989): 435–446.

37. Jorgensen, P. L., "Mechanism of the Na^+,K^+ pump: Protein structure and conformations of the pure $(Na^+ + K^+)$-ATPase," *Biochim. Biophys. Acta* **694** (1982): 27–68.

38. Shull, G. E., A. Schwartz, and J. B. Lingrel, "Amino-acid sequence of the catalytic subunit of the $(Na^+ + K^+)$ATPase deduced from a complementary DNA," *Nature (London)* **316** (1985): 691–695.

39. Kaplan, J. H., "Ion movements through the sodium pump," *Annu. Rev. Physiol.* **47** (1985): 535–544.

40. Post, R. L., "Structural aspects of Na,K-ATPase," *Curr. Top. Membr. Transp.* **19** (1983): 53–65.

41. Esmann, M., and D. Marsh, "Spin-label studies on the origin of the specificity of lipid–protein interactions in NaKATPase membranes from squalus acanthias," *Biochemistry* **24** (1985): 3572–3578.

42. Hediger, M. A., E. Turk, and E. M. Wright, "Homology of the human intestinal Na/glucose and *E. coli* Na/proline cotransporters," *Proc. Natl. Acad. Sci. U.S.A.* **86** (1989): 5748–5752.

43. Amzel, L. M., and P. L. Pederson, "Proton ATPases: Structure and mechanism," *Annu. Rev. Biochem.* **52** (1983): 801–824.

44. Forgac, M., and L. Cantley, "Characterization of the ATP-dependent proton pump of clathrin-coated vesicles," *J. Biol. Chem.* **259** (1984): 8101–8105.

45. Sachs, G., L. D. Faller, and E. Rabon, "Proton/hydroxyl transport in gastric and intaestinal epithelia," *J. Membr. Biol.* **64** (1982): 123–135.

46. Sigler, K., and M. Hofer, "Mechanisms of acid extrusion in yeast," *Biochim. Biophys. Acta* **1071** (1992): 375–379.

47. Nelson, N., "Evolution of organellar proton-ATPases," *Biochim. Biophys. Acta* **1100** (1992): 109–124.

48. Saier, M. H., "Mechanisms of carbohydrate transport in bacteria and comparisons with those in eukaryotes," *The Structure of Biological Membranes* (P. L. Yeagle, ed.) (Boca Raton: CRC Press, 1992), 833–891.

9

Membrane Fusion

Membranes are often considered stable, noninteracting structures that, among other things, form semipermeable barriers for cells and organelles. The lipid bilayer is the dominant structural feature of this picture for membranes. It is a continuous structure, which for the most part does not have any major structural defects that would otherwise destroy its semipermeable nature.

However, in biology, membranes are much more dynamic than the above description implies. A number of cellular processes require that two membranes become one membrane, or that one membrane segregate into two membranes. Central to these transformations of membrane structure is the process of membrane fusion. Figure 9.1 schematically represents some of the cellular processes that involve membrane fusion.

For example, membrane fusion is an essential step in transport of newly synthesized membrane constituents. Lipid and protein components of the plasma membrane are synthesized on the endoplasmic reticulum. To get those constituents from their site of synthesis to the plasma membrane requires transport of the newly synthesized material from the endoplasmic reticulum to the target membrane. The transport occurs by discrete membrane vesicles. These vesicles must pinch off of the endoplasmic reticulum and become part of the target membrane into which the newly synthesized material is to be inserted.

Transport can also occur from the surface of the cell to the inside. One of the major pathways of regulation of cholesterol biosynthesis is through uptake of low-density lipoproteins from serum. This uptake is mediated by a receptor. The lipoprotein binds to a receptor that subsequently becomes internalized through the process of receptor-mediated endocytosis. This process involves an endocytotic vesicle that pinches off from the plasma membrane (a fusion event) and subsequently fuses with an endosome.

Proteins that are to be secreted are initially synthesized at the ER and inserted into the lumen of the ER. To get from the ER to the exterior of

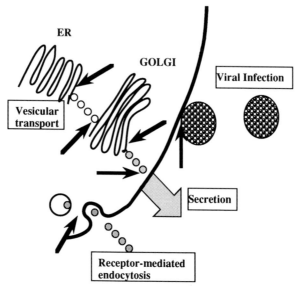

Fig. 9.1. Schematic representation of some biological processes involving membrane fusions.

the cell requires vesicular transport and a series of fusion events. Early fusion events involve vesicle transport between ER and Golgi. Late fusion events involve fusion of the transport vesicles with the plasma membrane of the cell.

Part of the pathway of conduction of nerve impulses involves the release of a neurotransmitter by one cell as a means of chemical communication with another cell. For this process, the neurotransmitter is packaged in secretory vesicles that reside near the presynaptic membrane. On receipt of the appropriate signal, these vesicles migrate to the plasma membrane and become one with that membrane. In so doing they expose their interior to the outside of the cell. Thus their contents are released to the exterior of the cell and into the synapse.

Many enveloped viruses infect cells by a pathway that has something in common with the above processes. After adhering to the target cell surface, these viruses cause their envelope, or membrane, to fuse with the target cell plasma membrane. The result is analogous to the secretion event just described. In this case, one can think of the virus as containing a nucleocapsid inside a vesicular membrane, the virus envelope. When the virus envelope becomes one with the plasma membrane of the target cell, its contents are released. However, in this case, the contents are released to the inside of the cell. By so doing, the virus injects its nucleocapsid containing the genetic material into the cell.

All of the above processes have one event in common. At some point, two membranes that were originally separate become continuous with each other, allowing some mixing of the components of the two original membranes. Or alternatively, one membrane entered a process that led to the separation of two distinguishable membranes as a product. These are all membrane fusion events.

To prepare for a discussion of the pathways for the fusion events mentioned above, the discussion will turn to fundamental considerations of membrane fusion, followed by a discussion of simple lipid vesicle fusion systems to gain an understanding of some molecular details that may be important to other fusion events. Then the discussion will return to the above-mentioned biological fusion events and a consideration of the mechanisms involved.

I. FUNDAMENTALS OF THE MEMBRANE FUSION PROCESS

The pathway leading to the fusion of two lipid bilayers or cell membranes has been suggested to include the following events:

1. Aggregation of the membranes that will fuse;
2. Close approach of the lipid bilayers of the membranes that are to fuse, leading to a removal of some of the water separating the membranes (partial dehydration);
3. Destabilization of the bilayer at the point of fusion (two bilayers closely opposed will not necessarily spontaneously fuse);
4. Mixing of the components of the bilayers and ultimate separation from the point of fusion into the new membrane structures.

Event 1 may involve random collisions in laboratory vesicle fusion experiments or targeting (via receptors) in cellular fusion events. A related problem is specificity. In a properly functioning cell, indiscriminate fusion cannot be permitted. Vesicles targeted to the Golgi should go to the Golgi and not to the plasma membrane. Endocytotic vesicles destined for lysosomes should not go to endoplasmic reticulum. Therefore there must be some mechanism within the cell to control which membranes will undergo fusion with each other. Event 2 is distinguished from event 1 in that close approach of biological membranes often encompasses rearrangements of the membrane components (proteins) that would otherwise keep the membranes sufficiently separated to inhibit fusion. Event 3 usually involves some other influence or component in the system that can promote sufficient bilayer destabilization to permit mixing of the lipid components of the two reacting membranes. This influence must be adequately effective to overcome the stability normally associated with the lipid bilayer. This event does not require, however, that the membrane bilayers actually contact each other to induce fusion.

A related problem is timing. For example with release of neurotransmitter, the fusion event is timed to occur on receipt of a signal. By so doing an orderly series of events can occur to ensure proper and accurate transmittal of information by the nervous system.

By examining details of fusion events, both in model systems and in biological membranes, the underlying principles of membrane fusion can be discerned. This discussion continues therefore with a consideration of model vesicle systems for membrane fusion.

II. MEASUREMENT OF FUSION

A number of fluorescence assays have been developed to measure quantitatively the progress of a membrane fusion event.[1] One type of assay is sensitive to the mixing of the components of two membranes. Consider the issue from the point of view of one of the membranes that is to fuse. When it fuses with another target membrane, its membrane components become diluted with the target membrane components in the product membrane. Thus if the dilution effect can be detected, one would have a measure of this feature of fusion. One means of detecting dilution is to include a fluorescent lipid in one of the membranes at a high enough concentration that the fluorescence of the fluorophore is partially self-quenched at the start of the assay. Then when this membrane fuses with another membrane that does not contain the fluorescent probe, the probe molecules are diluted and the quenching (which is related to the average intermolecular distance, or the frequency of collision, both of which depend on concentration of the probe in the membrane) is relieved (see Fig. 9.2).

Another type of fluorescence assay measures the mixing of vesicle contents. For example, imagine two chemical species that exhibit distinctly different fluorescent properties when combined than when separate. If these two species were then put in separate vesicles, the mixing of the vesicle contents could be detected from the changes in fluorescence. An example of this approach that has been used is the following: ANTS is a highly fluorescent compound; DPX is an effective quencher. Therefore, the mixing of vesicle contents can be monitored by incorporating ANTS in one vesicle population and DPX in another vesicle population before allowing them to undergo fusion. On mixing of the vesicle contents, quenching of the ANTS fluorescence occurs (see Fig. 9.3).

An alternate form of the experiment just described can be used to

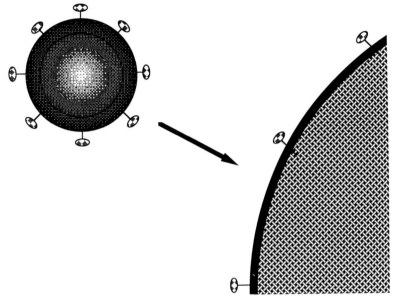

Fig. 9.2. Schematic representation of the effect of fusion on the relative concentration of a fluorescent probe in a membrane. This represents a lipid mixing assay for membrane fusion.

examine the leakage of vesicle contents during the fusion event. For example, consider the case when all the vesicles are initially loaded with a fluorescent probe at concentrations that lead to self-quenching. If the probe leaks out during the fusion process, the probe will enter a medium in which it is at relatively low concentration and is not quenched. Thus

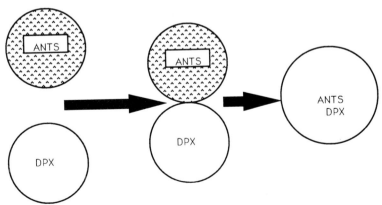

Fig. 9.3. Contents mixing membrane fusion assay described in text.

an increase in fluorescence will be observed if leakage occurs. Carboxy-fluorescein is an example of a fluorescent probe that is used to assay leakage.

III. FUSION OF PHOSPHOLIPID VESICLES

A. PS Vesicle Fusion Stimulated by Calcium

Some useful information about fusion of membranes has been obtained from relatively simple vesicle fusion studies. One system that has been studied in considerable depth is the fusion of phosphatidylserine (PS)-containing vesicles.[2] These vesicles obviously carry a negative charge on their surface, which tends to inhibit close approach of the vesicle surfaces. Mg^{2+}, as a cation, can screen this negative charge. Addition of Mg^{2+} to PS vesicles can therefore cause aggregation of the vesicles. The aggregation is manifest, for example, in an increase in the light-scattering properties of the system. However, little or no fusion occurs with Mg^{2+}.

Addition of Ca^{2+} to the medium surrounding the PS vesicles has a dramatically different effect from that of Mg^{2+}. Ca^{2+} causes a massive fusion of the vesicles. Fusion is followed by a substantial leakage of vesicle contents. Eventually cochleate structures are formed, which consist of dehydrated calcium–phosphatidylserine complexes. On removal of the calcium by chelation, for example, large vesicles are formed from the cochleates (see Fig. 9.4).

The most interesting part of this model fusion process is the early series of events. In this period, vesicle aggregation and fusion occur prior to release of vesicle contents. These early events most closely mimic the events of fusion of biological membranes. Interestingly, Mg^{2+} potentiates the effects of Ca^{2+}; fusion occurs at significantly lower Ca^{2+} concentrations in the presence of Mg^{2+}. This is presumably because Mg^{2+} promotes an essential step in the fusion process, namely the aggregation of the vesicle leading to close apposition of the membrane surfaces.

The next question is what does calcium do that magnesium does not, to promote fusion. When Ca^{2+} binds to a membrane surface, it can screen the surface charge as noted for Mg^{2+}. This screening would lead to an enhanced capability of the two membrane surfaces to approach each other closely. However, Ca^{2+} forms a specific complex with PS. It has been suggested that when the two membrane surfaces get close together, a "trans" complex forms, with calcium linking PS molecules in the opposing membranes.

What is the effect of the formation of such a complex? The binding of

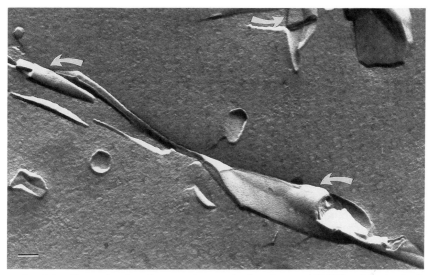

Fig. 9.4. Freeze-fraction electron micrograph of a mixture of DPPC and bovine brain PS at a molar ratio of 4 : 6, in the presence of 10 mM Ca^{2+}. Note that the PS fraction rolled into cochleates (arrows), whereas the PC fraction remains in vesicle form. Bar: 100 nm. Micrograph courtesy of S. W. Hui.

cations to the surface of a lipid bilayer can lead to an increase in the phase transition temperature. Therefore, the question arose whether calcium exerts its effect by creating patches of gel state lipid that promote fusion. However, it has been demonstrated that calcium has its effect when at least the bulk of the membrane is in the liquid crystal state. Therefore, one must look elsewhere for the basis for the calcium effect on fusion of PS vesicles.

The formation of this trans-PS–calcium complex does have another important effect on the membrane surface. Binding of the calcium leads to a decrease in the number of waters bound to the phospholipid, leading to a dehydration of the membrane surface. As was concluded previously, removal of water from between two membranes is one of the important barriers that must be surmounted for the two membranes to fuse. Therefore the dehydration of the surface of the membrane by calcium may be a factor in reducing the barriers to fusion.

B. Fusion of *N*-Methyl-DOPE LUV

A system that appears, at least superficially, to follow a different pathway for membrane fusion is the fusion of LUV made of *N*-methyldioleoylphosphatidylethanolamine (N-methyl-DOPE).[3] These vesicles made of a modified PE do not require calcium to fuse. They are initially formed by extrusion at high pH (pH 9.5), where the amino group is at least partially deprotonated. At pH 9.5 these membranes are stable phospholipid bilayers. To initiate fusion, the pH of the medium is dropped to pH 4.5. At pH 4.5 and the appropriate temperature, a relatively rapid fusion of the vesicles occurs, as measured by fluorescence assays of membrane fusion (contents mixing assay, for example).

The question that is stimulated by these results is what occurs at the low pH that allows fusion to proceed. Investigation has revealed that under fusion-permissive conditions, the bilayers of N-methyl-DOPE are unstable.[4] Nonlamellar structures can form. The presence of nonlamellar structures was detected by the techniques described in Chapter 4, including ^{31}P NMR and freeze-fracture electron microscopy. In particular, N-methyl-DOPE can form the kind of nonlamellar structures found in Fig. 4.9 under fusogenic conditions.

The relative incidence of formation of nonbilayer structures was found to be correlated with the initial rates of fusion of these LUV. In particular, there is a linear relationship between the percentage of the phospholipid in the nonlamellar structures and the initial rate of membrane fusion as measured by contents mixing assays. Companion studies revealed that membrane components such as diacylglycerol had a destabilizing effect on lipid bilayers. In N-methyl-DOPE nonlamellar structures appeared at significantly lower temperatures in the presence of diacylglycerol. Likewise, membrane fusion was accelerated at lower temperatures by the appearance of nonlamellar structures at those temperatures.

On the basis of correlations such as these, it was suggested that fusion in this system may proceed through nonbilayer intermediates (see Fig. 4.15). Nonbilayer intermediates provide a pathway for mixing bilayer components that is required by event 3 in the general fusion pathway outlined in Section I. Figure 9.5 shows freeze-fracture electron micrographs of fusion that occurred in a membrane system through nonbilayer intermediates.

The details of the structures of these fusion intermediates are not known with certainty. Some EM data have been obtained that are consistent with theoretical calculations on possible intermediates that involve interlamellar attachments.[5] Studies on the inhibition of membrane fusion by specific inhibitors have suggested that the structures of the fusion intermediates in the fusion of N-methyl-DOPE LUV have highly curved surfaces, i.e., with radii of curvature near that observed for SUV (see Chapter 3).[6] The interlamellar attachments have such curvature. This suggestion was

Fig. 9.5. Freeze-fracture electron micrograph showing interbilayer connections in the cross fracture. Lipidic particles are also seen. This represents fusion induced in PC/PE membranes by freeze-thawing. Bar: 100 nm. Micrograph courtesy of Dr. S. W. Hui.

derived from the observation that certain hydrophobic peptides that inhibit the fusion of N-methyl-DOPE LUV also inhibit the formation of highly curved phospholipid surfaces. These studies also had ramifications for the mechanism of enveloped virus fusion (see below).

C. Fusion Stimulated by Phospholipase C

Diacylglycerols have been shown to destabilize the lamellar phase of PE and its close structural relatives. In LUV of N-methyl-DOPE, the presence of small amounts of diacylglycerols also stimulate fusion, concomitant with the formation of nonlamellar structures. These observations raise the question whether activation of phospholipase C might also stimulate fusion. Experiments in phospholipid vesicles suggested that action of phospholipase C can stimulate membrane fusion.[7] However, it remains to be determined by what mechanism fusion is stimulated.

IV. VIRUS FUSION

The discussion now turns to the mechanism of membrane fusion facilitated by transmembrane glycoproteins of the enveloped viruses (viruses with a membrane surrounding the nucleocapsid containing the viral genome). Among these glycoproteins are a class of proteins, referred to as fusion proteins, that are responsible for promoting the membrane fusion event. This membrane fusion is a necessary part of the pathway of viral infection. Fusion of the viral envelope (membrane) with the target cell membrane will lead to injection of the nucleocapsid into the cell. Viral replication cannot occur until the nucleocapsid is inside the cell.

As mentioned in Chapter 6, the viral fusion proteins have an interesting organization of their primary structure. They contain a typical hydrophobic stretch of amino acids that form the transmembrane segment of the protein (typically only one transmembrane segment is found in these fusion proteins). In addition, a hydrophobic sequence is found in these fusion proteins that is not employed to anchor this membrane protein in the viral membrane. This additional hydrophobic sequence is required for the virus to be fusion competent and has been suggested to be directly involved in the fusion event. Some homology is evident from one fusion protein to another in this sequence. The sequence, Phe–X–Gly, is one consensus sequence for the amino terminus of a ''fusion peptide'' of the viral fusion protein among some classes of viruses (Fig. 6.12).

The enveloped viruses can be divided into two groups with respect to the pH at which their fusion proteins are active. One group is capable of fusion at neutral pH. These viruses are thus capable of fusing with the plasma membrane of the target cell. Sendai virus is a member of this group and will be used in the following discussion as a prototypical example. The other group requires acid pH to exhibit fusion. These viruses normally fuse only after endocytosis. Influenza will be used in the following discussion as a prototypical example of this latter group of viruses.

A. Sendai Virus

One of the best known of the fusion-promoting viruses is the Sendai virus.[8, 9] This is an enveloped virus with a negative strand of RNA. Sendai virus is prototypical for neutral pH-fusing viruses. Other viruses that can fuse at neutral pH include measles virus, herpes simplex virus, and human immunodeficiency virus. Sendai virus is capable of fusing to cells and vesicles with appropriate receptors and of inducing cell–cell fusion, as in the formation of hybridomas. It certain cases, Sendai is also capable of fusing with a vesicle without a receptor.

Figure 1.3 shows schematically the structural features of this virus important to membrane fusion. The envelope membrane contains two glycoproteins, which extend from the surface of the virion in the form of spikes as visualized in the electron microscope. The glycoproteins both have a portion of their mass buried in the membrane, and may be transmembrane proteins. A portion of the protein may be exposed on the inside of the envelope for interaction with the M protein, which may form a membrane skeleton for the structure of the virus. However, most of the mass of these viral glycoproteins, including the extensive carbohydrate, is on the outside of the virus.

One of the viral glycoproteins exhibits hemagglutinating activity and is referred to as the HN protein. This protein is responsible for the virus binding to the target cells. Thus HN plays a role in what was referred to in Section I as event 1 in the fusion process.

The other glycoprotein of the viral envelope is required for the virus to fuse with the target membrane and is consequently called the F protein. The structure of the F protein can be modeled, based on available indirect information on its structure. Figure 9.6 shows a schematic representation of this model. The F protein is activated by cleavage into F_1 and F_2 fragments, which are held together by disulfide bonds. The new amino terminal thus exposed contains a hydrophobic sequence that is apparently not used as the primary anchor for this protein in the envelope membrane. This hydrophobic sequence is referred to as the "fusion peptide" and appears to be required for the fusion of this virus.

Reference to the discussion of the fusion of N-methyl-DOPE LUV above provides an interesting framework within which to examine Sendai virus-mediated fusion. First, inhibitors of membrane fusion have been discovered that inhibit both Sendai virus fusion with cells and the fusion

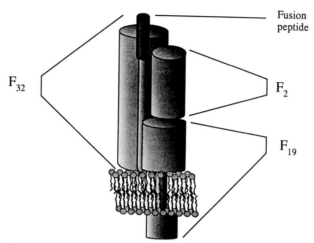

Fig. 9.6. Schematic representation of the fusion protein of Sendai virus. F_{19} and F_{32} refer to fragments of F_1 obtained by limited proteolysis. F_1 and F_2 are products of proteolytic activation of the Sendai F protein. The fusion peptide portion of the protein appears to be directly responsible for promotion of membrane fusion.

of N-methyl-DOPE LUV.[10] Since N-methyl-DOPE LUV fuse following a pathway involving nonlamellar intermediates and since Sendai virus can fuse with N-methyl-DOPE LUV without a glycolipid or protein receptor, the normal fusion pathway of Sendai likely involves nonlamellar intermediates. Second, the inhibitors just mentioned destabilize highly curved phospholipid assemblies. Furthermore, the "fusion peptide" of a similar virus, measles, was observed to destabilize bilayers and to facilitate the formation of nonbilayer structures, such as those putatively involved in the fusion of N-methyl-DOPE LUV.[11] These observations suggest that highly curved, nonlamellar intermediates may be involved in the fusion of Sendai virus with target membranes. Finally, inhibition by lysoPC of N-methyl-DOPE LUV fusion and Sendai viral fusion, and related evidence, suggests that the curvature of the intermediate is negative.

Therefore the role of the F protein may be to facilitate the formation of such intermediate non-bilayer structures sufficiently to catalyze membrane fusion. To do this some have suggested that the hydrophobic "fusion peptide" of the F protein may insert into the target membrane.[12] Model studies would suggest that this insertion could lead to a destabilization of the lipid bilayer and a formation of highly curved nonlamellar structures.

The overall pathway of fusion of Sendai virus can then be modeled as follows. The first step is binding of the virion to the target membrane, often enhanced by a ''receptor'' in the target membrane. The next step involves rearrangement of the surface of the virion and perhaps the surface of the target membrane. This rearrangement prepares the way for the action of the viral fusion protein. Step three is the facilitation of membrane destabilization by the viral fusion protein to form highly curved lipid assemblies as fusion intermediates. This allows the components of the two membranes to mix and fusion to proceed (see Fig. 9.7).

B. Influenza Virus

Influenza virus enters the cell by an alternate mechanism also involving membrane fusion.[13] Influenza is protypical for viruses that must experience acid pH to become fusion competent. This virus is taken up into the cell by receptor-mediated endocytosis. Binding initially occurs at the plasma membrane, mediated by cell surface receptors bearing sialic acid residues and by the HA envelope glycoprotein of influenza. Although the virus exhibits neuraminidase activity that can cleave off the sialic acid and thus the binding site for HA, the virus can nevertheless be endocytosed. The endocytotic vesicle fuses with an endosome. Proton pumps in the endosome membrane acidify the lumen of the endosome. It is after acidification that the fusion activity of the viral glycoprotein HA is manifest. At acid pH the HA glycoprotein of the viral envelope promotes membrane fusion. Entry of the viral nucleocapsid into the cell follows the fusion event.

The required pH change apparently leads to an alteration of the conformation of HA, changing it from an inactive form to an active form; HA

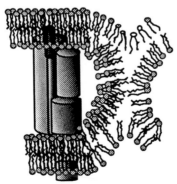

Fig. 9.7. Highly hypothetical induction of a fusion event by a Sendai fusion protein.

is a transmembrane glycoprotein. A crystal structure for the extramembraneous portion of the HA protein in the inactive form is available.[14] The structure describes a protein elongated along an axis approximately perpendicular to the membrane surface. No such structural information is available for the active form of HA.

In the current models, the HA protein forms trimers in a membrane. Oligomerization of HA is required for fusion competency.[15] The molecular details of the fusion pathway are not yet known, although the HA protein has been proposed to destabilize the membrane of the target cell.[16] The fusion peptide of the HA protein appears to interact with the target membrane.[17] As an alternative, stalks have been suggested to form between membranes as nonbilayer intermediates that facilitate membrane fusion.[18] Perhaps the HA protein plays a role in stabilizing the structures of such intermediates on the fusion pathway, thereby catalyzing the fusion event.

V. CALCIUM AND CELLULAR MEMBRANE FUSION

So far we have not discussed normal intracellular fusion events. Included in this category are fusion of coated vesicles with target membranes, endocytosis, release of neurotransmitters, and many other phenomena. For example, consider the role of calcium in fusion. In the model systems discussed above, calcium induced fusion at millimolar levels, or only slightly less. Inside cells, calcium is frequently present at three or four orders of magnitude lower concentration. Therefore, the systems discussed in this chapter do not yet model these cellular fusion events. Much remains to be learned. However, the protein synexin, a protein from adrenal medulla, can dramatically lower the calcium threshold for fusion.[19] With synexin, calcium concentrations near physiological levels are effective at promoting diffusion.

As particular biological examples, consider the following. Calcium plays a role in the fusion of synaptosomes with plasma membrane. The required concentrations for calcium are much lower than for the PS–calcium fusion systems. Likewise, rod outer segment disk membranes fuse with outer segment plasma membranes.[37] This fusion is dependent on micromolar calcium or less.

VI. INTRACELLULAR MEMBRANE FUSION

Available evidence suggests that intracellular fusions occur by different pathways with considerable opportunity for regulation. Regulation is more important in intracellular membrane fusion than in viral fusion, for example, since in the former indiscriminate fusion would compromise the integrity of the cell, whereas in the latter, indiscriminate fusion could enhance the incidence of viral replication.

A. Endocytosis

Consider first endocytosis. Endocytosis is one of the major pathways for specific uptake of extracellular materials. The first fusion event occurs when a coated vesicle pinches off the plasma membrane carrying the extracellular material. This vesicle is coated with a network of the protein clathrin, which is capable of forming baskets on its own or surrounding phospholipid vesicles.[20, 21] In this role, the self-assembly of clathrin likely plays a role in formation of the vesicle and may provide part of the energy required for the fusion process to occur. Clathrin consists of large (180 kDa) and small (25–29 kDa) subunits.[22] Three large and three small subunits assemble from a pool of soluble clathrin to form the basic unit of the clathrin basket.[23] Assembly of these basic units creates a spherical structure that can define the shape of the coated vesicle. Adapter proteins promote clathrin coat assembly[24] and interaction of the clathrin with the membrane.[25] Subsequently the coated vesicle loses its clathrin coat and fuses with endosomes (the clathrin basket around the vesicle probably sterically inhibits the membrane fusion event when the basket structure is complete by preventing close approach of the lipid bilayers). Uncoating of the vesicle requires ATP.[26] One of the smaller subunits of clathrin (LC_a) appears to regulate the disassembly of the clathrin coat through a conformational change, along with another associated protein that binds to a site on LC_a, hsc 70.[27] From the endosome, receptors can be recycled to the plasma membrane. Finally, endosome–lysozome fusion can take place, which brings the extracellular material to the compartment containing the hydrolytic enzymes for degradation. A good example of the process is the receptor-mediated endocytosis of human low-density lipoprotein (LDL) utilizing the LDL receptor (see Chapter 10).

B. Fusion in Intracellular Vesicular Transport

Conceptually, the opposite of this process occurs in secretion. Proteins for secretion are synthesized by a pathway that leaves them in the lumen of the ER. From there, vesicular transport involving sequential fusion steps carries the proteins to be secreted first to the Golgi and subsequently to the plasma membrane. Some of the steps of this process are now known.[28]

Consider the fusion of vesicles, derived from the ER, with the Golgi.[29] The process begins with formation of the vesicles from the ER. This is a regulated process in which the components of the transport vesicle are sorted from the components of the ER membranes. A nonclathrin coat forms around the budding vesicle and accompanies the vesicle when it leaves the ER. Adenosine triphosphate may be necessary to maintain this vesicle coat. This coat may have some similarity to the protein coats of clathrin-coated vesicles, although the proteins involved are different. Brefeldin A inhibits the assembly of this protein coat and thus blocks the whole pathway (one effect of which is to inhibit glycosylation of proteins).[30] This inhibition is expressed through an interaction of Brefeldin A with one of the coat proteins, β-COP, that leads to a dissociation of this coat protein from the vesicle.[31] The transport vesicle moves to the Golgi through some ill-defined, but directed (perhaps involving cytoskeletal elements), transport process. At the Golgi membrane, there is a G-protein-dependent uncoating of the vesicle and association of that vesicle with the Golgi membrane; GTP must be hydrolyzed for this process to proceed. Some evidence suggests that one of the G proteins involved consists of three subunits,[32] α, β, γ, similar to the G proteins involved in signal transduction (see Chapter 10). An activated $G_{\alpha i}$ subunit appears to inhibit uncoating when in the activated form. Conversion to the inactive form (thus permitting uncoating) is achieved by hydrolysis of GTP bound to the G_α subunit. Thus a nonhydrolyzable form of GTP (for example, GTPγS) inhibits the uncoating process.

Then a factor, referred to as NSF, is required. The acronym refers to N-ethylmaleimide (NEM)-sensitive factor; NEM can inhibit this process by chemically modifying the NSF protein required for this process.[33] NSF appears to be involved in establishing a necessary prefusion complex (although NSF is not likely the fusion protein). Other factors involved include ATP, acyl-CoA, proteins called SNAPS, and an integral membrane protein in the target membrane.[34] After uncoating the vesicle and establishment of the prefusion complex, disassembly of the prefusion complex and membrane fusion occur. At this point the components of the transport vesicle are incorporated into the Golgi membrane. Figure 9.1 schematically represents the events of this fusion pathway.

The fusion process used for fusion of ER-derived vesicles with Golgi membranes is widespread within the cell. Fusion of transport vesicles among the elements of the Golgi (cis, medial, and trans) utilizes the same NSF component.[35] Endocytosis and fusion of endocytotic vesicles with endosomes may also involve the same or similar factors. Exocytosis may also proceed by a similar pathway.[36] Specificity (or targeting to particular membranes of transport vesicles) may be directed by specific receptors in the target membrane for SNAPS. [38]

VII. SUMMARY

Membrane fusion is a phenomenon important to many cellular processes. It involves the mixing of two sets of membrane components, such that what was originally two membranes ends up as one membrane. Processes of endocytosis and formation of intracellular transport vesicles involve the reverse of this process, in which two separate membranes are formed from what was originally one membrane.

Studies of model membrane systems have led to the identification of important steps in the fusion of membranes. A close apposition of the two membranes that are to fuse is a preliminary process. This process requires removal of the barriers that prevent two membranes from getting close together. A catalytic event is likely required for the components of the two original membranes to mix as one membrane (or for one membrane to separate into two). Some destabilization of the lipid bilayers of the membrane(s) involved in the fusion process or some stabilization of intermediates on the fusion pathway is likely involved.

Fusion has been studied in lipid vesicle systems. These can be divided into a class that requires a negatively charged lipid and calcium and a class that requires flexibility in the morphology that the lipid assembly can adopt.

One of the best studied biological fusions is that involving enveloped viruses. Glycoproteins of the viral envelope, which are integral membrane proteins, promote aggregation of the virus with the target cell by interaction with a receptor on the target membrane and facilitate the membrane fusion event. Some viral fusion proteins can be active at neutral pH (and thus fuse with the plasma membrane) and some require acidic pH (and thus fuse after endocytosis and acidification of the endosomal compartment). Viral membrane fusion may follow a pathway involving highly curved nonbilayer intermediates, such as stalks, between the membranes that are to fuse.

Intracellular membrane fusions, for example, involving transport vesicles, require a greater array of factors for fusion to occur than in the viral fusion. Such vesicles must be denuded of their protein coat, associate with a target membrane, and assemble a fusion complex that involves several soluble proteins and at least one integral membrane protein. This process requires GTP (and G proteins) and ATP and can be inhibited by NEM.

REFERENCES

1. Bentz, J., and H. Ellens, "Membrane fusion: Kinetics and mechanisms," *Colloids Surfaces* **30** (1988): 65–112.

2. Wilschut, J., N. Düzgüneş, D. Hoekstra, and D. Papahadjopoulos, "Modulation of membrane fusion by membrane fluidity: Temperature dependence of divalent cation fusion of phosphatidylserine vesicles," *Biochemistry* **24** (1985): 8–14.
3. Ellens, H., D. P. Siegel, D. Alford, P. L. Yeagle, L. Boni, L. J. Lis, P. J. Quinn, and J. Bentz, "Membrane fusion and inverted phases," *Biochemistry* **28** (1989): 3692–3703.
4. Gagne, J., L. Stamatatos, T. Diacovo, S. W. Hui, P. L. Yeagle, and J. Silvius, "Physical properties and surface interactions of bilayer membranes containing N-methylated phosphatidylethanolamines," *Biochemistry* **24** (1985): 4400–4408.
5. Siegel, D. P., J. L. Burns, M. H. Chestnut, and Y. Talmon, "Intermediates in membrane fusion and bilayer/non-bilayer phase transitions imaged by time-resolved cryo-transmission electron microscopy," *Biophys. J.* **56** (1989): 161–169.
6. Yeagle, P. L., J. Young, S. W. Hui, and R. M. Epand, "On the mechanism of inhibition of viral and vesicle membrane fusion by carbobenzoxy-D-phenylalanyl-L-phenylalanyl-glycine," *Biochemistry* **31** (1992): 3177–3183.
7. Nieva, J.-L., F. M. Goni, and A. Alonso, "Liposome fusion catalytically induced by phospholipase C," *Biochemistry* **28** (1989): 7364–7367.
8. Hoekstra, D., "Membrane fusion of enveloped viruses: Especially a matter of proteins," *J. Bioenerg. Biomembr.* **22** (1990): 121–155.
9. Ohnishi, S., "Fusion of viral envelopes with cellular membranes," In *Membrane Fusion in Fertilization, Cellular transport and Viral Fusion* (N. Duzgunes, and F. Bronner, eds.) *Curr. Topics Membr. Transp.* (San Diego: Academic Press, 1988). **32**: 257–296.
10. Kelsey, D. R., T. D. Flanagan, J. Young, and P. L. Yeagle, "Peptide inhibitors of enveloped virus infection inhibit phospholipid vesicle fusion and Sendai virus fusion with phospholipid vesicles," *J. Biol. Chem.* **265** (1990): 12178–12183.
11. Yeagle, P. L., R. M. Epand, C. D. Richardson, and T. D. Flanagan, "Effects of the "fusion peptide" from measles virus on the structure of N-methyl dioleoylphosphatidylethanolamine membranes and their fusion with Sendai virus," *Biochim. Biophys. Acta* **1065** (1991): 49–53.
12. Novick, S. L., and D. Hoekstra, "Membrane penetration of Sendai virus glycoproteins during the early stages of fusion with liposomes as determined by hydrophobic photoaffinity labeling," *Proc. Natl. Acad. Sci. U.S.A.* **85** (1988): 7433–7437.
13. Bentz, J., H. Ellens, and D. Alford, "Liposomes, membrane fusion and cytoplasmic delivery," In *The Structure of Biological Membranes* (P. L. Yeagle ed.) (Boca Raton: CRC Press, 1992).
14. Wilson, I. A., J. J. Skehel, and D. C. Wiley, "Structure of the haemagglutinin membrane glycoprotein of influenza virus at 3 Å resolution," *Nature (London)* **289** (1981): 366–373.
15. Bentz, J., H. Ellens, and D. Alford, "An architecture for the fusion site of influenza hemagglutinin," *FEBS Lett.* **276** (1990): 1–5.
16. Stegmann, T., R. W. Doms, and A. Helenius, "Protein-mediated membrane fusion," *Annu. Rev. Biophys. Chem.* **18** (1989): 187–211.
17. Brunner, J., "Testing topological models for the membrane penetration of the fusion peptide of influenza virus hemagluttinin," *FEBS Lett.* **257** (1989): 369–372.
18. Siegel, D. P., *Viral fusion mechanisms* (J. Bentz, ed.) (Boca Raton: CRC Press, 1992).
19. Pollard, H. B., E. Rojas, and A. L. Burns, "Synexin and chromaffin granule membrane fusion," *Ann. N.Y. Acad. Sci.*: 524–541.
20. Keen, J. H., "Clathrin and associated assembly and disassembly proteins," *Annu. Rev. Biochem.* **59** (1990): 415–438.
21. Pearse, B. M. F., and M. S. Robinson, "Clathrin, adaptors, and sorting," *Annu. Rev. Cell Biol.* **6** (1990): 151–171.
22. Brodsky, F. M., "Living with clathrin: Its role in intracellular membrane traffic," *Science* **242** (1988): 1396–1402.

23. Goud, B., C. Huet, and D. Louvard, "Assembled and unassembled pools of clathrin: A quantitative study using an enzyme immunoassay," *J. Cell Biol.* **100** (1985): 521–527.

24. Zaremba, S., and J. H. Keen, "Assembly polypeptides from coated vesicles mediate reassembly of unique clathrin coats," *J. Cell. Biol.* **28** (1983): 47–58.

25. Virshup, D. M., and V. Bennett, "Clathrin coated vesicle assembly polypeptides: Physical properties and reconstitution studies with brain membranes," *J. Cell Biol.* **106** (1988): 39–50.

26. Schmid, S. L., and L. L. Carter "ATP is required for receptor-mediated endocytosis in intact cells," *J. Cell Biol.* **111** (1990): 2307–2318.

27. Rothman, J. E., and S. L., Schmid, "Enzymatic recycling of clathrin from coated vesicles," *Cell* (*Cambridge, Mass.*) **46** (1986): 5–9.

28. Wattenberg, B. W., "Vesicular traffic in eukaryotic cells," In *The Structure of Biological Membranes* (P. L. Yeagle, ed.) (Boca Raton: CRC Press, 1992), 997–1046.

29. Wilson, D. W., S. W. Whiteheart, L. Orci, and J. E. Rothman, "Intracellular membrane fusion," *Trends Biochem. Sci.* **16** (1991): 334–337.

30. Orci, L., M. Tagaya, M. Amherdt, A. Perrelet, J. G. Donaldson, J. Lippencott-Schwartz, R. D. Klausner, and J. E. Rothman, "Brefeldin A: A drug that blocks secretion, prevents the assembly of non-clathrin coated buds on Golgi cisternae," *Cell* (*Cambridge, Mass.*) **64** (1991): 1183–1195.

31. Donaldson, J. E., J. Lippencott-Schwartz, G. S. Bloom, T. E. Kreis, and R. D. Klausner, "Dissociation of a 110-kDa peripheral membrane protein from the Golgi apparatus is an early event in brefeldin A action," *J. Cell Biol.* **111** (1990); 2295–2306.

32. Donaldson, J. G., R. A. Kahn, J. Lippencott-Schwartz, and R. D. Klausner, "Binding of ARF and β-cop to Golgi membranes: Possible regulation by a trumeric g protein," *Science* **254** (1991): 1197–1199.

33. Glick, B. S., and J. E. Rothman, "Possible role for fatty acyl-Coenzyme A in intracellular protein transport," *Nature* (*London*) **326** (1987): 309–312.

34. Clary, D. O., I. C. Griff, and J. E. Rothman, "SNAPs: A family of NSF attachment proteins involved in intracellular membrane fusion in animals and yeast," *Cell* (*Cambridge, Mass.*) **61** (1990): 709–721.

35. Rothman, J. E., and L. Orsi, "Movement of proteins through the Golgi stack: A molecular dissection of vesicular transport," *FASEB J.* **4** (1990): 1460–1468.

36. Edwardson, J. M., and P. U. Daniels-Holgate, "Reconstitution in vitro of a membrane-fusion event involved in constitutive exocytosis," *Biochem. J.* **285** (1992): 383–385.

37. Boesze-Battaglia, K., A. D. Albert, and P. L. Yeagle, "Fusion between disk membranes and plasma membrane of bovine photoreceptor cells is calcium dependent," *Biochemistry* **31** (1992): 3733–3738.

38. Söllner, T., S. W. Whiteheart, M. Brunner, H. Erdjument-Bromage, S. Geromanos, P. Tempst, and J. Rothman, "SNAP receptors implicated in vesicle targetting and fusion," *Nature* (*London*) **362** (1993): 318–322.

10

Membrane Receptors

The plasma membranes of cells separate the inside of the cells from the outside. Therefore if there is to be any communication between the outside of the cell (that is, from the organism as a whole) and the inside of the cell, that communication must be mediated by the plasma membrane.

The process whereby signals external to the cell can alter intracellular behavior is called signal transduction. The signals can be hormones or other molecular species that encounter the plasma membrane of the cell, often sent from other parts of the same organism. In one specialized case, the signal can be light. The plasma membrane component that mediates the signal transduction is called a receptor.

Receptors of the plasma membrane are transmembrane proteins, usually with binding sites on the portion of the protein exposed to the exterior of the cell. In general, occupation of these binding sites by agonists (molecules that will elicit a response from the receptor) will cause a conformational change in the receptor protein that is propagated to the portion of this transmembrane protein that is exposed to the interior of the cell. On the cytoplasmic face of the plasma membrane, interactions can occur between the receptor and other proteins that lead to the subsequent stages of the signal transduction process.

This chapter describes several different kinds of membrane receptors that function in signal transduction. First to be described is the ever-expanding field of receptors that utilize G proteins as part of their means to communicate with the interior of the cell. Next, a receptor that operates a cation channel will be examined (the acetylcholine receptor). This will be followed by a look at the insulin receptor. Finally, the LDL receptor will be used as an example of receptors that operate by receptor-mediated endocytosis.

I. G-PROTEIN RECEPTORS

In this section we will examine general patterns in receptor function with respect to receptors utilizing G proteins. A variety of signal transduction systems utilize this class of receptor. Among them are the rhodopsins of the retinal photoreceptors (cones and rods), the β-adrenergic receptor involved in hormonal (catecholamine) signaling to cellular metabolism, α_1-adrenergic receptor, muscarinic cholinergic receptors, vasopressin receptor, angiotensin receptor, and the olfactory sensory systems.

With respect to this group of receptors, transmembrane signaling involves three different kinds of proteins. The first kind of protein is the receptor, a transmembrane protein. This receptor will usually be found in the plasma membrane, and thus located on the cell surface, like the β-adrenergic receptor.[1] Alternatively, this protein may be located in an intracellular membrane, as in the visual transduction system of retinal rod cells.[2] The receptor is the recognition component of the signal transduction system. It is specific for the particular signal to be recognized. Signals include catecholamines, serotonin, purines, peptides, and light.

The second kind of protein is the coupling factor, often referred to in these systems as G protein.[3] This protein may be membrane bound, or may be soluble, or may shuttle between the two forms. The G protein binds GTP and has hydrolytic activity, turning over GTP to GDP + P_i.

The G protein, when activated by the receptor-mediated response, in turn activates the third kind of protein in this signal transduction system, an enzyme that controls the level of an intracellular second messenger. In this general way, a signal is transduced to an intracellular biochemical event with serious functional consequences for the cell in question.

The following two examples of receptor systems that utilize G proteins both lead to second messenger production. One is the β-adrenergic receptor. This receptor system contains a plasma membrane-bound receptor protein. When the specific agonist binds to this receptor, the receptor couples to the G protein, which in turn can activate adenylate cyclase and raise intracellular cAMP levels. The increase in cAMP can lead to activation of cAMP-dependent protein kinase, which in turn can lead to protein phosphorylation and specific metabolic consequences.

The visual pigment, rhodopsin (38 kDa), of vertebrate retinal rod cells also functions as a receptor. Light-activated rhodopsin is capable of activating G protein, sometimes called transducin in this system, which in turn can activate phosphodiesterase. In the rod outer segment, this leads to a reduction in cGMP, which may result, in one current hypothesis, in the closing of Na^+ channels in the plasma membrane.

With this overview, one can now examine in more detail each of the components of these signal transduction systems. Begin with the receptor, the primary recognition point of signal transduction and the site for specificity with respect to the signal given. In each case of signal transduction systems utilizing G proteins, the receptor is a membrane bound protein. However, the similarity in these receptors goes far beyond that. The amino acid sequences of a β-adrenergic receptor, a rhodopsin, and a muscarinic receptor, among others, have been determined. This was accomplished through cloning cDNA's for each of these receptors, sequencing the cDNA and translating the triplet code for the amino acids in the appropriate reading frame. When the sequences are compared pairwise for these three receptors, there is 20–30% identity among the amino acid sequences. When taking into account conservative replacements, the structural homology becomes striking.

Figure 10.1 shows a deduced arrangement in the membrane for the amino acid sequence of rhodopsin. The regions of the polypeptide that are found in the membrane are those regions of the amino acid sequence that are hydrophobic. In most cases they correspond to linear sequences of hydrophobic amino acids.

This structure applies in general to both the β_2-adrenergic receptor and rhodopsin. In each case the amino terminus is located in a morphologically analogous position. In the β_2-adrenergic receptor, the amino terminus is located on the exterior of the plasma membrane, or in an extracellular position. The amino terminus of rhodopsin is located in the intradiskal space, which is morphologically identical to the extracellular surface, considering the morphology of the formation of the disk membranes. Furthermore, the amino terminus is glycosylated at two sites in both receptors.

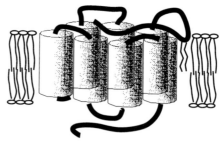

Fig. 10.1. Schematic representation of the photoreceptor, rhodopsin, in a membrane. The acylation of the two cysteines on the carboxyl terminus is represented by the two wavy lines on that portion of the protein. The seven transmembrane α helices are represented by cylinders.

The carboxyl terminus is located on the opposite side of the membrane. This is a consequence of the seven, or odd number, transmembrane segments in these proteins.

As was mentioned earlier, these receptors, when activated, interact with the next protein in the sequence of signal transduction, the G protein. Some information is available on the likely sites of such interaction from studies on rhodopsin. Two regions have been implicated in interactions of rhodopsin with G_t, or transducin. One is the third cytoplasmic loop. The other is the carboxyl terminus on the cytoplasmic surface of the membrane. Presumably, interaction with agonist (or light) causes a conformational change in the protein such that the cytoplasmic surface of the protein has a conformation suitable for binding the G protein. As will be discussed further later, the G proteins from the two systems can be interchanged *in vitro* and function retained. Therefore, similar regions on the β_2-adrenergic receptor are likely involved in binding the G protein. The structures represented here show that such similarities are likely.

In the case of the visual pigment rhodopsin, the receptor is activated by absorption of one photon of light. The absorption of light leads to an isomerization of the chromophore 11-*cis*-retinal. This chromophore is part of a Schiff's base with a lysine on one of the transmembrane helices of rhodopsin. On absorption of one photon of light, the 11-*cis*-retinal isomerizes to all-*trans*-retinal. The change in conformation of the retinal forces an alteration in the disposition of the transmembrane helices and thus a conformational change in the receptor protein. The conformational changes can be monitored by their characteristic absorption. The state known as Metarhodopsin II is the conformational state characterized by absorption spectroscopy that appears to be capable of activating the G_t protein.

The structures of both receptors offer possible mechanisms of desensitization. On the carboxyl terminal region of the protein are serines and threonines that are potential sites for protein phosphorylation.[4] Kinase activity has been found for both receptor systems that leads to phosphorylation of these sites. For the β_2-adrenergic receptor, a specific receptor kinase (80 kDa) phosphorylates the receptor only when agonist is bound.[5] Correspondingly for rhodopsin, there is a light-activated kinase (63 kDa) that phosphorylates rhodopsin.[6] The analogy is that rhodopsin is a substrate for the kinase only when it is activated, i.e., when it absorbs a photon of light. Therefore it would appear that the particular protein conformation that results when the receptor is activated is the only form of the receptor that functions as substrate for the receptor kinase. The result in both cases is that the receptor, when phosphorylated, is inhibited from interacting with G protein to activate the G protein. [Phosphorylation of rhodopsin also enhances the binding of the regulatory protein arrestin (48 kDa), which further inhibits signal transduction,[7] whereas the analogous reaction enhances binding of β-arrestin to the β-adrenergic receptor.] Therefore, signal transduction is inhibited. The extent of the inhibition may be controlled by the level of phosphorylation of the receptor. These receptors can be multiply phosphorylated. With phosphorylation, the system is desensitized, a well-known phenomenon in receptor physiology, resulting from prolonged stimulation of the receptor system. Long-term desensitization of the β-adrenergic receptor, for example, can also be achieved through increased degradation of the receptor.

Again at this level, crossover in functionality has been demonstrated in these systems. The kinase from the β_2-adrenergic receptor will phosphorylate rhodopsin, but only when rhodopsin has been activated by light. The kinase from the disk system will phosphorylate the β_2-adrenergic receptor, but only when the receptor is agonist bound.

That description summarizes the functional state of the receptor in these G-protein receptor systems. The next level in these signal transduction systems is the G protein. The G proteins consist of three subunits, α, β, and γ. The activated receptor interacts with the heterotrimer and leads to an exchange of GTP for GDP on the α subunit. This is not a hydrolysis reaction; the GDP leaves the binding site on the protein and is replaced by GTP. At the same time, the α subunit dissociates from the $\beta\gamma$ dimer. This activated G_α subunit protein can then diffuse laterally in the plane of the membrane and bind to a target (effector) enzyme. Binding to the effector enzyme by the G_α subunit modulates the activity of that enzyme. Depending on the system, the G protein may stimulate the effector enzyme (G_s) or may inhibit the effector enzyme (G_i). The effector enzyme produces the change in the cytoplasmic level of the second messenger, and thus communicates a signal to the cell, which originated from a stimulus that was exterior to the cell. The cycle of the G protein is completed when the GTPase activity of the G protein is manifest, turning over GTP to GDP. With GDP bound, the α subunit can no longer activate the target (effector) enzyme modulating the second messenger.

Once again it is instructive to compare the primary structure of the G proteins from various receptor systems. In particular, consider the β subunit. The β subunit has been sequenced from liver receptor systems by obtaining the cDNA clone for the β subunit. The open reading frame of the cDNA clone codes for a sequence identical to transducin, the G protein from the retinal rod cell photoreceptor system. The liver G proteins in question here are the G_s, the G protein that is capable of stimulating adenylate cyclase, and the G_i, the G protein that is capable of inhibiting adenylate cyclase. The specificity of the G protein action then likely arises from the α subunit, which must interact with, and modulate the activity of, the effector enzyme that controls the second messenger levels.

Although the similarities in the G proteins are striking, there are differences. In the β_2-adrenergic receptor system, the G protein is membrane bound. In the photoreceptor system, the G protein can be found in soluble form, as well as membrane bound. For the G_α subunit of these G proteins, membrane association may be promoted by myristoylation on the amino terminus. Association of the $G_{\beta\gamma}$ subunit with the membrane can be promoted by a geranylgeranyl moiety in a thioester linkage to a serine at the carboxyl terminus.

The G proteins are the target of several bacterial toxins, including pertussis toxin and cholera toxin. The catalytic subunits of these toxins are capable of modifying the G protein and thus modulating its activity. For example, cholera toxin leads to ADP-ribosylation of the $G_{s\alpha}$ subunit, freezing it in an active conformation. This leads to uncontrolled stimulation of the effector enzymes (adenylate cyclase) and an extensive secretion (loss) of intracellular fluids. Pertussis toxin leads to similar modification of $G_{i\alpha}$, which limits the ability to inhibit effector enzymes.

Finally this discussion can turn to the target of these signal transduction systems. These are the effector enzymes that modulate the intracellular second messenger levels. There are several classes of such enzymes. One class is the adenylate cyclase of the plasma membrane. This enzyme synthesizes cAMP, a second messenger that regulates, among other things, the cAMP-dependent protein kinase. This in turn phosphorylates, among other things, metabolic enzymes and regulates their activity accordingly. The adenylate cyclase was one of the first receptor-regulated systems described.

A second class of receptor regulated enzymes is phosphodiesterase. This enzyme figures prominently in the photoreceptor system. The phosphodiesterase contains a regulatory subunit. When this subunit is bound, the enzyme is largely inactive. When the α subunit of G_t interacts with the phosphodiesterase, the regulatory subunit of the PDE is released, and the enzyme becomes activated. When the GTPase activity hydrolyzes the GTP to GDP, the α subunit of G_t is no longer capable of dissociating the regulatory subunit of phosphodiesterase, and the phosphodiesterase becomes inactive.

The function of the PDE in the rod photoreceptor cell is to regulate the levels of cGMP, which in turn regulate the Na^+ channels in the plasma membrane. In particular, the sodium channels are sensitive to the occupancy of a cGMP binding site on the channel. When cGMP is high, these sites are occupied and the sodium channel is open. When phosphodiesterase is activated, the levels of cGMP in the rod outer segment decrease, the cGMP site on the sodium channel becomes vacated, and the sodium channel closes. Closure of the sodium channel inhibits a current that is normally sustained by an inward flux of sodium through the channel. This leads to a hyperpolarization of the plasma membrane and an electrical signal that is transmitted to the synapse on the other end of the rod cell. Signal transduction in the rod cell through G-protein receptors is schematically represented in Fig. 10.2.

A third class of receptor-regulated enzymes is phospholipase. Phospholipases cleave phospholipids as described in Chapter 2. One receptor-stimulated phospholipase is phospholipase A_2. Phospholipase A_2 cleaves phospholipids to release one fatty acid from the 2' position of the glycerol, leaving lysophospholipid. One example is found in the retinal rod cell. Often the fatty acid released is arachidonic acid, which is a precurser for prostaglandin biosynthesis.

Another phospholipase, which is activated by receptors and which has caused considerable interest recently, is phospholipase C. This releases the headgroup of the phospholipid and leaves diacylglycerol. For example, PI can be phosphorylated to form PIP and PIP_2. When phospholipase C acts on these phosphorylated phospholipid substrates, in addition to re-

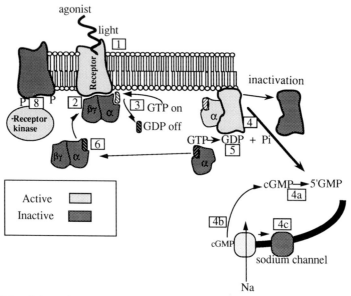

Fig. 10.2. Schematic representation of signal transduction in the rod cell through G-protein receptors.

lease of diacylglycerol, IP$_3$ is released. Both of these products have important intracellular effects.

Diacylglycerol is capable of activating protein kinase C, along with other cofactors, and thus beginning another cascade of metabolic regulation through phosphorylation of protein substrates. The other product is IP$_3$ if the substrate for the reaction is PIP$_2$; IP$_3$ in some of its isomers has been suggested to cause release of calcium from intracellular stores such as endoplasmic reticulum. Release corresponds to the binding of IP$_3$ to a receptor operating a calcium channel in the ER.[8] This leads to an increase in intracellular calcium levels in response to an extracellular signal. These activities are summarized schematically in Fig. 10.3.

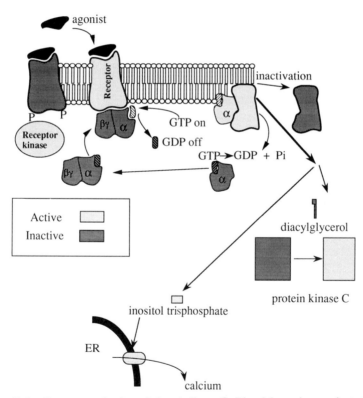

Fig. 10.3. Receptor activation of phospholipase C. Diacylglycerol, a product of phospholipase C, participates in activatioin of protein kinase C. Inositol trisphosphate binds to a receptor on the ER and opens a calcium channel.

II. RECEPTOR-COUPLED CHANNELS

Some receptors are coupled to ion channels. The example to be used here, the nicotinic acetylcholine receptor, is a receptor coupled to a cation channel.[9] The nicotinic acetylcholine receptor is a transmembrane protein of the plasma membrane. This receptor has a binding site specific for the chemical structure of acetylcholine. Occupancy of that site by acetylcholine opens a sodium channel in the same protein (see Fig. 10.4). Some analogies exist between this receptor and receptors for δ-aminobutyric acid (GABA), glutamate, and glycine. Possibly, the calcium channel that responds to IP_3 (Section I) is an analogous receptor-coupled channel.

Four different kinds of subunits are found in the nicotinic acetylcholine receptor; α, β, γ, and δ. The composition of a receptor is $\alpha_2\beta\gamma\delta$. These proteins are glycosylated. The α and β subunits were acylated in some forms. Oligomeric forms of the receptor have been found.

The high-affinity binding site for acetylcholine is on the extracellular portion of the α subunit of the receptor. In its most simple form, this acetylcholine receptor operates in a straightforward fashion. Occupancy of the ligand binding site for acetylcholine by this agonist leads to an open state of the Na^+ channel, which is an integral part of the same protein. Apparently the energy of binding of acetylcholine is utilized for a conformational change in the protein that opens the Na^+ channel. The depolarization of the plasma membrane that occurs from the opening of this channel triggers other receptor-independent sodium channels to open in a manner that transmits an action potential along the plasma membrane.

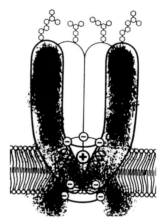

Fig. 10.4. Schematic representation of the acetylcholine receptor. Reprinted with permission from Hucho, F. and R. Hilgenfeld, *FEBS Lett.* **257** (1989): 17.

The activity of the acetylcholine receptor can be modulated. Over longer term, the bound ligand can induce a desensitized form of the receptor, which corresponds to a high affinity (for acetylcholine) binding site and a closed channel conformation. Furthermore, each of the subunits can be phosphorylated. Some evidence has been interpreted in terms of receptor desensitization on phosphorylation.

The function of the receptor is dependent on the lipid composition of the membrane. Both negatively charged lipids and cholesterol are components of the plasma membrane that play a role in maintaining the activity of this receptor. Thus the acetylcholine receptor may be another example of a plasma membrane protein that is activated by the cholesterol in that membrane (see Chapter 5).

III. INSULIN RECEPTOR

The insulin receptor is one of the most important and perhaps one of the most complex of the plasma membrane receptors. It is involved in extensive regulation of intracellular metabolism, with both short-term effects and long-term effects. Short-term effects include influences on carbohydrate and fat metabolism. Long-term effects include influences on gene expression and on protein synthesis. The ligand for this receptor is the peptide hormone, insulin. Insulin consists of two peptides joined by disulfide bonds. Originally synthesized as a single polypeptide precurser in the pancreas, it undergoes post-translational modification involving a loss of some of the peptide due to action of proteases. This cleaves the insulin into two smaller peptides that remain covalently attached.

As in the case of the other receptors examined so far in this chapter, it is a transmembrane protein. Insulin acts as an extracellular hormone, influencing cellular metabolism by occupancy of a binding site on the insulin receptor. Influences include a growth-factor-like response to insulin. As a transmembrane protein, the insulin receptor can then be involved in signal transduction carrying the extracellular signal of the hormone to the inside of the cell. The intracellular conduction of the signal has not yet been shown to involve classic second messengers.

The critical importance of the insulin receptor to humans can be observed in the family of diseases referred to as diabetes. The insulin receptor regulates glucose homeostasis in humans. A malfunctioning receptor will lead to uncontrolled serum glucose levels. This loss of regulation can result in the problems in glucose metabolism, including high serum glucose levels, that characterize diabetes.

The cellular response to insulin is rapid and dramatic. In a matter of minutes, glucose transport can increase by over an order of magnitude, particularly in responsive cells such as muscle and fat cells.

The insulin receptor protein consists of a dimer of dimers ($\alpha_2\beta_2$) held together by disulfide bonds. The β subunit is the transmembrane portion of the receptor with a molecular weight near 70,000. In the carboxyl terminal region of the β subunit is found the tyrosine kinase activity of the insulin receptor. The α subunit is entirely extramembraneous and the polypeptide has a molecular weight near 85,000. The α subunit contains the insulin binding domain. Interestingly the two subunits are synthesized from the same mRNA.

The tyrosine kinase activity of the receptor is stimulated by occupancy of the ligand binding site on the α subunit by insulin. This tyrosine kinase activity autophosphorylates the insulin receptor as well as phosphorylating other substrates in the cell. Other receptors may be in the same family because they also exhibit tyrosine kinase activity on binding ligand. Included in this group is the receptor for epidermal growth factor.

One of the functions of the insulin receptor must be to effect the classical response of a cell to insulin, an increase in glucose uptake. This response is promoted, at least in part, by a recruitment of glucose transporters to the plasma membrane to increase the flux of glucose across the plasma membrane. These transporters facilitate glucose diffusion across the plasma membrane. (The active glucose transport process described in Chapter 8 that is dependent on the sodium gradient is not the process stimulated by insulin.) These glucose transporters facilitate passive diffusion that responds to a concentration gradient of glucose. At least two different kinds of glucose transporters, which have been termed GLUT1 and GLUT4, can be expressed.[10]

The most simple picture is that excess glucose transporters are "stored" near the plasma membrane and that some of these transporters are inserted into the plasma membrane to increase the flux of glucose into the cell.[11] New protein synthesis is not required for this immediate response to insulin. "Storage" in this model would involve membrane vesicles since the glucose transporter is a transmembrane protein and must reside in a membrane. The GLUT4 form of the transporter exhibits the greatest increase in number in the plasma membrane on insulin stimulation.[12] Insertion into the plasma membrane of extra glucose transporters is a regulated event. Insertion likely involves membrane fusion because of the membranous nature of the storage form of the transporters. Insertion could involve a G-protein-regulated event that is similar to that used for other intracellular fusions (see Chapter 9).

Another mechanism for insulin regulation of the uptake of glucose is through modulation of the intrinsic ability of the transporter to facilitate glucose transport. It is possible that effector proteins may mediate between the insulin receptor and the glucose transporter. Perhaps even the kinase activity of the receptor is important, either directly, or indirectly, in this regulation.

IV. RECEPTOR-MEDIATED ENDOCYTOSIS

The last class of receptor that will be considered in this chapter contains receptors that operate by the mechanism of receptor-mediated endocytosis. Among the receptors that utilize this mechanism are the LDL receptor[13] and the transferrin receptor (for uptake of iron into the cell).[14] The example to be examined here will be the LDL receptor.

The LDL receptor is responsible for the uptake of LDL from serum. A major role is played by this receptor in uptake of LDL into liver in humans. This receptor-mediated process is a crucial part of the much larger process of cholesterol homeostasis. Cholesterol and lipids are transported in serum via the serum lipoproteins. These lipoproteins include (from largest to smallest) chylomicrons, derived from intestinal mucosa; very low density lipoproteins (VLDL), derived from liver; LDL; and high density lipoproteins (HDL), also derived from liver. A schematic picture of a serum lipoprotein (LDL) appears in Fig. 10.5. The serum lipoproteins contain triglyceride, cholesterol ester, cholesterol, phospholipids, and protein. The phospholipids (PC and sphingomyelin) form a monolayer of amphipathic lipids on the surface of the lipoprotein, with the headgroups of these lipids facing the aqueous medium. The nonpolar lipids, triglycerides and cholesterol esters, reside mostly in the core of the LDL. The unesterified cholesterol is located mostly in the surface monolayer with the PC and the sphingomyelin. Low density lipoprotein is the major serum carrier of cholesterol. Therefore LDL uptake by the liver is the major means by which cholesterol is removed from the circulation. It is for this reason that LDL receptor plays a major role in cholesterol homeostasis in humans.

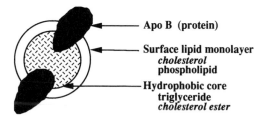

Fig. 10.5.   Schematic representation of the structure of LDL.

The critical importance of the LDL receptor can be dramatically identified in the genetic disease familial hypercholesterolemia. This family of
diseases is characterized by a deficiency or a complete absence of function
of the LDL receptor in the plasma membrane of liver cells. The result is
a decrease in LDL uptake by the liver and a corresponding increase in
serum LDL level (which results clinically in an increase in serum cholesterol levels). The serum cholesterol levels that result can be several times
higher than normal for the human population and fatal atherosclerosis
results, often rapidly (at an early age).

There is another connection between the LDL receptor and the disease
atherosclerosis. There is a receptor in macrophages that operates in much
the same fashion as the liver LDL receptor in LDL uptake. However,
the macrophage LDL receptor does not recognize normal LDL. Instead
uptake is of oxidized LDL by this macrophage receptor. Oxidation can
be the result of exposure of LDL to cells that are producing oxidizing
compounds, such as H_2O_2, as a result of the cellular metabolism. This
oxidized LDL is taken up by receptor-mediated endocytosis and can lead
to a massive intracellular accumulation of the components of LDL. This
process can lead to the foam cells that may be part of the pathogenesis
of atherosclerotic plaques.

The LDL receptor is a transmembrane protein of the plasma membrane,
like the other receptors that have already been examined. The polypeptide
has a molecular weight near 96,000 and in addition is glycosylated. There
is a binding site on the extracellular portion of the LDL receptor that
recognizes the protein component of LDL. In particular, LDL has a B
peptide that is the major protein component of the LDL particle. The
LDL receptor has a binding site that recognizes a portion of the B peptide.
It also can bind the E peptide, which can be found on LDL and on some
other serum lipoproteins.

Occupancy of an LDL receptor by LDL can stimulate the series of
events known as receptor-mediated endocytosis. Ligand–receptor complexes aggregate in an invagination of the plasma membrane called a
coated pit. The coated pit is lined (on the cytoplasmic surface of
the plasma membrane) by a protein coat of clathrin (see Chapter 9).
The cytoplasmic domains of the receptor may interact with adapter proteins that are involved in the interaction between the clathrin coat
and the membrane of the coated pit/endocytic vesicle.[15,16] In the process
of receptor-mediated endocytosis, basket formation is coincident with the
formation of a vesicle the coated vesicle.[17] The LDL is on the inside of
this coated vesicle, bound to its receptor. At about this point in the cycle,
a sorting process occurs. The receptor is separated from the LDL and
the receptor is recycled to the plasma membrane. Thus this receptor can

be reused. The LDL is transported to the lysosome and catabolized. This process is schematically represented in Fig. 10.6.

V. SUMMARY

Membrane receptors mediate signal transduction for cellular responses to extracellular stimuli. Membrane receptors are usually transmembrane proteins; this allows receptors to communicate with both sides of the plasma membrane of a cell. The extracellular portion of the receptor often has a binding site for a ligand. Ligands can be, for example, hormones or neurotransmitters. An alternative to ligand binding is found in the photoreceptor, in which a photon of light substitutes for a ligand for rhodopsin.

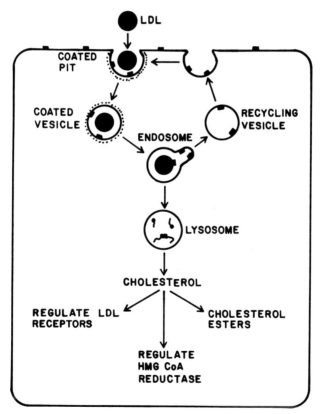

Fig. 10.6. Receptor-mediated endocytosis of LDL.

The receptors discussed in this chapter can be grouped into four groups: G-protein receptors, receptors coupled to ion channels, receptors with kinase activity, and receptors involved in receptor-mediated endocytosis. The mechanisms of signal transduction by each of these receptor groups is different.

G proteins mediate the initial response of receptors to which they are coupled by modulating the activity of target enzymes. G proteins bind to activated receptors and themselves become activated through an exchange of GTP (on the protein) for GDP (off the protein) on the α subunit. It is the α subunit that is involved in modulating (inhibiting or stimulating) the activity of the target enzyme. The G protein is then recycled through a hydrolysis of GTP to GDP, inactivation of the G_α subunit, and reassociation of the G protein subunits. Phosphorylation of the receptor desensitizes the receptor to further stimulus. Target enzymes include phospholipases, adenylate cyclase, and phosphodiesterases. The levels of second messengers are modulated by these target enzymes and they include diacylglycerols, phosphorylated inositols, cAMP,and cGMP.

The nicotinic acetylcholine receptor is coupled to a sodium channel. Occupancy of the ligand binding site by acetylcholine leads to an opening of the sodium channel and a depolarization of the membrane.

The insulin receptor modulates the response of cells to insulin. This receptor has kinase activity. Also, this receptor leads to a recruitment of glucose transporters into the plasma membrane, which permits an increase in the flux of glucose into the cell.

The LDL receptor is an example of receptors that are involved in receptor-mediated endocytosis. Occupancy of the ligand binding site on this receptor by LDL starts the endocytosis process. Coated vesicles transport the receptor–ligand complex inside the cell. After sorting, the LDL is catabolized and the receptor can be recycled to the plasma membrane.

REFERENCES

1. Caron, M. G., and R. J. Lefkowitz, "Structure and regulation of G protein-coupled receptors: the beta-2-adrenergic receptor as a model," *Vitamin Horm.* **46** (1991): 1–39.
2. Hargrave, P. A., and J. H. McDowell, "Rhodopsin and phototransduction: A model system for G protein-linked receptors," *FASEB J.* **6** (1992): 2323–2331.
3. Lefkowitz, R. J., "G proteins: The subunit story thickens," *Nature (London)* **358** (1992): 372.
4. Dohlman, H. G., M. Bouvier, J. L. Benovic, M. G. Caron, and R. J. Lefkowitz, "The multiple membrane spanning topography of the β_2-adrenergic receptor:Localization of the sites of binding, glycosylation and regulatory phosphorylation by limited proteolysis," *J. Biol. Chem.* **262** (1987): 14282–14288.

5. Benovic, J. L., R. H. Strasser, M. G. Caron, and R. J. Lefkowitz, "β-adrenergic receptor kinase: Identification of a novel protein kinase that phosphorylates the agonist-occupied form of the receptor," *Proc. Natl. Acad. Sci. U.S.A.* **83**, (1986): 2797–2801.

6. Palczewski, K., J. H. McDowell, and P. A. Hargrave, "Rhodopsin kinase: Substrate specificity and factors that influence activity," *Biochemistry* **27** (1988): 2306–2313.

7. Wilden, U., S. W. Hall, and H. Kuhn, "Phosphatidiesterase activated by photoexited rhodopsin is quenched when rhodopsin is phosphorylated and binds the intrinsic 48-kDa protein of rod outer segments," *Proc. Natl. Acad. Sci. U.S.A.* **83** (1986): 1174–1178.

8. Ferris, C. D., and S. H. Snyder, "Inositol 1,4,5-triphosphate-activated calcium channels," *Annu. Rev. Physiol.* **54** (1992): 469–488.

9. Pradier, L., and M. G. McNamee, "The nicotinic acetylcholine receptor," In *The Structure of Cell Membranes*, (P. L. Yeagle, ed.) (Boca Raton: CRC Press, 1992), 1047–1106.

10. Gould, G. W., and G. I. Bell, "Facilitative glucose transporters: An expanding family," *Trends Biochem. Sci.* **15** (1990): 18–23.

11. Simpson, I. A., and S. W. Cushman, "Hormonal regulation of mammalian glucose transport," *Annu. Rev. Biochem.* **55** (1986): 1059–1089.

12. Slot, J. W., "Immuno-localization of the insulin regulatable glucose transporter in brown adipose tissue of the rat," *J. Cell. Biol.* **113** (1991): 123–135.

13. Bradley, W. A., and S. H. Gianturco, "Lipoprotein receptors in cholesterol metabolism," In *Biology of Cholesterol* (P. L. Yeagle, ed.) (Boca Raton: CRC Press, 1988), 95–120.

14. Klausner, R. D., J. van Renswoude, G. Ashwell, C. Kempf, A. N. Schechster, A. Dean, and K. R. Bridges, "Receptor-mediated endocytosis of transferrin in K562 cells," *J. Biol. Chem.* **258** (1983): 4715–4724.

15. Pearse, B. M. F., "Receptors compete for adaptors found in plasma membrane coated pits?," *EMBO J.* **7** (1988): 3331–3336.

16. Glickman, J. N., E. Conibear, and B. M. F. Pearse, "Specificity of binding of clathrin adaptors to signals on the mannose-6-phosphate/insulin-like growth factor II receptor," *EMBO J.* **8** (1989): 1041–1047.

17. Pearse, B. M. F., and R. A. Crowther, "Structure and assembly of coated vesicles," *Annu. Rev. Biophys. Biophys. Chem.* **16** (1987): 49–68.

11

The Metabolism of Membrane Lipids

The biosynthesis of the lipid components of membranes is a vast and complicated process because of the wide variety of lipid structures found in biological membranes (for further reading and detailed references, the reader is referred to Ref. 1). The structures of lipids vary in both the composition of their hydrocarbon chains and the structure of their polar headgroups. Taking into account the various possible combinations, biological membranes contain literally thousands of individual lipids species. The following discussion will emphasize the synthesis of some of the more common lipids and their components.

The discussion will begin with the biosynthesis of fatty acids, since most classes of lipids and some membrane proteins contain fatty acids in their structures. Closely linked with the synthesis of the fatty acids is the oxidative degradation of the fatty acids, which will also be reviewed. The discussion will then turn to the biosynthesis of the major phospholipid classes of cellular membranes of prokaryotes and eukaryotes. This will be followed by a consideration of the biosynthesis of sphingolipids, including some glycolipids. The chapter will then end with a discussion of the biosynthesis of cholesterol.

There are differences between the pathways of biosynthesis of these components in eukaryotes and prokaryotes. At several points during the discussion, comparisons between the systems will be described. However, it should be noted that there is variation in the biosynthetic pathways among eukaryotes and among prokaryotes, as well as between eukaryotes and prokaryotes, and the discussion presented is not meant to be comprehensive.

I. FATTY ACID BIOSYNTHESIS

Although some cells can utilize exogenous fatty acids for the synthesis of their membrane lipids, the intracellular synthesis of fatty acids is an important step in the construction of cellular membranes. Although it is not part of the subject of this book, fatty acid biosynthesis is also essential for the formation of triglyceride, a major form of energy storage for many organisms.

Fatty acid biosynthesis is common to many cells and tissues. It is particularly active in the liver of fed animals and is largely turned off in starved animals. This regulation results because insulin stimulates fatty acid biosynthesis whereas glucagon and free fatty acids from the diet inhibit fatty acid biosynthesis. Furthermore, the synthesis of fatty acids starts with products of glycolysis and the tricarboxylic acid cycle, and both cycles are fed from glucose, which can be derived from the diet. (Hence in times of plenty, the excess intake of food can in part be turned into the useful storage molecules, the triglycerides.) Citrate, which can be transported via a carrier out of the mitochondria in exchange for malate, is the source for acetyl-CoA and is a regulator of the pathway. Acetyl-CoA is formed from citrate by citrate lyase outside the mitochondria. Acetyl-CoA is the starting point for this consideration of the biosynthesis of fatty acids.

Initially, the discussion will be limited to describing the steps involved in the biosynthesis of fatty acids and then will be expanded to consider regulation of these pathways. The primary fatty acid produced in many systems is palmitate, so it is with the biosynthesis of palmitate that this examination of fatty acid biosynthesis begins.

A. Synthesis of Palmitate

1. ACETYL-COA CARBOXYLASE

In both bacteria and in animals, the synthesis of malonyl-CoA by the acetyl-CoA carboxylase system is one of the early key steps. The reaction scheme is

$$E - biotin + HCO_3^- + ATP \rightarrow E - biotin\text{-}CO_2 + ADP + P_i$$
$$E - biotin\text{-}CO_2 + acetyl\text{-}CoA \rightarrow E\text{-}biotin + malonyl\text{-}CoA$$

Three individual enzymes are required to carry out the synthesis of malonyl-CoA in *Escherichia coli,* whereas one protein (molecular weight about 250,000) is required in yeast and animals. The provision of acetyl-CoA and malonyl-CoA makes possible the initiation of fatty acid synthesis. Thus, this is the first committed step in this pathway.

It would be expected, therefore, that acetyl-CoA carboxylase is a key regulatory point in the synthesis of fatty acids. The reaction is largely irreversible, which is expected of a regulatory step. Furthermore, the rate of this reaction is slower than the other reactions in the pathway.

2. FATTY ACID SYNTHASE

The linking of two carbon units to yield a fatty acyl-CoA or a free fatty acid product is carried out by the fatty acid synthase complex. In animal cells, the growing product is covalently attached to the protein complex via the prosthetic group, 4'-phosphopantetheine. Two gene products are involved in the fatty acid synthase complex in yeast and one gene product in animal cells.

The reactions involved in mammalian fatty acid biosynthesis are the following:

Acetyl-CoA + E-pan-SH → acetyl-S-pan-E + CoA (acyltransferase)

Malonyl-CoA + E-pan-SH → malonyl-S-pan-E + CoA (malonyltransferase)

Acetyl-S-pan-E + E-cys-SH → acetyl-S-cys-E + E-pan-SH (β-ketoacyl synthase)

$$\text{Acetyl-S-cys-E} + \text{malonyl-pan-E} \rightarrow \underset{\substack{\|\\O}}{CH_3C}\underset{\substack{\|\\O}}{CH_2C}\text{-S-pan-E} + \text{E-cys-SH} + CO_2 \ (\beta\text{-ketoacyl synthase})$$

$$\underset{\substack{\|\\O}}{CH_3C}\underset{\substack{\|\\O}}{CH_2C}\text{-S-pan-E} + \text{NADPH} + H^+ \rightarrow \underset{\substack{|\\OH}}{CH_3CH}\underset{\substack{\|\\O}}{CH_2C}\text{-S-pan-E} + \text{NADP}^+ \ (\beta\text{-ketoacyl reductase})$$

$$\underset{\substack{|\\OH}}{CH_3CH}\underset{\substack{\|\\O}}{CH_2C}\text{-S-pan-E} \rightarrow CH_3CH=\underset{\substack{\|\\O}}{CHC}\text{-S-pan-E} + H_2O \ (\beta\text{-hydroxyacyl dehydrase})$$

$$CH_3CH=\underset{\substack{\|\\O}}{CHC}\text{-S-pan-E} + \text{NADPH} + H^+ \rightarrow CH_3CH_2CH_2\underset{\substack{\|\\O}}{C}\text{-S-pan-E} + \text{NADP}^+ \ (\text{enoyl reductase})$$

$$CH_3CH_2CH_2\underset{\substack{\|\\O}}{C}\text{-S-pan-E} + \text{E-cys-SH} \rightarrow CH_3CH_2CH_2\underset{\substack{\|\\O}}{C}\text{-S-cys-E} + \text{E-pan-SH} \ (\text{ketoacyl synthase})$$

$$CH_3CH_2CH_2\underset{\substack{\|\\O}}{C}\text{-S-cys-E} + \text{malonyl-S-pan-E} \rightarrow CH_3(CH_2)_2\underset{\substack{\|\\O}}{C}\underset{\substack{\|\\O}}{CH_2C}\text{-S-pan-E} + \text{E-cys-SH} + CO_2 \ (\text{ketoacyl synthase})$$

Cycling these reactions to produce palmitoyl-pan-E is then followed by

$$\text{Palmitoyl-pan-E} + H_2O \rightarrow \text{palmitate} + \text{E-pan-SH}$$

The overall reaction for the production of palmitate is

$$\text{Acetyl-CoA} + 7 \text{ malonyl-CoA} + 14 \text{ NADPH} + 14 \text{ H}^+ \rightarrow \text{palmitate} \\ + CO_2 + 14 \text{ NADP}^+ + 8 \text{ CoASH} + 6 \text{ H}_2O$$

In *E. coli*, the growing product is covalently attached to a distinct acyl carrier protein (ACP) via 4′-phosphopantetheine. This is a small protein, of molecular weight just under 9000. The phosphopantetheine prosthetic group is attached to a serine on the protein and serves as the site of attachment of the growing acyl-CoA product. In *E. coli*, the individual reactions of the elongation of the acyl chain are carried out by separate enzymes.

In animal cells, two carbon units are successively added until palmitate is achieved. In *E. coli*, there is a branch point in this reaction scheme. When a 10-carbon chain is produced, a specific enzyme, 3-hydroxy-decenoyl-ACP dehydrase, comes into play. This enzyme is capable of producing the *cis*-3-decanoyl-ACP intermediate rather than *trans*-2-decanoyl-ACP intermediate. The former will become the 16 : 1 product in successive stages of elongation. The latter will become the 16 : 0 product. These, plus *cis*-vaccenic acid which is derived from the 16 : 1 product, are the main fatty acids of *E. coli*. Eukaryotic cells have a more complicated and capable system for producing a wider range of fatty acids for membrane lipid components.

Fatty acid synthase from animal cells appears to exhibit all enzymatic functions on a single polypeptide. In its active form, the enzyme is a dimer of identical subunits of molecular weight about a quarter of a million.

A model for the organization of fatty acid synthase has been proposed. In this model the various enzymatic activities are organized in a linear fashion among domains on each of the subunits. The acetyltransferase, malonyltransferase, and ketoacyl synthase are located in one domain; the dehydrase, enoyl reductase, ketoacyl reductase, and 4-phosphopante-theine-containing peptide (functionally equivalent to the ACP of *E. coli*) are organized into another domain. The thioesterase is contained in the third domain of molecular weight about 35,000. The subunits are then organized into a dimer with the subunits oriented in opposite directions. This allows the close apposition of donors and acceptors in the acyl transfer reactions.

B. Regulation of Fatty Acid Biosynthesis

Regulation of the biosynthesis of fatty acids is largely at the level of the acetyl-CoA carboxylase. Regulation is accomplished through at least four different mechanisms.

1. The first mechanism is regulation by citrate. Citrate is capable of activating an inactive enzyme by binding to the protein and causing a conformation change. As a result of this change in conformation, the protein may aggregate into dimers and higher oligomers. This activation is reasonable since high citrate will result from the extensive degradation of dietary carbohydrate during a well-fed period for the animal. Activation of fatty acid biosynthesis can lead to a storage of dietary energy in the form of triglyceride.

2. Regulation can be achieved by covalent modification. One pathway involves phosphorylation–dephosphorylation for control. Acetyl-CoA carboxylase is a substrate for cAMP-dependent protein kinase. Increases in intracellular cAMP will lead to phosphorylation of the acetyl-CoA carboxylase, which decreases the activity of the enzyme. This inhibition by phosphorylation is stimulated in liver by glucagon, for example. Glucagon leads to a reduction in fatty acid biosynthesis by inhibiting the first committed step in the reaction pathway through phosphorylation of the acetyl-CoA carboxylase. In contrast, insulin stimulates the same enzyme, apparently through phosphorylation at a different site, which is a substrate for a cAMP-independent protein kinase. Another form of covalent modification is removal of a 30,000-Da fragment that activates the enzyme five-fold.

3. The product of the pathway is also inhibitory. High levels of acyl-CoA, such as palmitoyl-CoA, have been shown *in vitro* to inhibit acetyl-CoA carboxylase. However, it is not clear whether this mechanism functions *in vivo*.

4. Long-term regulation of the enzymes of the fatty acid biosynthetic pathway can be achieved by controlling the number of enzyme units available. This is likely done by controlling the availability to the protein synthesis apparatus of message (mRNA) for these enzymes. Feeding a diet high in carbohydrate or a fat-free diet stimulates the production of mRNA for the lipogenic enzymes, whereas starvation or a high-fat diet inhibits the production of message. This suggests that glucagon and insulin have long-term effects on the activity of lipogenic enzymes by controlling mRNA levels, glucagon inhibiting production of message, and insulin stimulating production of message.

C. Elongation

The pathway just described provides palmitate as a fatty acid for incorporation into the lipids of membranes (as well as other uses such as for the synthesis of triglyceride). The reaction pathway described is limited in that this pathway does not, in general, produce fatty acids longer than palmitate. Yet such longer fatty acids are found esterified to membrane lipids, and complex patterns of unsaturation are also found in the fatty acids of native membrane lipids. Therefore, there must be additional enzyme-catalyzed reactions for the elongation of palmitic acid and for the introduction of carbon–carbon double bonds into the fatty acids. Elongation will be considered first.

An important elongation step involves the formation of stearoyl-CoA from palmitoyl-CoA, so this will be used as an example. One general feature of this reaction is that the fatty acid is lengthened by two carbons. Thus this reaction pathway involves the addition of a two-carbon unit to the chain, just as in the formation of palmitate. However, the elongation pathway utilizes different enzymes than those employed in the synthesis of palmitate. Figure 11.1 shows the reaction sequence using palmitoyl-

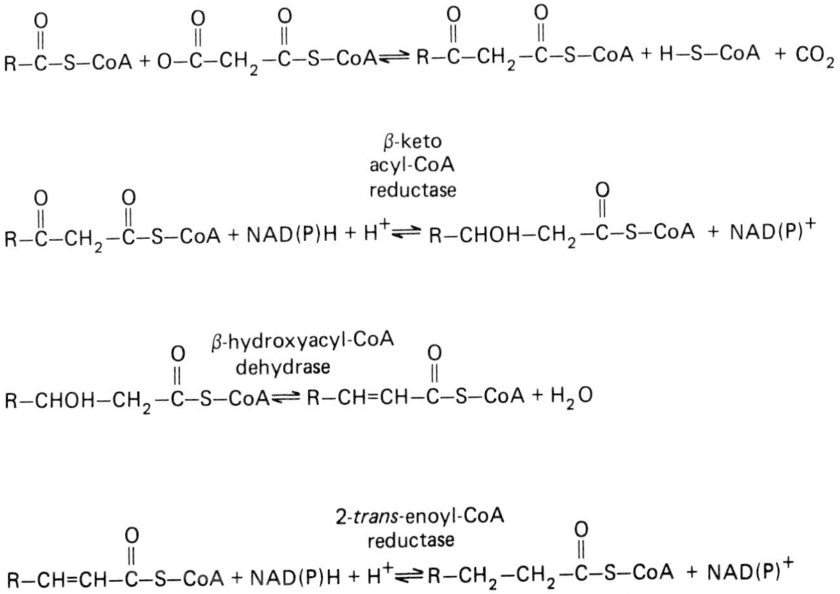

$$R-\overset{O}{\overset{\|}{C}}-S-CoA + O-\overset{O}{\overset{\|}{C}}-CH_2-\overset{O}{\overset{\|}{C}}-S-CoA \rightleftharpoons R-\overset{O}{\overset{\|}{C}}-CH_2-\overset{O}{\overset{\|}{C}}-S-CoA + H-S-CoA + CO_2$$

$$R-\overset{O}{\overset{\|}{C}}-CH_2-\overset{O}{\overset{\|}{C}}-S-CoA + NAD(P)H + H^+ \overset{\beta\text{-keto} \atop \text{acyl-CoA} \atop \text{reductase}}{\rightleftharpoons} R-CHOH-CH_2-\overset{O}{\overset{\|}{C}}-S-CoA + NAD(P)^+$$

$$R-CHOH-CH_2-\overset{O}{\overset{\|}{C}}-S-CoA \overset{\beta\text{-hydroxyacyl-CoA} \atop \text{dehydrase}}{\rightleftharpoons} R-CH=CH-\overset{O}{\overset{\|}{C}}-S-CoA + H_2O$$

$$R-CH=CH-\overset{O}{\overset{\|}{C}}-S-CoA + NAD(P)H + H^+ \overset{2\text{-}trans\text{-enoyl-CoA} \atop \text{reductase}}{\rightleftharpoons} R-CH_2-CH_2-\overset{O}{\overset{\|}{C}}-S-CoA + NAD(P)^+$$

Fig. 11.1. Elongation pathway.

CoA as the substrate. The additional two carbons come from malonyl-CoA. Although only one addition of a two-carbon unit is described in Fig. 11.1, this process can be repeated to produce the $20:0$, $22:0$, and $24:0$ fatty acids. These latter fatty acids are important for sphingomyelin formation, for example, in myelination of nerve, and thus are important in development. The elongation pathway can also be used to elongate unsaturated fatty acids.

For the lipids of cellular membranes, the endoplasmic reticulum provides much of the necessary fatty acids longer than $16:0$. However, elongation is also carried out in the mitochondria from acetyl-CoA. The role of the mitochondrial reaction pathway is not clear.

D. Desaturation of Fatty Acids

Although the reactions just discussed can provide the long-chain fatty acids often found in membrane lipids, only the synthesis of fully saturated fatty acids has so far been described. A large portion of the fatty acids found in cellular membrane lipids are unsaturated. In fact, although $18:1$ is common, $20:4$ and $22:6$ are also found and are important for the function of a number of tissues. Therefore, the aim of this section is the description of the desaturase system that produces these unsaturated fatty acids.

The reactions of desaturation of fatty acyl-CoA's can be seen in Fig. 11.2. Stearoyl-CoA is used as the example, since this is a common substrate for this pathway. The product, oleoyl-CoA, is the result of electron transfer reactions, and the pathway further requires molecular oxygen.

The three main proteins involved are the NADH–cytochrome-b_5 reductase, cytochrome b_5, and the desaturase, which in case of the example above is the Δ_9 desaturase. These are all integral membrane proteins of the smooth ER. Recall that cytochrome b_5 has a relatively large hydrophilic headgroup that contains the heme moiety and is attached to the membrane by a much smaller, hydrophobic segment of the protein. Similarly,

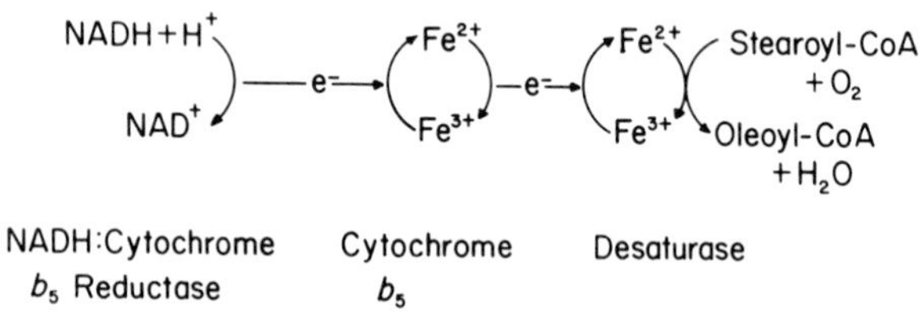

Fig. 11.2.   Desaturation reaction sequence.

NADH–cytochrome-b_5 reductase also consists of a large head, containing a flavin cofactor, and a hydrophobic tail, anchoring the protein to the membrane of the endoplasmic reticulum. The proteins involved in desaturation diffuse freely in the plane of the membrane. The electron transfer reactions occur by collision of these proteins through lateral diffusion. This reaction mechanism is in contrast to the cytochrome oxidase complex, where the constituent proteins exist as an enzyme complex in the membrane (see Chapter 6).

1. REGULATION OF DESATURATION

The desaturase contains an iron, but not the heme iron found in cytochrome b_5. The desaturase is the largest of the three proteins with a molecular weight about 53,000. Each desaturase is apparently specific for the position of action. In contrast, the cytochrome b_5 and its reductase can be shared among the desaturases. The Δ_9 desaturase can be regulated extensively. Starvation, for example, leads to a dramatic reduction of desaturase activity in liver. Enhancement in the use of the product of the reaction, oleoyl-CoA, such as in the synthesis of phospholipids, stimulates the enzyme. Insulin also appears to stimulate the enzyme. In many cases, regulation is via control of protein synthesis, which alters the number of copies of the enzyme in the membrane.

One of the interesting questions concerning the mechanism of desaturation is the physical location of the active site of the enzyme. It would seem likely that the active site is located in the correct position relative to a cross section of the membrane to have contact with the acyl-CoA (or phospholipid) substrate at the appropriate carbon atoms in the bond to be oxidized. Unless the enzyme was to be disruptive to the endoplasmic reticulum membrane (and thus impair its permeability barrier), the active site of the enzyme can be expected to be buried in the hydrophobic interior of the membrane. Most active sites, although perhaps containing hydrophobic pockets, are thought to be exposed directly to the aqueous media. Therefore, in the cases of the desaturases, some interesting "hydrophobic phase" chemistry may be going on. This will be an important subject for future study.

2. PATHWAY OF DESATURATION

In animal systems, the first step in desaturation is the production of oleoyl-CoA from stearoyl-CoA. Further elongation and desaturation can take place, but it is interesting to note that most animal systems cannot insert double bonds lower in the chain (further from the carboxyl of the fatty acid) than position 9. Thus, in animal systems there are desaturases for positions 4, 5, and 6, but not for 12 or 15. Yet remember that many of the fatty acids found in the lipids of animal membranes are unsaturated at those latter positions. For example, consider the three common fatty acids, 18:2 ($\Delta_{9, 12}$), 20:4 ($\Delta_{5, 8, 11, 14}$), and 22:6 ($\Delta_{4, 7, 13, 16, 19}$). These fatty acids cannot be made *de novo* in animal systems.

The key to the synthesis of these latter fatty acids is the utilization of plant fatty acid precursors derived from the diet. The fatty acid linoleic acid is an essential precursor of the synthesis of arachidonic acid, for example. Plants have the capability of desaturating oleic acid to linoleic acid, whereas animals do not. Therefore, linoleic acid is termed an essential fatty acid for animals, since it is needed for the synthesis of important products and cannot be made in the animal.

As an example, consider the pathway of synthesis of arachidonic acid, which is a precursor to prostaglandin synthesis. Figure 11.3 shows the steps involved. Note that a combination of desaturation and elongation steps is used to synthesize the arachidonic acid. Similar combinations are used to synthesize the wide variety of fatty acids found in the lipids of biological membranes.

The substrate specificity of the desaturases, as well as the availability of various desaturases, determines the lack of conjugation of carbon–carbon double bonds in the naturally occurring fatty acids. Only rarely is this pattern violated.

The primary substrate for the desaturation reaction is likely the acyl-CoA derivative of the fatty acid. However, this does not rule out the action of desaturases on phospholipid substrates. There is some evidence that phospholipids can serve as substrate in some systems, for example, in *Tetrahymena pyriformis*. In *E. coli,* the conversion of the unsaturated fatty acids to cyclopropane containing fatty acids during the stationary phase is apparently performed with phospholipid as substrate.

DIET

18:3

20:3

20:4

Fig. 11.3.   Steps in the synthesis of arachidonic acid.

E. Degradation of Fatty Acids

Cellular components are always undergoing turnover by synthesis and degradation. This section will be concerned with the metabolism of the fatty acids whose synthesis was just reviewed.

The source of these fatty acids may be endogenous triglyceride, membrane lipids, or dietary fat. The fatty acids are transported in the circulation in the form of serum lipoproteins or bound to serum albumin. However, for degradation to take place, the fatty acids must be transported into the cell. If the fatty acids are esterified to triglyceride or phospholipid they must first be hydrolyzed to fatty acid. Then the fatty acids can be transported to the endoplasmic reticulum and the mitochondria, where the first step of the degradation pathway can take place.

Generally, for the fatty acids to be suitable for further metabolism, conversion to the acyl-CoA derivative must take place first. Acyl-CoA synthetases are found in the mitochondria and in the endoplasmic reticulum. This reaction is ATP dependent:

$$RCOO^- + ATP + CoASH \rightarrow Acyl\text{-}CoA + AMP + PP_i$$

When long-chain acyl-CoAs are formed, they can be the substrate for several reactions, including desaturation, phospholipid synthesis, and oxidative degradation. The latter is the subject of interest here.

There are at least three classes of acyl-CoA synthetases. They are classed according to their substrate specificity. One class operates primarily on acetate, another on short to medium length fatty acid chains, and a third on long chain fatty acids. The first two are appropriately water-soluble enzymes, whereas the last class is membrane bound to accommodate the substrates that are bound to either membranes or protein. The latter enzyme is located in the outer mitochondrial membrane and in the endoplasmic reticulum membrane. The molecular weight of the enzyme in both locations is close to 80,000.

The oxidative degradation pathway is largely located in the interior of the mitochondria, but the substrate, acyl-CoA, is not capable of penetrating the inner mitochondrial membrane. However, the carnitine derivative is permeable. Therefore, the enzyme carnitine acyltransferase is employed to derivatize the acyl-CoA to permit it to enter the mitochondria via the carnitine:acylcarnitine translocase in the inner mitochondrial membrane. Once inside, this derivative is returned to the acyl-CoA form.

If the acyl-CoA contains an even number of carbon atoms, it can enter the oxidation pathway described in Fig. 11.4 and, by a cyclic application of that pathway, be successively degraded by two carbon units until the

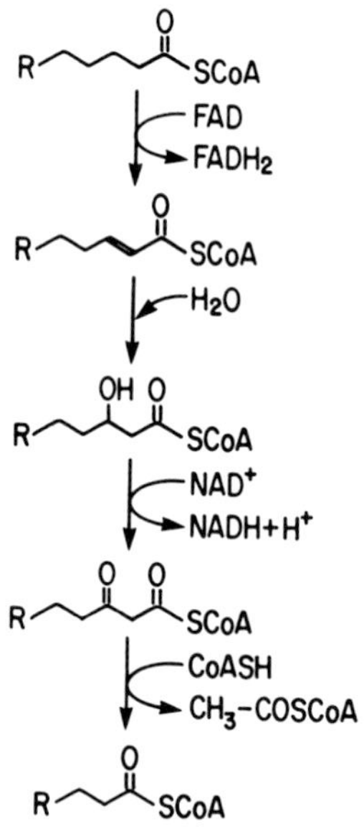

Fig. 11.4.   Fatty acid oxidation.

entire fatty acid has been oxidized to the appropriate number of two-carbon acetyl-CoA units. It is common to find that the enzymes catalyzing the reactions can be divided into the same general classes as for the acyl-CoA synthetases. This is reasonable in that the solubility properties of substrates are dependent on the length of the acyl chain. Initially, long-chain acyl-CoA must be acted on by enzymes that have access to the dominant location of such species, which is in the membrane due to the hydrophobic nature of the substrate. Hence, such enzymes must be integral membrane proteins. As the fatty acid is degraded sequentially to short-chain species, water-soluble enzymes are satisfactory.

The pathway just described is effective for fatty acids with an even number of carbon atoms in the molecule. For fatty acids with an odd number of carbons in the molecule, some additional reactions must take place. Oxidation proceeds as for fatty acids with an even number of carbon atoms, successively producing acetyl-CoA, until the final three carbons remain in proparionyl-CoA. This is subsequently converted to succinyl-CoA, which can enter the tricarboxylic acid cycle in the mitochondria.

Unsaturated fatty acids require even more perturbations of the oxidative pathway. These are exemplified in Fig. 11.5 for linoleoyl-CoA. Ultimately, it too is catabolized to acetyl-CoA two-carbon units. These acetyl-CoA molecules can then be used for a variety of reaction pathways.

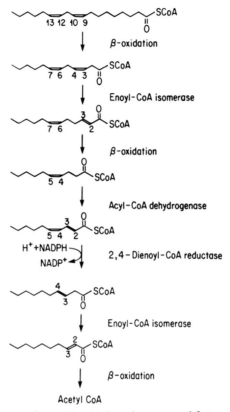

Fig. 11.5.   Fatty acid oxidation of unsaturated fatty acid.

An analogous series of reactions are carried out in *E. coli* for the oxidation of fatty acids. In *E. coli*, the reactions appear to be carried out on an enzyme complex consisting of two different subunits.

II. BIOSYNTHESIS OF PHOSPHATIDATE

One of the important intermediates in phospholipid biosynthesis is phosphatidate. One of the precursors of phosphatidate biosynthesis is fatty acyl-CoA. Another precursor is glycerol-3-P derived from glycolysis. A fatty acid in the form of the acyl-CoA can be esterified to the glycerol-3-P by glycerol-phosphate acyltransferase. This reaction can take place both in the endoplasmic reticulum and in the mitochondria. However, only in the liver is the mitochondrial reaction of significant proportions in mammalian systems. The product of this reaction is lysophosphatidate (see Fig. 11.6).

The lysophosphatidate can also be synthesized from dihydroxyacetone-P, which, as in the case of glycerol-3-P, is derived from glycolysis. The glycerol-phosphate acyltransferase can also use the dihydroxyacetone phosphate as substrate, making acyldihydroxyacetone-P. This compound has two fates. One is conversion to lysophosphatidate. The other is conversion to alkyldihydroxyacetone-P, which is on the pathway to the synthesis of the alkyl and alkenyl phospholipids.

Lysophosphatidate is the substrate for monoacylglycerol-phosphate (lysophosphatidate) acyltransferase, and the product formed is phosphatidate. This enzyme is found both in the endoplasmic reticulum and in mitochondria. The enzymes in these two organelles are different. In endoplasmic reticulum, the enzyme will use both saturated and unsaturated acyl-CoA, whereas in mitochondria the enzyme preferentially employs saturated acyl-CoA. Control of phosphatidate biosynthesis lies in part with the availability of fatty acyl-CoA. The other major user of fatty acyl-CoA is the fatty acid oxidation pathway described earlier. Therefore, the control of the biosynthesis of phosphatidate derives from a competition between synthesis of the phosphatidate from acyl-CoA and the oxidation of acyl-CoA.

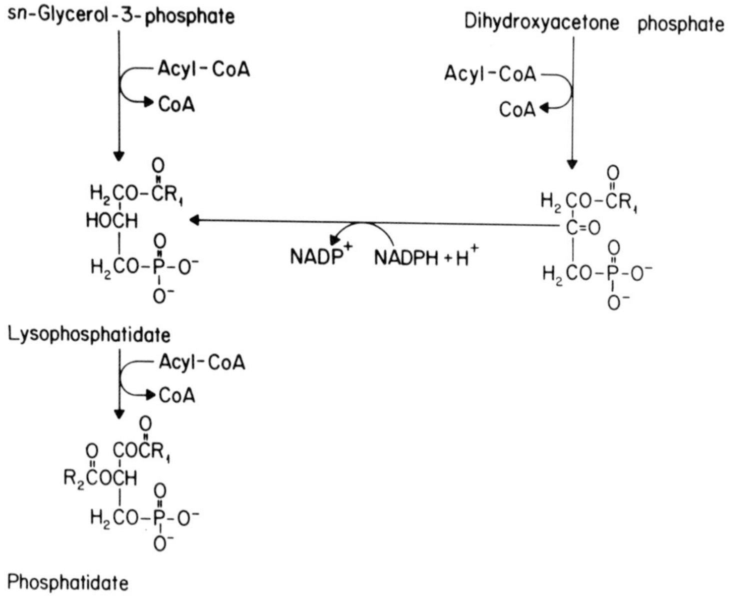

Fig. 11.6. Synthesis of phosphatidate.

Phosphatidate sits at a pivotal point in the lipid biosynthesis pathways. It can be considered to have at least three important fates. First, it serves as a precursor for the formation of diacylglycerol, which can be used for triglyceride biosynthesis or for the biosynthesis of phosphatidylcholine and phosphatidylethanolamine. Second, phosphatidate is a substrate for the reaction forming CDP diacylglycerol, which is a precursor for the synthesis of phosphatidylinositol, phosphatidylglycerol, and diphosphatidylglycerol. Finally, phosphatidate can be metabolized to fatty acids and glycerolphosphate by phospholipases.

III. BIOSYNTHESIS OF DIACYLGLYCEROL FROM PHOSPHATIDATE

The production of diacylglycerol will be considered first, since it will lead directly into the biosynthesis of phosphatidylcholine and phosphatidylethanolamine, the next subjects for discussion. The conversion of phosphatidate to diacylglycerol is carried out by the enzyme phosphatidate phosphohydrolase. This enzyme is found in the endoplasmic reticulum membrane and in a soluble fraction. The activity of this enzyme can be rapidly regulated over a wide range. Both the number of copies and the percentage of active enzyme can be regulated. The former is regulated hormonally.

An interesting model has been proposed for the regulation of the percentage of active phosphohydrolase. The soluble fraction of the phosphatidate phosphohydrolase is thought to be of low activity, whereas the membrane-bound fraction exhibits high activity. This organization of the enzyme follows the partitioning of the phosphatidate, which will be found almost exclusively in the membrane because of its highly hydrophobic hydrocarbon chains. One method for increasing the number of active enzymes is to increase the number of membrane-bound enzymes at the expense of the soluble fraction. An increase in the number of membrane-bound enzymes is apparently stimulated by an increase in fatty acid concentration. The increase in fatty acid levels corresponds to the concomitant increase in triglyceride synthesis subsequent to the phosphatidate-phosphohydrolase-catalyzed step. cAMP has an opposing regulatory effect. cAMP apparently causes a displacement of the phosphohydrolase enzyme from the membrane, thereby decreasing the total activity of the overall enzyme-catalyzed reaction.

By this process, the diacylglycerol intermediate is formed. The diacylglycerol species can be destined for triglyceride synthesis or for phospholipid synthesis. Triglyceride synthesis is primarily a storage function, designed to operate when an excess of fatty acids is present. In contrast, synthesis of phosphatidylcholine and phosphatidylethanolamine is essential to the assembly of the cellular membranes.

IV. BIOSYNTHESIS OF PHOSPHATIDYLCHOLINE

Phosphatidylcholine is one of the most abundant of the membrane phospholipids. It will be the first phospholipid whose biosynthesis will be reviewed.

There is more than one pathway for the synthesis of phosphatidylcholine, but one pathway is dominant in many eukaryotic cells. This involves the condensation of diacylglycerol with CDP-choline. The diacylglycerol comes from the phosphatidate whose synthesis was just described. CDP-choline is synthesized from choline using the following reactions:

Choline + ATP → phosphocholine + ADP (choline kinase)
Phosphocholine + CTP → CDP-choline + PP$_i$ (CTP:phosphocholine cytidylyltransferase)

The first reaction is carried out in solution since the substrates and the enzyme, choline kinase, are soluble. Phosphorylation of choline activates the choline for further reactions and is the normal mode of entry of choline into biosynthetic pathways. The same enzyme may be capable of catalyzing the phosphorylation of ethanolamine.

The cytidylyltransferase appears to be active primarily in its membrane-bound form, since phospholipids are required to activate the enzyme. At the same time, there is a considerable pool of soluble enzyme. This situation sets the stage for regulation of the total activity of this enzyme by regulating its distribution between the soluble fraction and the membrane-bound fraction. This regulatory scheme would allow for a rapid increase (or decrease) in the total tissue activity of the enzyme, in much the same manner as for the phosphatidate phosphohydrolase discussed earlier in this chapter. As in the case of this latter enzyme, the cytidyltransferase is activated by fatty acids and fatty acyl-CoAs, which favor the partitioning of the enzyme into the membrane-bound form. Also available for modulation of the activity of cytidyltransferase is cAMP-dependent protein kinase-catalyzed phosphorylation, which inactivates the enzyme. Finally, it is the cytidyltransferase that appears to be rate limiting in the biosynthesis of phosphatidylcholine. Therefore, the regulation of the activity of this enzyme just discussed is important in the overall rate of phosphatidylcholine synthesis. A corollary is that a restriction in the availability of fatty acyl-CoA will restrict the synthesis of membrane phospholipids, although as mentioned earlier, phospholipid biosynthesis takes precedence over the other major constructive use of fatty acids, triglyceride synthesis. Here one is beginning to see some of the important factors in the control of cell membrane biogenesis.

Returning to the main pathway for biosynthesis of phosphatidylcholine, only one step remains to complete the synthesis. That is the reaction of CDP-choline with diacylglycerol:

CDP-choline + diacylglycerol → phosphatidylcholine + CMP
(CDP-choline:1,2-diacylglycerol phosphocholinetransferase)

As one might expect given the nature of the one of the substrates and one of the products, this phosphocholinetransferase is a membrane-bound enzyme. It does not share activity with the corresponding ethanolamine species. It does show some substrate specificity, favoring 1-palmitoyl-2-linoleoyl diacylglycerol.

Phosphatidylcholine biosynthesis can occur by methylation of phosphatidylethanolamine. The only structural difference between phosphatidylcholine and phosphatidylethanolamine is the methylation of the nitrogen of phosphatidylcholine. Methylation of phosphatidylethanolamine is also a way to synthesize phosphatidylcholine in the laboratory. Long before the organic chemist, cells developed their own method for doing the same synthesis.

The methylation of phosphatidylethanolamine occurs stepwise, adding one methyl group at a time. The methyl group donor in each case is S-adenosylmethionine. It is a minor pathway in most tissues; only in liver does the action make a significant overall contribution to phosphatidylcholine biosynthesis.

The third pathway of phosphatidylcholine biosynthesis to be considered here is the reacylation of lysophosphatidylcholine. Endogenous phospholipase A_2 promotes continual turnover of membrane phospholipids by hydrolysis at position 2' on the glycerol. The products are lysophosphatidylcholine and fatty acid, when phosphatidylcholine is the substrate. The lysophosphatidylcholine can then be reacylated, using acyl-CoA. There is an interesting by-product of this otherwise apparently futile reaction. The reacylation reaction selects for unsaturated acyl-CoA's. Thus, this reaction is responsible in part for the observation that most common phospholipid species in animals contain a saturated fatty acid on position 1' and an unsaturated fatty acid on position 2' of the glycerol. Furthermore, as can be seen by comparing substrate specificities in the steps in the major pathway of phosphatidylcholine biosynthesis, this reacylation reaction is an effective means for the introduction of highly unsaturated fatty acids into the phospholipid pool, such as arachidonic acid. Since the latter is a substrate for prostaglandin biosynthesis, introduction into the phospholipid pool is obviously an important effect.

V. PHOSPHATIDYLETHANOLAMINE BIOSYNTHESIS

Two of the four major pathways for phosphatidylethanolamine biosynthesis are analogous to two of the three pathways just described for phosphatidylcholine biosynthesis. One of those is synthesis via CDP-ethanolamine, analogous to the pathway involving CDP-choline, and the other is synthesis via reacylation of lysophosphatidylethanolamine.

For the first of these, as in the case of choline, one begins with the synthesis of CDP-ethanolamine, which proceeds by the following reactions:

$$\text{Ethanolamine} + \text{ATP} \rightarrow \text{phosphoethanolamine}$$
$$+ \text{ADP (ethanolamine kinase)}$$
$$\text{Phosphoethanolamine} + \text{CTP} \rightarrow \text{CDP-ethanolamine} + \text{PP}_i$$
$$\text{(CTP:phosphoethanolamine cytidylyltransferase)}$$

In general, the enzymes catalyzing these reactions involving ethanolamine are distinct from the enzymes catalyzing the reactions already discussed involving choline. In fact, the cytidyltransferase in this case does not appear to be a membrane protein or to require phospholipids for activation.

The final step in this pathway is the formation of phosphatidylethanolamine:

$$\text{CDP-ethanolamine} + \text{diacylglycerol} \rightarrow \text{phosphatidylethanolamine}$$
$$+ \text{CMP} \quad \text{(CDP-ethanolamine:1,2-diacylglycerol}$$
$$\text{phosphoethanolaminetransferase)}$$

The species of diacylglycerol favored in this last step is a species containing a saturated fatty acid on position 1' and the unsaturated 22 : 6 fatty acid at position 2'. An excellent example of the consequences of this specificity is observed in the rod outer-segment disk membrane. In this membrane, phosphatidylethanolamine is one of the two major phospholipids. The phosphatidylethanolamine of this membrane is rich in the fatty acid, 22 : 6.

The second pathway that is in common between phosphatidylcholine synthesis and phosphatidylethanolamine synthesis is reacylation of the lyso derivative. Because this is basically the same reaction presented previously, this mechanism will not be examined further.

Two other pathways have been described for the biosynthesis of phosphatidylethanolamine. They both involve another phospholipid, phosphatidylserine. In one, serine of phosphatidylserine can be exchanged for ethanolamine in a base exchange reaction. The reaction is calcium dependent and likely takes place on endoplasmic reticulum membranes. An analogous reaction is carried out in the laboratory using calcium-dependent phospholipase D, which is capable of base exchange. In the other, the enzyme phosphatidylserine decarboxylase, catalyzes the structural change in the phospholipid headgroup required to convert the serine headgroup to the ethanolamine headgroup. This reaction apparently takes place in mitochondria.

The reactions just examined outline the biosynthetic pathways for phosphatidylethanolamine in animal cell membranes. Somewhat different pathways are utilized in *E. coli*. As in the case of the animal cells, phosphatidate is an important intermediate. It is synthesized by the acylation of glycerophosphate. The substrates for the reaction are fatty acids esterified to acyl carrier protein, rather than CoA. However, the phosphatidate resulting is similar to that synthesized in animal cells in that position one is primarily saturated fatty acids, whereas position two is primarily occupied by unsaturated fatty acids.

This phosphatidate is then the source for the formation of CDP-diacylglycerol. This reaction was referred to earlier and is catalyzed by the enzyme phosphatidate cytidyltransferase. Phosphatidylserine synthase then acts on the CDP-diacylglycerol in the presence of serine and forms phosphatidylserine. This phospholipid is subsequently decarboxylated by phosphatidylserine decarboxylase to form the phosphatidylethanolamine commonly found in the *E. coli* membranes.

VI. PHOSPHATIDYLSERINE BIOSYNTHESIS

The predominant pathway of phosphatidylserine synthesis in animals is the base exchange reaction with phosphatidylethanolamine just described. In bacteria, phosphatidylserine is made by a pathway involving CDP-diacylglycerol as described above in the biosynthesis of phosphatidylethanolamine.

VII. PHOSPHATIDYLGLYCEROL AND DIPHOSPHATIDYLGLYCEROL SYNTHESIS

Phosphatidylglycerol synthesis begins with CDP-diacylglycerol. A reaction between glycerol phosphate and CDP-diacylglycerol, catalyzed by phosphatidylglycerolphosphate synthase, forms phosphatidylglycerolphosphate. The latter product is dephosphorylated to form phosphatidylglycerol with phosphatidylglycerolphosphate phosphatase. This reaction pathway provides the second common phospholipid of the *E. coli* membranes and is also the pathway in animals (where phosphatidylglycerol is a minor phospholipid).

In animals, diphosphatidylglycerol is formed in mitochondria according to the following reaction:

$$PG + CDP\text{-diacylglycerol} \rightarrow disphosphatidylglycerol + CMP$$

However, in *E. coli* (where phosphatidylglycerol is a major phospholipid), two molecules of phosphatidylglycerol are condensed to form diphosphatidylglycerol, with the release of glycerol.

VIII. PHOSPHATIDYLINOSITOL BIOSYNTHESIS

Phosphatidylinositol is perhaps one of the most interesting phospholipids metabolically. This is because metabolism of this phospholipid is stimulated by some receptors in the plasma membrane. These reactions are most important when the metabolism involves phosphorylated derivatives of phosphatidylinositol. In that case, diacylglycerol and phosphorylated inositols are released, apparently in response to the activation of a phospholipase C, which in turn is activated by a G-protein receptor (see Chapter 10). Both these compounds appear to act as powerful second messengers. The diacylglycerol released binds in a 1 : 1 stoichiometry to protein kinase C, activating this enzyme. Inositol 1,4,5-triphosphate is apparently capable of stimulating the endoplasmic reticulum system to release calcium into the cytoplasm, with all the concomitant metabolic effects of increased intracellular calcium.

The synthesis of phosphatidylinositol is carried out by the following set of reactions, using the CDP-diacylglycerol pathway:

$$Phosphatidate + CTP \rightarrow CDP\text{-diacylglycerol} + PP_i$$
$$CDP\text{-diacylglycerol} + inositol \rightarrow phosphatidylinositol + CMP$$

Phosphatidylinositol is then subject to further phosphorylation by a kinase on the inositol moiety, with ATP as the phosphate donor:

$$Phosphatidylinositol + ATP \rightarrow phosphatidylinositol\text{-}4\text{-}P + ADP$$
$$Phosphatidylinositol\text{-}4\text{-}P + ATP \rightarrow phosphatidylinositol$$
$$- 4,5,\text{-biphosphate} + ADP$$

After the breakdown of phosphatidylinositol into diacylglycerol and the inositol, the diacylglycerol can be recycled back into phosphatidylinositol. Thus, the series of reactions is referred to as the phosphatidylinositol metabolic cycle.

IX. BIOSYNTHESIS OF THE SPHINGOLIPIDS

The sphingolipids exhibit considerably greater variety in structure than is seen in the phospholipids. One of the species is sphingomyelin, which is closely related in its structure to the phospholipids just discussed. However, the class of sphingolipids also includes many glycolipids, in which one can find an amazing variety of structures and properties that depend on the sequence of sugars in the polar headgroups, as well as on the structure of the hydrocarbon region. Thus, a discussion of the biosynthesis of these lipids is a formidable task. Here some representative examples will be described.

A. Biosynthesis of Sphingomyelin

An appropriate lipid to begin with is sphingomyelin, which is most closely related to the phospholipids that were just examined. It is likely that two different pathways may be used for the synthesis of sphingomyelin. Both require the ceramide base, whose synthesis will be examined shortly. The following two reactions represent the two likely pathways:

Ceramide + phosphatidylcholine → sphingomyelin + acylglycerol

This first reaction is essentially a transfer mechanism, using a phospholipid as a donor. Second,

Ceramide + CDP-choline → sphingomyelin + CMP

If this last pathway is operative, it would be analogous to the synthesis of phosphatidylcholine discussed earlier.

Clearly, both of these pathways require ceramide. Likewise, other sphingolipids require ceramide for their synthesis. The first step in the biosynthesis of ceramide involves palmitoyl-CoA and serine:

Palmitoyl-CoA + serine → 3-oxosphinganine

The next compound to be made, sphinganine, is the precursor to the three most common bases for this class of sphingolipids.

3-Oxosphinganine + NADPH + H^+ → sphinganine + $NADP^+$

The structure of sphinganine appears in Fig. 11.7. Sphinganine acts as one of the bases for glycosphingolipid biosynthesis. It can also be converted by dehydrogenation into another of the common bases, 4-sphinganine (see Fig. 11.7 for the structure). Furthermore, sphinganine can be oxidized to a third base, 4-hydroxysphinganine.

From the structure of these sphinganine bases, one can see that there is available in these structures a site for derivatization, which is the amino

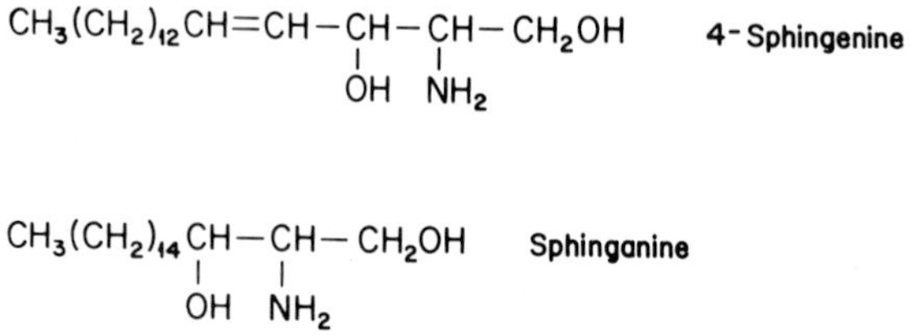

$$CH_3(CH_2)_{12}CH{=}CH{-}CH{-}CH{-}CH_2OH \qquad \text{4-Sphingenine}$$
$$\underset{\displaystyle OH}{|} \quad \underset{\displaystyle NH_2}{|}$$

$$CH_3(CH_2)_{14}CH{-}CH{-}CH_2OH \qquad \text{Sphinganine}$$
$$\underset{\displaystyle OH}{|} \quad \underset{\displaystyle NH_2}{|}$$

Fig. 11.7. Chemical structure of 4-sphingenine and sphinganine.

group. It is in fact derivatization of this amino group that leads to the formation of ceramide. Two different reactions may be possible.

$$\text{Fatty acid } + \text{ sphinganine} \rightarrow \text{ceramide}$$
$$\text{Acyl-CoA } + \text{ sphinganine} \rightarrow \text{ceramide}$$

A generic structure for the ceramide appears in Fig. 11.8.

The fatty acid used for the synthesis of the ceramide may be of several different classes, but in general these fatty acids are distinctly different from the fatty acids commonly used in the biosynthetic pathways for the phospholipids examined earlier. Both moderate and long-chain saturated fatty acids are employed in this synthesis, including 16:0, 18:0, 22:0, and 24:0. Monounsaturated fatty acids are also employed, including 18:1 and 24:1. Furthermore, α-hydroxy derivatives of these fatty acids are used, which are rarely found in the phospholipids.

Because of the very long chain fatty acids and because of the relatively high percentage of saturated fatty acids attached to the amino group of the ceramide, naturally derived sphingolipids like sphingomyelin exhibit gel to liquid crystalline phase transitions for the pure lipid at relatively high temperatures. Because of this property, it is interesting to ponder what roles in membrane structure and function a sphingolipid with a high-phase transition temperature might have. For example, it is not unreasonable that such lipids may phase separate, to some extent, at physiological temperature to form patches on the cell surface with quite different properties than exhibited by the membrane as a whole.

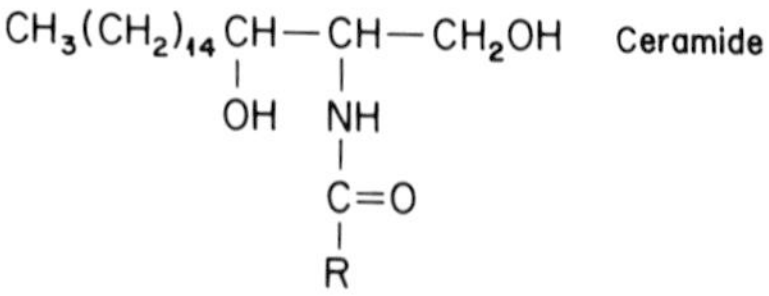

$$CH_3(CH_2)_{14}CH{-}CH{-}CH_2OH \qquad \text{Ceramide}$$
$$\underset{\displaystyle OH}{|} \quad \underset{\displaystyle NH}{|}$$
$$\underset{\displaystyle C{=}O}{|}$$
$$\underset{\displaystyle R}{|}$$

Fig. 11.8. Chemical structure of ceramide.

B. Biosynthesis of Glycosphingolipids

The glycosphingolipids also use ceramide as the starting material. Consider, for example, the biosynthesis of galactosylceramide. The synthesis is carried out by a transfer of the sugar to the ceramide. However, for this transfer to take place, the sugar must be activated.

Galactose-1-P + UTP → UDP-galactose + PP$_i$ (UDP-galactose pyrophosphorylase)

This activated sugar can then be transferred:

UDP-galactose + ceramide → galactosylceramide + UDP (UDP-galactose:ceramide galactosyltransferase)

Interestingly, although most of the lipid biosynthesis reactions occur in endoplasmic reticulum membranes, some of the glycosyltransferase activities are found in the Golgi.

Glucosylceramide is one of the reactants in the pathway that produces lactosylceramide, which is an important precursor to the synthesis of many neutral glycolipids. This synthesis is described in the following reaction:

UDP-galactose + glucosylceramide → UDP + galactosyl-β_1→4-glucosylceramide (UDP-galactosyl:glycosylceramide galactosyltransferase)

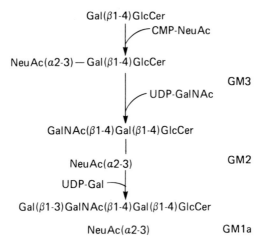

Fig. 11.9. Biosynthesis of a ganglioside.

With this starting material, the synthesis of many glycolipids can proceed. The detailed discussion of the many reaction pathways is beyond the scope of this book, and the reader is referred to the references listed at the end of the chapter for further details. See Fig. 11.9 for an example.

X. METABOLISM OF SPHINGOLIPIDS

The pathways for the degradation of the glycosphingolipids are as complex as their structures suggest. Some limited specific examples will be described. These degradation pathways are important clinically, because genetic deficiencies in these pathways often lead to serious disease states.

First, consider the metabolism of sphingomyelin. This lipid is a substrate for the enzyme sphingomyelinase, which catalyzes the following reaction:

$$\text{Sphingomyelin} \rightarrow \text{phosphorylcholine} + \text{ceramide}$$

The sphingomyelinase activity can be found in the lysosome. A deficiency in this enzyme leads to Niemann–Pick disease, in which there is an abnormal accumulation of sphingomyelin in some cells.

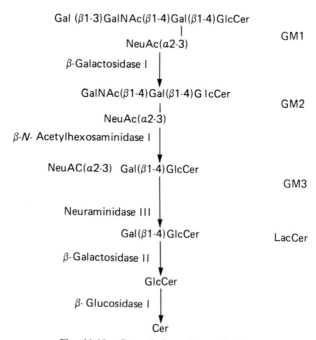

Fig. 11.10. Degradation of a ganglioside.

The metabolic pathway can be carried further by the degradation of ceramide:

Ceramide + water → base (i.e., sphinginine) + fatty acid (ceramidase)

The base can be further degraded to along-chain aldehyde and phospho-ethanolamine.

An entirely different set of enzymes is required for the breakdown of other more complex glycosphingolipids. Before the process produces ceramide, the complex carbohydrate must be broken down into simple sugars that are cleaved from the lipid. Deficiencies in these pathways can also lead to diseases. One of the best known is Tay–Sachs disease. This results in the accumulation of the ganglioside, GM_2, due to the deficiency of an acetylhexosaminidase required in the pathway of degradation of gangliosides (see Fig. 11.10).

XI. CHOLESTEROL BIOSYNTHESIS

The discovery of the pathways of biosynthesis and degradation of the important membrane lipid, cholesterol, is due in large part to the efforts of Konrad Bloch and colleagues.[2]

The biosynthesis of cholesterol begins with the production of mevalonate. Mevalonate is synthesized from acetyl-CoA. The first step is the condensation of two acetyl-CoA units to acetoacetyl-CoA. The addition of one more unit of acetyl-CoA yields hydroxymethylglutaryl-CoA. Finally, that intermediate is reduced to mevalonate:

$$\text{Acetyl-CoA} + \text{acetyl-CoA} \rightarrow \text{acetoacetyl-CoA}$$

Acetoacetyl-CoA + acetyl-CoA → hydroxymethylglutaryl-CoA (HMG-CoA synthase)

Hydroxymethylglutaryl-CoA + 2 NADPH + $2H^+$ → mevalonate
→ CoASH + 2 $NADP^+$ (HMGCoA reductase)

HMG-CoA reductase is an integral membrane protein of the endoplasmic reticulum membrane system. The reaction catalyzed by HMG-CoA reductase is the rate-limiting step in cholesterol biosynthesis. The regulation of the HMG-CoA reductase is important to both the cholesterol content of the cell and the rate of cholesterol ester export from the cell if, for example, it is a liver cell making serum lipoproteins.

HMG-CoA reductase can be regulated in several ways. One is a phosphorylation–dephosphorylation cycle. The enzyme is a substrate for a protein kinase. When the HMG-CoA reductase is phosphorylated it loses activity, and when it is acted on by a phosphatase to dephosphorylate it, the HMG-CoA reductase regains activity. The protein kinase itself is apparently subject to regulation by phosphorylation such that it is most active when phosphorylated. Thus, there is amplification in this system of regulation of HMG-CoA reductase.

As will be described in more detail shortly, HMG-CoA reductase activity is regulated by cholesterol delivered to the cell by serum low-density lipoprotein. Two possible mechanisms for the regulation by cholesterol have been examined. One is direct cholesterol modulation of the activity of the HMG-CoA reductase. This could occur by direct sterol interaction with the protein or by the characteristic cholesterol-mediated modulation of membrane properties. Although considerable attention has been paid to these possible mechanisms, the issue is not clear at this time. Regulation of activity is also achieved by oxygenated sterols.

A second possible mechanism for regulation of HMG-CoA reductase activity by cholesterol is through regulation of the number of copies of the enzyme in the endoplasmic reticulum. This may be achieved through controlling the rate of synthesis and the rate of degradation of the enzyme. It would appear that the regulation of synthesis is at least as far back as the message level. However, much remains to be examined in this area of regulation.

The synthesis of squalene is the next milestone in the biosynthesis of cholesterol. The relevant series of reactions is given in Fig. 11.11.

The third stage of cholesterol biosynthesis is the synthesis of lanosterol from squalene. Squalene is first converted to its oxide. Then it undergoes an amazing reaction in which the fused ring system of the sterol is created in the form of lanosterol. The molecular mechanism of this reaction is not completely understood but probably will provide one of the most interesting reaction mechanisms yet described, given the complexity of the reaction catalyzed.

It is interesting to note that the biosynthesis of the predominant membrane sterol of mammalian systems does not stop with lanosterol. At this point the fused ring system of the sterol is present, as noted above. However, several stages of demethylation and isomerization must take place (involving 19 reaction steps) before cholesterol results (see Fig. 11.12).[3]

In the process of the conversion of lanosterol to cholesterol, several demethylation reactions are required. The first methyl group to be removed is at position 14. Subsequently, the methyl groups at C-4 are removed. After several other steps a carbon–carbon double bond is formed at the characteristic position of cholesterol, between carbons 5 and 6.

One might well wonder whether the process is worth all the effort.[4] Assuming that such a complicated series of reactions is important to the biology of the cell, one can identify in the available literature a likely reason for the complete synthesis of cholesterol. Particular sterol structures are

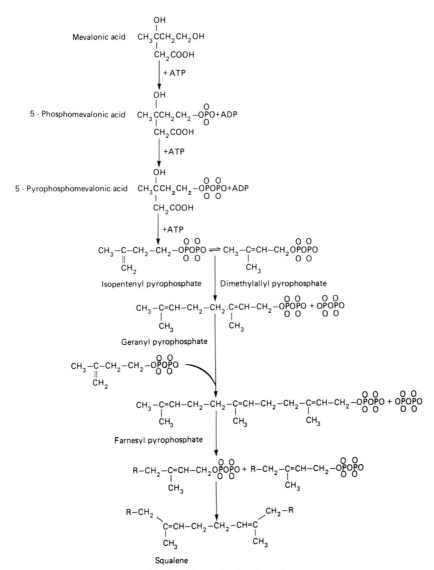

Fig. 11.11. Biosynthesis of squalene.

required for the modulation of crucial enzymes involved in membrane biogenesis. Where cholesterol is required, for example, lanosterol is not an effective substitute. The mechanism for this effect is likely recognition of the sterol by an enzyme or enzymes at a binding site. The recognition sites will be structurally specific, so that the differences in structure of lanosterol from cholesterol will render lanosterol less competent to modu-

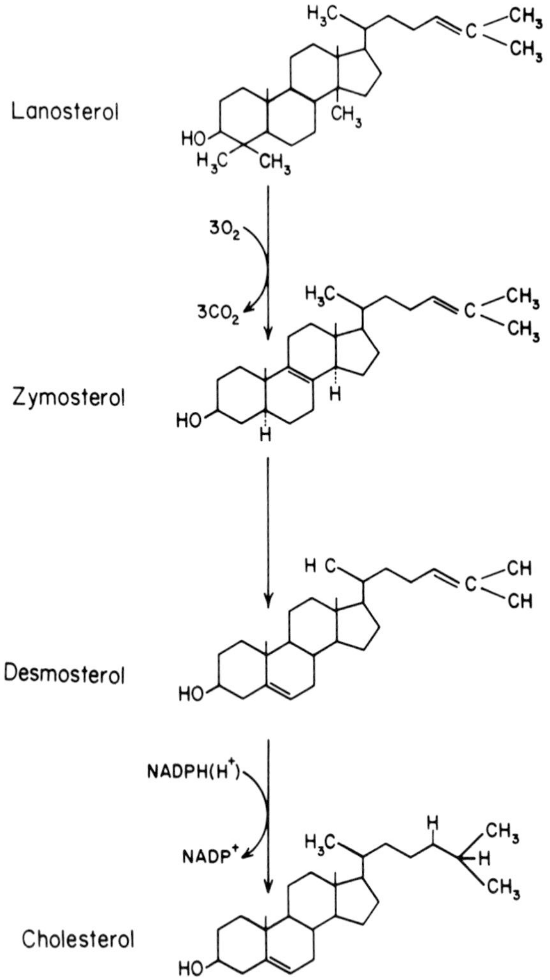

Fig. 11.12. Outline of the pathway from lanosterol to cholesterol.

late the activity of the enzyme. Much remains to be elucidated in this area (see Chapter 5).

XII. REGULATION OF CHOLESTEROL LEVELS IN THE CELL

Several different pathways are available to maintain proper cholesterol levels in the cell. Considerable effort is expended by a mammalian cell to maintain its cholesterol levels in the various membranes. In particular, cholesterol levels in the plasma membrane are important, since that is where 90% or more of the cellular cholesterol resides. Furthermore, the much lower amounts of cholesterol found in the endoplasmic reticulum appear to be important to further membrane biogenesis and so must be maintained. Seven different pathways directly impinge on cholesterol homeostasis.

First is cellular synthesis of cholesterol. The rate of this synthesis appears dependent on the other pathways of control of cholesterol levels. If the other means fail to maintain sufficient cholesterol, the cell responds by synthesizing more, as described in the previous section.

Second are the catabolic pathways that lead to the production of bile salts or hormones. Bile salt synthesis begins with the production of 7α-hydroxycholesterol. This step, catalyzed by 7α-hydroxylase, is a control point for the synthesis of bile salts. After this committed step, several more steps are required for the formation of the bile salts. These reactions are summarized in Fig. 11.13. The formation of bile salts is important in the liver. A number of steroid hormones are synthesized, starting with cholesterol, but that subject is large and beyond the scope of this book. More cholesterol is required for the synthesis of bile salts than for steroid hormones, but together these pathways account for the majority of catabolism of cholesterol. Clearly, the rate of cholesterol loss is important to the overall maintenance of cellular cholesterol levels.

Third, cholesterol levels in the cell can be reduced by the esterification of cholesterol to cholesterol esters, catalyzed by acyl-CoA cholesterol acyltransferase. This process is stimulated by an influx of free cholesterol into the cell.

Fourth, hydrolysis of cholesterol esters by cholesteryl ester hydrolase will modulate cholesterol levels. This will result in an increase of free cholesterol in the cell.

Fifth, cellular cholesterol can be reduced by loss of cholesterol from the cell. This may occur by transfer to serum lipoproteins via the plasma membrane. It may also occur in liver cells by packaging into very low density lipoproteins for export to the plasma or in intestinal cells, where lipids are packaged into chylomicrons for export through the portal vein.

Sixth, cellular cholesterol can be increased by equilibration of the plasma membrane of the cell with cholesterol-rich plasma lipoproteins. The plasma membrane can also bring cholesterol into the cell through a receptor-independent endocytosis of plasma.

Seventh is receptor-mediated endocytosis of low density lipoprotein by the LDL receptor, as described in Chapter 10.

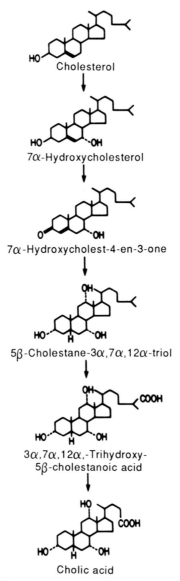

Cholesterol

7α-Hydroxycholesterol

7α-Hydroxycholest-4-en-3-one

5β-Cholestane-3α,7α,12α-triol

3α,7α,12α,-Trihydroxy-
5β-cholestanoic acid

Cholic acid

Fig. 11.13.   Production of a bile salt from cholesterol.

XIII. SUMMARY

This chapter has examined, in overview, the metabolism of the lipids of
the membranes of cells. The important stages of biosynthesis of membrane

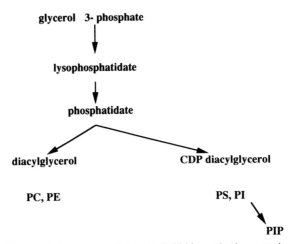

glycerol 3- phosphate

↓

lysophosphatidate

↓

phosphatidate

diacylglycerol   CDP diacylglycerol

PC, PE   PS, PI

PIP

Fig. 11.14. Outline of the pattern of phospholipid biosynthesis, centering on phosphatidate.

lipids include synthesis of the fatty acid components and synthesis of the lipids from the fatty acids and the polar headgroups. Fatty acids are synthesized by elongation of acyl-CoA, until palmitate is made. Longer chain fatty acids and unsaturated fatty acids are made from palmitate by elongation and desaturation. Some of the longest chain unsaturated fatty acids found in mammalian cells require precursors from the diet. Phosphatidate is a focal point for the major pathways of biosynthesis of the phospholipids found in mammalian cells. Phosphatidylethanolamine and phosphatidylcholine have parallel pathways for synthesis, whereas phosphatidylserine and phosphatidylinositol are synthesized on a separate pathway (see Fig. 11.14). Sphingolipids and sterols have independent pathways for biosynthesis. A great deal of metabolic effort goes into making cholesterol, suggesting an important biochemical purpose for this sterol in cholesterol-requiring cells (i.e., mammalian cells).

REFERENCES

1. Vance, D., and J. Vance, *The Biochemistry of Lipids* (New York: Benjamin, 1985).
2. Bloch, K., "On the evolution of a biosynthetic pathway," In *Reflections in Biochemistry* (A. Kornberg *et al.*, eds.) (Oxford: Pergamon Press, 1976), 143.
3. Faust, J. R., J. M. Trzaskos, and J. L. Gaylor, "Cholesterol biosynthesis," In *Biology of Cholesterol* (P. L. Yeagle, ed.) (Boca Raton: CRC Press, 1988), 19–38.
4. Bloch, K., "Sterol structure and membrane function," *Crit. Rev. Biochem.* **14** (1983): 47–92.

12

Membrane Biogenesis

The assembly of membrane components into the membranes of cells is a complex process utilizing both newly synthesized membrane lipids and newly synthesized membrane proteins. This chapter will first consider the incorporation of newly synthesized lipid into membranes. Next, this chapter will consider the biosynthesis of membrane proteins and their incorporation into membranes. The last problem in the area of membrane biogenesis to be discussed will be the sorting of membrane components to create and maintain functionally and compositionally distinct membrane-bound organelles or plasma membranes in the eukaryotic cell.

I. ASSEMBLY OF LIPIDS INTO MEMBRANES

There are three important problems to consider concerning the assembly of newly synthesized lipid into membrane.[1] The first is the initial incorporation of newly synthesized lipid into an acceptor membrane. The second is the generation of the proper transmembrane distribution of lipids. The third is the maintenance of the proper lipid composition in the various target membranes for which the new lipid is being synthesized.

The first problem is a relatively simple one to solve, at least in a topological sense. The late steps in synthesis of the lipids that were examined in Chapter 11 involve membrane-bound enzymes. These enzymes act on substrates that themselves are incorporated into the membrane due to their hydrophobicity (i.e., phosphatidate, or ceramide or lanosterol). Therefore the final stages of lipid biosynthesis lead directly to incorporation of newly synthesized lipid into the endoplasmic reticulum membrane at the location of the enzymes catalyzing lipid synthesis.

The second problem is more difficult. The various classes of lipids are not distributed symmetrically across biological membranes. For example, in human erythrocyte membrane, the outside surface of the plasma membrane is enriched in phosphatidylcholine and sphingomyelin, whereas the inside surface of the plasma membrane is enriched in phosphatidylethanolamine and phosphatidylserine, as described in Chapter 4. Thus there is a substantial asymmetry in the transmembrane distribution of these major lipids of the erythrocyte membrane.[2] Furthermore, the fatty acid composition in these lipids is not the same on each side of the membrane either. Other cellular membranes also exhibit different lipid compositions on either leaflet of the membrane, although the asymmetry is frequently not as pronounced as in the erythrocyte membrane.

The other fact important to the problem of transmembrane distribution of lipids is that the enzymes for synthesis of the membrane lipids for the most part display their active sites on the cytoplasmic surface of the endoplasmic reticulum (this does not include the glycosyltransferases of the Golgi).[3] Such a disposition of the active sites, and thus the site of synthesis, solves a problem directly for phosphatidylethanolamine, which tends to be preferentially distributed on the cytoplasmic side of the endoplasmic reticulum and of the plasma membrane (although the asymmetry appears to be less severe on the intracellular membrane than on the plasma membrane). However, phosphatidylcholine is synthesized from CDP-choline and diacylglycerol on the cytoplasmic surface of the endoplasmic reticulum, and yet, phosphatidylcholine is found preferentially on the lumenal surface of the same membrane and largely on the extracellular surface of the plasma membrane (with which the lumenal surface of the endoplasmic reticulum is topologically related). Flip-flop of phospholipids across a pure lipid membrane is hindered, although in biological membranes the rate of phospholipid translocation may be accelerated (see Chapter 4).[4] In particular, lipid transmembrane movement in membranes involved in lipid biosynthesis is relatively rapid.[5,6] Thus, one key question in this field is the mechanism of transport of newly synthesized phospholipids to the correct side of the endoplasmic reticulum.

The question has, in part, been solved for the endoplasmic reticulum. It is known that proteins in membranes can accelerate the process of transmembrane movement of phospholipids. In particular, proteins exist in the endoplasmic reticulum that are capable of facilitating the transmembrane movement of short-chain phospholipids in an energy-independent manner.[7,8] Furthermore, ATP-driven pumps for phospholipid translocation have been found in the plasma membrane (Chapter 4). How then is membrane asymmetry established or maintained in the ER?

Membrane curvature influences lipid asymmetry in model membrane systems (see Chapter 4). Given the size of the transport vesicles for phospoholipid transport, membrane curvature may contribute to phospholipid asymmetry in biological membranes if the phospholipid asymmetry in the vesicles can be maintained during fusion of the transport vesicles with the target membrane. It is possible that membrane proteins might also influence the transmembrane distribution of membrane lipids. Specific lipid–protein interactions have not been well characterized in the endoplasmic reticulum. However, electrostatic interactions between lipid headgroups and membrane proteins are a potential factor in determining the transmembrane distribution of lipids. For example, peripheral membrane proteins might play a role in phosphatidylserine distribution across the ER. Another interesting speculation is that the transport proteins mentioned above might have enough specificity to enhance transmembrane movement of some species over another. Another time at which transmembrane movement of lipids might occur is during the membrane fusion process that occurs as a part of the separation of a transport vesicle from the ER that is destined to, for example, the Golgi. There is a lipid sorting process that occurs during this fusion and separation. Since the membrane fusion event likely involves some disruption of the integrity of the lipid bilayer, at least transiently, the transmembrane movement of lipids could be facilitated (see event 3 in the fusion process described in Chapter 9).

An interesting enzymatic mechanism for the transmembrane movement of a subclass of membrane lipids has been suggested in one of the biosynthetic pathways for phosphatidylcholine. The pathway involves methylation of phosphatidylethanolamine by a series of sequential steps (see Fig. 12.1). Some experiments have suggested that the first enzyme in this pathway is located on the cytoplasmic surface, whereas the final enzyme in the pathway is located on the opposite side of the membrane. Thus, translocation of the lipid across the membrane during its synthesis is built into the biosynthetic mechanism. One major problem with this mechanism is that in most cases, the methylation pathway is a minor source for the

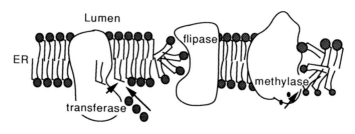

Fig. 12.1. Phospholipid biosynthesis and transmembrane phospholipid movement. On the left, the phospholipid biosynthesis on the cytoplasmic face of the ER is followed by membrane protein-facilitated transmembrane lipid movement. On the right, sequential methylation of PE leads to exposure of PC on the opposite surface.

synthesis of phosphatidylcholine. Clearly, much remains to be delineated in this area.

Concerning the control of the different lipid compositions of the various cellular membranes even less is known. Present evidence favors transport of newly synthesized membrane material via vesicular transport. Different vesicles transport newly synthesized protein and newly synthesized lipid even though both derive from the transitional elements of the ER.[9] The kinetics of transport of newly synthesized membrane lipid components are, in general, different from each other. Transport of cholesterol to the plasma membrane occurs with a half-time in the range 10–60 min.[10,11] Transport of newly synthesized phosphatidylcholine to the plasma membrane is much faster (half-time of about 1 min).[12] Transport of newly synthesized phospholipid to the mitochondria apparently is much slower.[13] This vesicular transport of lipids operates in the following way.

Morphological studies on the organelles in eukaryotic cells suggest that the structures of, for example, ER and Golgi, are separate identities. That is, they are not directly connected by membraneous structures. If no direct connections exist, then transport of membrane components, at least the strongly hydrophobic ones, must be by a vesicular transport mechanism (direct phospholipid exchange between membranes, such as can be observed with cholesterol, has been shown to be unlikely by studies with simple lipid vesicles). Such a transport process can operate by the separation (reverse of the fusion process) of vesicles containing the newly synthesized material, which then move to the target membrane and fuse with it, thereby depositing the newly synthesized material in the next organelle in the transport pathway (see Chapter 9).

Newly synthesized lipid is transported from the ER to the Golgi, at least partially, in 500- to 700-Å vesicles by a process that requires ATP and is sensitive to N-ethylmaleimide.[14] That sensitivity suggests comparison to the vesicular transport pathway utilized for glycoprotein transfer between ER and Golgi or within elements of the Golgi (see below and Chapter 9). These transport vesicles are enriched in phosphatidylcholine and cholesterol. That observation suggests a phospholipid sorting process in vesicle formation from the transitional elements of the ER. The sorting of the PC to enrich the transport vesicles by itself would be sufficient to enrich the vesicles in cholesterol through thermodynamics of cholesterol partitioning (see Chapter 5).

Although vesicular transport can explain the movement of lipids from one intracellular organelle to another, the maintenance of proper lipid composition in the various organelles is not readily explained by this mechanism. How do the various membranes maintain their identity with respect to lipid composition, when most of the cellular lipids are synthesized on the endoplasmic reticulum and must be separated and transported from the site of the synthesis to the appropriate target membrane? Above, sorting of PC was observed in the formation of some transport vesicles. In the case of diphosphatidylglycerol, synthesis of DPG is largely confined to the mitochondria, so this phospholipid does not appear in other cellular membranes. However, control of the distribution of the other membrane lipids among the membranes of the cell remains a mystery. For example, it is not clear why the vast majority of cellular sphingomyelin ends up in the plasma membrane of the cell, even though synthesis takes place in the endoplasmic reticulum. Transport occurs by vesicles to the Golgi and thence to the plasma membrane, but how the sorting process works is not known.[15] Another example can be found in the formation of new disks in the retinal rod outer segment. The phospholipids are immediately sorted upon new disk formation. Cholesterol sorting then occurs by gradual thermodynamic equilibration during the process of disk aging. Table 12.1 shows some of these results.[29] More questions than answers remain in this field, and many challenges await future research efforts of membrane biochemists.

II. BIOSYNTHESIS OF MEMBRANE PROTEINS AND THEIR ASSEMBLY INTO MEMBRANES

Having examined what is known about the synthesis and assembly of membrane lipids into cell membranes, one can now turn to the biosynthesis and incorporation into cell membranes of their protein constituents. Many studies have been performed to decipher the detailed mechanisms involved in this process. The problems encountered are considerably more complex

TABLE 12.1
Comparison of Phospholipid Composition[a] of Bovine
Rod Outer Segment Disk and Plasma Membranes[b]

Phospholipid	Disk (%)	Plasma Membrane (%)
PE	42	11
PS	14	24
PI	3	<1
PC	45	65

[a] Percentage of total phospholipid by headgroup.
[b] Data from K. Boesze-Battaglia and A. D. Albert, *Exp. Eye Res.* **54** (1992): 821–823.

than those involved in the synthesis of water-soluble proteins because of the topology of the membrane proteins. Membrane proteins, as in the case of membrane lipids, are amphipathic. The portions of their mass outside the membrane are similar in structure to water-soluble proteins. These portions therefore have hydrophilic amino acids on their surface and hydrophobic amino acids on the protein interior.

Portions of integral membrane proteins are buried in the hydrophobic interior of the membrane. These portions are often α-helical and almost exclusively contain hydrophobic amino acids. It is this part of the membrane protein that makes it insoluble in water. It is this topology of the membrane protein that presents a considerably different set of problems to solve in biosynthesis and incorporation into the membrane.

Two important themes need to be followed to gain a complete picture of the pathways membrane proteins follow in the biosynthesis and assembly of cellular membranes. One area is the mechanism(s) of biosynthesis and insertion of the newly synthesized protein into the membrane (the endoplasmic reticulum, for example, where the membrane-bound ribosomes are found). The other major theme is the sorting of membrane proteins, a distribution process that must successfully send membrane proteins to the correct target membrane from their site of synthesis (likely the endoplasmic reticulum). The discussion to follow will examine the current state of knowledge in each of these two areas.

Three different experimental directions have led to the current hypotheses concerning membrane protein biosynthesis and incorporation into membranes. One was the discovery of the pathway of biosynthesis of proteins that are secreted by the cell. The second was the investigation of the pathway of biosynthesis of a viral glycoprotein, the vesicular stomatitis G protein.[16] And the third was the description of the biosynthesis of another viral protein, the M13 coat protein.[17] It is worth noting at this point that none of these proteins are natural cellular integral membrane proteins. Therefore, extrapolations from these systems to the biosynthesis of native cellular membrane proteins must be made with caution.

A. Protein Secretion

The pathway of protein secretion, largely worked out by Palade, begins with the synthesis of the protein (to be secreted) on a ribosome (see Fig. 12.2). The morphology of the system immediately distinguishes between the synthesis of the protein to be secreted and the synthesis of the protein to be placed in the cytoplasm of the cell. The latter proteins are synthesized on unbound ribosomes, whereas the former are synthesized on membrane-bound ribosomes. To get a better picture of this differ-

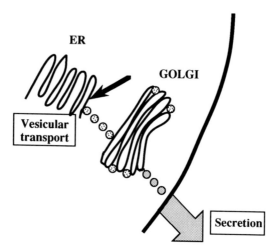

Fig. 12.2. The secretion pathway from synthesis in the ER.

ence, it is worthwhile to review briefly the process of membrane protein synthesis.

Ribosomes function as a factory for the biosynthesis of new proteins. Ribosomes contain both protein and RNA. The ribosome is capable of translation of the message provided by mRNA into a sequence of covalently linked amino acids that result in a functional protein. The fundamental translation process is carried out by the interaction of tRNA with mRNA and the ribosomal enzymes, all operating as a complex. The codons on the mRNA are recognized by the anticodon of the tRNA bearing a specific amino acid matching its anticodon. The tRNA binds to the ribosome in part via the interaction of its anticodon with the codon on the mRNA. This is the translation process. Subsequently, enzymes of the ribosome catalyze the covalent attachment of the amino acid on the tRNA to the existing polypeptide chain by formation of a peptide bond. Thus, the protein grows by addition of the next amino acid in the primary sequence of the protein encoded by the mRNA. This process is repeated, with the departure of the tRNA, which no longer contains an amino acid, and the binding to the ribosomal complex of another aminoacylated tRNA, whose anticodon will correctly pair with the next codon on the mRNA. The growing polypeptide chain, when it gets long enough, then starts to emerge from the ribosomal complex. This is the protein synthesis process that appears at this level to be the same for cytoplasmic, membrane-bound, and secreted proteins.

The ribosomes involved in protein synthesis can be found either free or bound to the endoplasmic reticulum. There is little apparent difference between the structure or composition of the membrane bound versus free ribosomes. It has been suggested that the controlling factor in the binding of ribosomes to the endoplasmic reticulum is contained in the protein to be synthesized and its mRNA. Furthermore, morphologists have suggested that ribosomes directed to bind to the endoplasmic reticulum do so in a particular region of that membrane, hence, the observation of smooth (no ribosomal binding) endoplasmic reticulum and of rough (ribosomes bound) endoplasmic reticulum in electron micrographs of cells.

How then are some ribosomes directed to bind to the endoplasmic reticulum? Some have suggested that proteins to be secreted have what has been termed a signal sequence as the first portion of the peptide to be synthesized. This sequence may or may not appear in the mature protein. The signal sequence is about 20 ± 3 amino acids. When this sequence is translated and synthesized, along with sufficient additional length of peptide to expose the signal sequence outside the ribosome, the signal sequence directs the ribosome to bind to the endoplasmic reticulum. Therefore, according to this hypothesis, secreted proteins (and most membrane proteins) are synthesized initially on free ribosomes, and it is the translated product that causes the ribosomal complex to bind to the membrane.

The signal sequence is usually hydrophobic, which may play a role in its binding to membranes. Furthermore, structure prediction algorithms suggest that the signal sequence is favored to be in an α-helical conformation, again enhancing its membrane-binding properties. Finally, signal peptides have been observed to be disruptive to the bilayer structure of simple lipid membranes. Thus the signal sequence may also be involved in the penetration of membranes by newly synthesized membrane proteins.

As stated above, the nascent polypeptide chain with its amino terminal signal sequence directs the binding to the membrane. The binding of the ribosome with the nascent polypeptide to the endoplasmic reticulum is apparently assisted by a complex containing protein and RNA, called the signal recognition particle (SRP). The SRP consists of six proteins, ranging in molecular weight from 9000 to 72,000 and a 7 S RNA. The signal recognition particle binds with high affinity to ribosomes synthesizing a polypeptide with a signal sequence. The SRP binds with only low affinity to other ribosomes. Interestingly, the binding of the SRP to the ribosomal complex with the signal sequence inhibits further rounds of protein synthesis (addition of further amino acids from tRNA). The growth of that particular polypeptide chain is therefore blocked.

The blockage of protein synthesis can be released by the binding of the SRP–ribosomal complex just described to what is apparently a receptor in the endoplasmic reticulum membrane. This binding is competitive with ribosomes synthesizing other proteins with signal sequences. Experimentation in this area has been done in cell-free systems in which all the ingredients for protein synthesis are present, including tRNA, the acylating system for tRNA, ribosomes, and other required factors. A system commonly used is derived from rabbit reticulocyte lysate. To this can be added microsomal membranes and a particular mRNA. If two different mRNAs are added, each coding for a signal sequence, they will compete with each other for binding to the microsomal membranes. Furthermore, the binding to the membrane can be inhibited by protease treatment of the membranes prior to the binding assay. Therefore, it has been suggested that a protein receptor is present in the endoplasmic reticulum for the binding of a ribosome-nascent polypeptide with a leader sequence–SRP complex.

One of the proteins that has been suggested to be involved in the binding to the receptor is the docking protein, of molecular weight 73,000. This protein, along with two other integral membrane proteins, the ribophorins, apparently participate in the binding of the ribosomal complex and participate in the continued synthesis of the protein to be secreted.

Once the binding has taken place, the synthesis of the nascent polypeptide is renewed. Through a process that is poorly understood, this polypeptide begins to appear on the lumenal side of the endoplasmic reticulum as its synthesis continues. Concurrently, a protease called the signal peptidases acts to cleave the signal sequence off the growing peptide as the signal appears on the lumenal side (although this does not seem to happen in all cases). The polypeptide chain must be long enough to span the membrane and to expose all the signal sequence on the lumenal side before this cleavage takes place. However, the cleavage occurs before the completion of the translation of the secreted proteins. This constitutes a committed step in the synthesis of the secreted protein.

Finally, the secreted protein is complete and appears totally within the lumen of the endoplasmic reticulum, ready for the export process from the cell. At this time, the ribosomal complex, having finished translating the mRNA, disassembles and is available for another round of protein synthesis.

This system of synthesis for proteins to be secreted has been offered as a method for the synthesis of membrane-bound proteins. This idea will be considered for the case of the vesicular stomatitis virus G protein, which is the most studied example of a simple membrane protein that may be synthesized by a process similar to that for secreted proteins.

B. Biosynthesis of Vesicular Stomatitis G Protein

Vesicular stomatitis virus is a small rhabdovirus, which is coated by a membrane, referred to as the viral envelope. The envelope surrounds a nucleocapsid, which is a complex of RNA and proteins. The envelope consists of a lipid bilayer in which is inserted a glycoprotein, called the viral G protein.

Outlined below is the pathway followed during a normal infection process. The virus infects the cell by fusing the viral envelope with the plasma membrane and injecting its nucleocapsid into the cell. The normal transcription and translation processes of the cell are then commandeered by the foreign RNA, and the cell begins to make viral proteins. The G proteins are transported to the cell plasma membrane. There, association of the G protein with the M protein of the virus triggers the assembly process, whereby the newly synthesized nucleocapsids become associated with a patch of plasma membrane enriched with viral G protein. This complex then buds out, forming a new virion.

The biosynthesis of the viral G protein has been worked out in large part by experiments in cell-free extracts and in whole cell systems.[18] It is a simple protein with only one crossing of the hydrophobic interior of the membrane by hydrophobic sequence of amino acids. Its synthesis was therefore modeled on the secretion pathway just described.

Experiments have shown that the protein exists in three different forms along the pathway of biosynthesis and viral assembly. The first product, G_0, is made on membrane-bound ribosomes and cotranslationally inserted into the endoplasmic reticulum membranes. This part of the pathway derives from observations made in cell-free protein synthesis systems. If membranes are added to the synthesis system at the time or shortly after initiation of translation of mRNA for the G protein, the protein will normally end up inserted into the membrane. However, if membranes are added later, but still before synthesis is complete, the protein will not be inserted into the membranes. Therefore, insertion must take place during synthesis of the G protein on a membrane-bound ribosome. The binding of the ribosomal complex and the insertion of the protein into the membrane appear to occur at discrete sites in the membrane. Addition to the system of message for a secreted protein that also has a signal sequence,[19] as does the mRNA for the viral G protein, will reduce the production of membrane-bound G protein, as if the two signal sequences were competing for the same sites on the endoplasmic reticulum for binding.

The G_0 form contains the signal sequence. It is not a membrane-bound form and is not normally seen as a product in the replication of viruses. When microsomes are included during synthesis, another form of the protein, G_1, is found. G_1 is membrane bound, does not contain the signal sequence, and is glycosylated. All these processes take place in the endoplasmic reticulum. Apparently, cleavage of the signal sequence as well as insertion into the membrane occurs cotranslationally. Glycosylation takes place with the addition of the mannose-rich core carbohydrate.

Subsequently, the G_1 protein is transported via vesicles to the Golgi.[20,21] There the carbohydrate is extensively modified. Some of the mannose-rich core carbohydrate is removed and additional carbohydrate residues are added, making a carbohydrate that is more complex in composition and structure.

Another interesting process takes place in the Golgi. The G_1 protein is acylated with fatty acid, frequently palmitic acid, at a cysteine. However, this acylation is apparently not required for transport of the G protein to the plasma membrane, whereas the glycosylation is required for the production of virus. This acylated G protein represents the G_2 form of the viral envelope protein. The G_2 form is eventually transported to the plasma membrane by coated vesicles for the viral assembly process. Figure 12.3 shows a schematic representation of this process.

The transmembrane segment of the G protein is required for proper development of the protein and virus. If this section is deleted from the

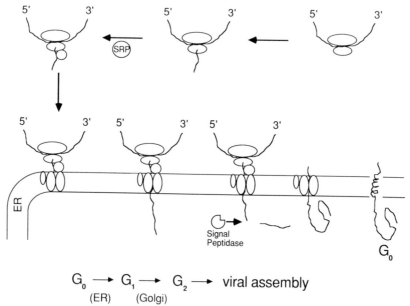

$$G_0 \longrightarrow G_1 \longrightarrow G_2 \longrightarrow \text{viral assembly}$$
$$\quad\;\; (ER) \quad\; (Golgi)$$

Fig. 12.3. Schematic representation of the pathway for biosynthesis of vesicular stomatitis G envelope protein.

message, a protein is synthesized that is secreted. Thus, the analogy of this pathway with the secretion pathway has a basis in experiment.

C. Biosynthesis of M13 Coat Protein

An extensive study of the biosynthesis of the coat protein from bacteriophage M13 in *Escherichia coli* has revealed an alternate pathway for insertion of newly synthesized membrane protein into membranes.[22] In contrast to the vesicular stomatitis virus, the M13 bacteriophage does not have an envelope membrane surrounding a nucleocapsid. Rather, the many copies of the coat protein assemble into a capsule surrounding the viral DNA inside. During synthesis in the host cell, this coat protein passes through a phase in which it is in residence in the *E. coli* inner membrane. While there, it appears to have the typical topography of a transmembrane protein, with a hydrophobic transmembrane segment of about 20 hydrophobic amino acids and additional portions of the protein exposed on both the cytoplasmic surface and the periplasmic surface of the inner *E. coli* membrane. Therefore, the synthesis of this protein has also been used as a model for the synthesis of cellular membrane proteins.

Evidence has been reported that the coat protein is synthesized by a different pathway than was just described for vesicular stomatitis G protein biosynthesis. The data suggested that the coat protein is made on polysomes not bound to membranes. Furthermore, the data suggested that the assembly into membranes occurs via a transient existence as an intermediate consisting of newly synthesized protein free in the cytoplasm.

Data such as these led to the suggestion of the trigger hypothesis.[22] According to this hypothesis, membrane proteins could be synthesized on ribosomes not bound to membranes and the protein released from the ribosome without cotranslational insertion into a membrane. In this model, the trigger for insertion into the membrane is provided by the signal sequence. Once the protein encounters a membrane, it changes conformation and inserts its hydrophobic signal sequence into the membrane, based on the hydrophobic effect. This is followed by cleavage of the signal sequence by the signal peptidase, as in the model based on protein secretion, which prevents the protein from leaving the membrane again (see Fig. 12.4).

D. Membrane Protein Insertion into Membranes in Eukaryotic Cells

As mentioned at the beginning of the discussion of membrane protein biosynthesis, the two protein systems studied in greatest detail, namely vesicular stomatitis virus G protein and M13 coat protein, are not normal

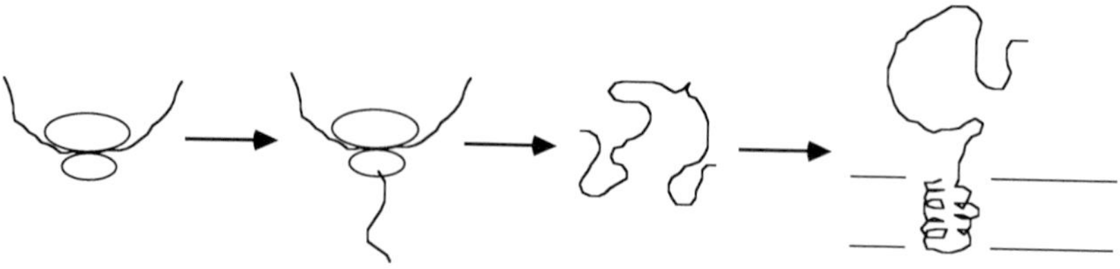

MEMBRANE

Fig. 12.4. Schematic representation of post-translational insertion.

cellular membrane proteins. These studies provide us interesting models for the synthesis of cellular membrane proteins, but it still remains to be determined how cellular membrane proteins are actually synthesized.[23] Furthermore, considerable controversy still exists on the details of the models presented for the synthesis of the two viral proteins.

At this point, three major questions concerning cellular membrane proteins should be examined. One is whether any evidence exists linking the synthesis of cellular membrane proteins with the pathways for these viral proteins. The second question is how to model the synthesis and insertion into membranes of membrane proteins whose amino acid sequence contains more than one transmembrane hydrophobic domain. The third question is how the newly synthesized proteins are sorted and transported to the appropriate target membranes.

Regarding the first question, there is considerable evidence that the mRNA for some integral membrane proteins contains a signal sequence. Examples include the HLA histocompatibility antigens, the acetylcholine receptor, and members of the oxidative phosphorylation system in mitochondria. However, a number of membrane proteins do not have precursor forms with an amino terminal signal sequence. Examples include the Ca^{2+}-ATPase, cytochrome b_5, band 3, rhodopsin, and the asialoglycoprotein receptor.

A further difficulty with application of some of the models for cellular membrane protein synthesis and insertion is exemplified by band 3. The process described for the vesicular stomatitis virus G protein calls for the amino terminus of the protein to be exposed on the lumenal side of the endoplasmic reticulum. Topologically, this corresponds to the extracellular surface of the plasma membrane. However, band 3 exposes the carboxyl terminus to the extracellular surface of the plasma membrane.

Therefore, it remains to be seen how generally applicable the model for vesicular stomatitis virus is. Are all these membrane proteins cotranslationally inserted into the endoplasmic reticulum by membrane-bound ribosomes? Apparently, some proteins do not have the amino terminal signal sequence required for binding of the ribosome to the membrane. Furthermore, some proteins, if they were to be cotranslationally inserted, would not expose their carboxyl terminus on the lumenal side of the endoplasmic reticulum, yet, in their ultimate expression in the membrane, are found in that orientation.

Rhodopsin exemplifies the second question concerning the models for cellular membrane protein synthesis and insertion. Not only does the message for rhodopsin not contain a signal sequence, but the polypeptide of rhodopsin crosses the membrane seven times. In the most strict application of the model based on the vesicular stomatitis virus, this topology of rhodopsin would be achieved by the leader sequence snaking in and out of the membrane, sequentially producing the set of transmembrane helices that constitute the intramembraneous structure of rhodopsin. This type of insertion is clearly in violation of the hydrophobic effect.

Several researchers have proposed models to surmount this topological problem.[24] One is the hairpin loop hypothesis.[25,26] This model proposes that two hydrophobic domains of a membrane protein are connected by amino acids that allow or prescribe a sharp turn in the protein structure. Thus, as the protein emerges from the ribosome, the amino acid sequence directs the formation of a loop structure whose hydrophobic nature causes it to be inserted into the membrane spontaneously, based on a favorable free energy change for the insertion process. This model is attractive because it readily accommodates the topology of proteins, like rhodopsin, whose polypeptide traverses the membrane more than one time. It also does not require that the amino terminus be exposed to the lumenal side of the endoplasmic reticulum.

The last question of this section is the sorting of newly synthesized proteins and their targeting to the appropriate destination. For example, both HMG-CoA reductase and band 3 are made initially in the endoplasmic reticulum. However, HMG-CoA reductase stays in the endoplasmic reticulum as its destination, and band 3 must be glycosylated and transported to the plasma membrane. How are these two proteins sorted such that only one is transported to other membranes? Related to this question is the pathway of receptor-mediated endocytosis. Some plasma membrane receptors are recycled after endocytosis and transported to a lysosome. How is the patch of membrane containing the receptor recycled to the plasma membrane and the other proteins of the lysosome not also sent to the plasma membrane?

Finally some proteins of the mitochondria are made from nuclear DNA and some from mitochondrial DNA. How is this synthesis coordinated? And how are the nuclear DNA-coded proteins imported into the mitochondria?[27] The import process is beginning to be understood. This process involves proteins in the membrane that facilitate the transport of newly synthesized protein. Import into the inner membrane or matrix of the mitochondria is likely through a point of contact between the inner an the outer mitochondrial membranes.[28]

An important aspect of membrane biogenesis is how the membrane material is sorted. The sequences of proteins that end up in the plasma membrane are all different, yet they all end up there after synthesis. If a receptor is involved, it must recognize some aspect of secondary or tertiary structure of the proteins and not their primary structure. At the level of the endoplasmic reticulum, vesicles that form or transport must do so from a patch of that membrane that is laterally separated and compositionally distinct from the remainder of the endoplasmic reticulum. Therefore, the sorting process for membrane proteins likely takes place as a lateral phase separation in the endoplasmic reticulum before the formation of the transport vesicles, with the targeting of the vesicles in terms of their destination as a separate process. Directional transport of the transport vesicles likely involves at least two factors. One is perhaps some sort of receptor process that recognizes some structural aspect of the transport vesicles specific for the target membrane, as mentioned above. The other is the involvement of the cell cytoskeleton. Directional transport is possible by elements of the cytoskeleton, supported by an ATP energy source. Vesicles therefore could move along paths predetermined by cytoskeletal elements that are already attached to the target membrane. All that need happen then is, ultimately, fusion with the target membrane to insert the newly synthesized protein into that membrane. This would confine the sorting process to the attachment of a particular transport vesicle to the proper cytoskeletal element. Perhaps even this can be accomplished by a lateral phase separation in the donor membrane, with vesicles pinching off next to the appropriate cytoskeletal component.

III. SUMMARY

Membrane biogenesis involves the biosynthesis and insertion into membranes of newly synthesized lipid and protein components. Newly synthesized lipid components must be both inserted into membranes and positioned on the proper side of the membrane. The former goal is achieved through the membrane localization of the terminal enzymes of lipid biosynthesis.

The latter goal, of correct transmembrane placement of the newly synthesized lipid, has not been adequately described. Presumably, it involves membrane protein but not the biosynthetic enzymes that are located on the cytoplasmic surface of the endoplasmic reticulum.

The synthesis and insertion of membrane proteins into membranes can be modeled after three related systems that have been described in some detail. One is the pathway of protein secretion, a second is the biosynthesis of vesicular stomatitis virus G protein, and a third is the synthesis of the M13 coat protein.

One consensus pathway starts with the synthesis of the protein on the membrane-bound ribosome. Apparently, the ribosomes becomes bound to the endoplasmic reticulum through the signal sequence, a specially coded sequence on the newly synthesized protein. This process results in cotranslational insertion of the protein. A second consensus pathway synthesizes the membrane protein on a free ribosome. Post-transitionally, this protein encounters a membrane, alters its conformation, and, consequently, inserts itself into the membrane. It is to be expected that mixtures of these pathways may operate in the insertion of normal cellular membrane proteins into the membranes of cells.

REFERENCES

1. Bishop, W. R., and R. M. Bell, "Assembly of phospholipids into cellular membranes," *Annu. Rev. Cell Biol.* **4** (1988): 579–610.
2. Bretscher, M. S., "Asymmetrical lipid bilayer structure for biological membranes," *Nature N. Biol.* **236** (1972): 11–12.
3. Ballas, L. M., and R. M. Bell, "Topography of phosphatidycholine, phosphatidyethanolamine and tracylglycerol biosynthetic enzymes in rat liver microsomes," *Biochim. Biophys. Acta* **602** (1980): 578–590.
4. Donohue-Rolfe, A. M., and M. Schaechter, "Translocation of phospholipids from the inner to the outer membrane of *E. coli.*," *Proc. Natl. Acad. Sci. U.S.A.* **77** (1980): 1867–1871.
5. Zilversmit, D. B., and M. E. Hughes, "Extensive exchange of rat liver microsomal phospholipids," *Biochim. Biophys. Acta* **469** (1977): 99–110.
6. Hutson, J. L., and J. A. Higgins, "Asymmetric synthesis and transmembrane movement of phosphatidylethanolamine synthesized by base-exchange in rat liver endoplasmic reticulum," *Biochim. Biophys. Acta* **835** (1985): 236–243.
7. Bishop, W. R., and R. M. Bell, "Assembly of the endoplasmic reticulum phospholipid bilayer: The phosphatidylcholilne transporter," *Cell (Cambridge, Mass.)* **42** (1985): 51–60.
8. Backer, J. M., and E. A. Dawidowicz, "Reconstitution of a phospholipid flippase from rat liver microsomes," *Nature (London)* **327** (1987): 342–343.
9. Urbani, L., and R. D. Simoni, "Cholesterol and vesicular stomatitis virus G protein take separate routes from the endoplasmic reticulum to the plasma membrane," *J. Biol. Chem.* **265** (1990): 1919–1923.
10. Kaplan, M. R., and R. D. Simoni, "Transport of cholesterol from the endoplasmic reticulum to the plasma membrane," *J. Cell Biol.* **101** (1985): 446–453.
11. Lange, Y., and H. J. G. Matthies, "Transfer of cholesterol from its site of synthesis to the plasma membrane," *J. Biol. Chem.* **259** (1984): 14624–14630.

12. Kaplan, M. R., and R. D. Simoni, "Intracellular transport of phosphatidylcholine to the plasma membrane," *J. Cell Biol.* **101** (1985): 441–445.
13. Yaffe, M. P., and E. P. Kennedy, "Intracellular phospholipid movement and the role of phospholipid transfer proteins in animal cells," *Biochemistry* **22** (1983): 1497–1507.
14. Moreau, P., M. Rodriguez, C. Cassagne, D. M. Morré, and D. J. Morré, "Trafficking of lipids from the endoplasmic reticulum to the Golgi apparatus in a cell-free system from rat liver," *J. Biol. Chem.* **266** (1991): 4322–4328.
15. Lipsky, N. G., and R. E., Pagano, "Intracellular translocation of fluorescent sphingolipids in cultured fibroblasts," *J. Cell Biol.* **100** (1985): 27–34.
16. Katz, F. N., J. E. Rothman, V. R. Lingappa, G. Blobel, and H. F. Lodish, "Membrane assembly in vitro: Synthesis, glysylation, and asymmetric insertion of transmembrane protein," *Proc. Natl. Acad. Sci. U.S.A.* **74** (1977): 3278–3282.
17. Ito, K., T. Date, and W. Wickner, "Synthesis, assembly into the cytoplasmic membrane and proteolytic processing of the precursor of coliphage M13 coat protein," *J. Biol. Chem.* **255** (1980): 2123–2130.
18. Katz, F., J. E. Rothman, D. M. Knipe, and H. F. Lodish, "Membrane assembly: Synthesis and intracellular processing of the vesicular stomatitis viral glycoprotein," *J. Supramol. Struct.* **7** (1977): 353–370.
19. Garnier, J., P. Gaye, J. C. Mercier, and B. Robson, "Structural properties of signal peptides and their membrane insertion," *Biochimie* **62** (1980): 231–239.
20. Rothman, J. E., and R. E. Fine, "Coated vesicles transport newly synthesized membrane glycoproteins from the endoplasmic reticulum to plasma membrane in two successive stages," *Proc. Natl. Acad. Sci. U.S.A.* **77** (1980): 780–784.
21. Rothman, J. E., "Transport of the vesicular stomatitis glycoprotein to trans Golgi membranes in a cell-free system," *J. Biol. Chem.* **262** (1987): 12502–12510.
22. Wickner, W., "The assembly of proteins into biological membranes: The membrane trigger hypothesis," *Annu. Rev. Biochem.* **48** (1979): 23–43.
23. Jokinen, M., C. G. Gahmberg, and L. C. Anderson, "Biosynthesis of the major human red blood cell silaoglycoprotein, glycophorin A, in a continuous cell line," *Nature (London)* **279** (1979): 604–607.
24. Wickner, W. T., and H. F. Lodish, "Multiple mechanisms of protein insertion into and across membranes," *Science* **230** (1985): 400–407.
25. Engelman, D. M., and T. A. Steitz, "The spontaneous insertion of proteins into and across membranes: The helical hairpin hypothesis," *Cell (Cambridge, Mass.)* **23**, (1981): 411–422.
26. Steitz, T., and D. Engelman, "Quantitative application of the helical hairpin hypothesis to membrane proteins," *Biophys. J.* **37** (1982): 124–125.
27. Hartl, F. U., N. Pfanner, D. W. Nicholson, and W. Neupert, "Mitochondrial protein import," *Biochim. Biophys. Acta* **988** (1989): 1–45.
28. Söllner, T., G. Griffiths, R. Pfaller, N. Pfanner, and W. Neupert, "MOM19, an import receptor for mitochondrial precursor proteins," *Cell (Cambridge, Mass.)* **59** (1989): 1061–1070.
29. K. Boesze-Battaglia and A. D. Albert, "Phospholipid distribution in bovine rod outer segment membranes," *Exp. Eye Res.* **54** (1992): 821–823.

Index

A